T0325109

Marketing Fresh Fruits and Vegetables

Marketing Fresh Fruits and Vegetables

R. BRIAN HOW
Cornell University

Springer Science+Business Media, LLC

An AVI Book
(AVI is an imprint of Van Nostrand Reinhold)

Originally published by Van Nostrand Reinhold in 1991
Library of Congress Catalog Card Number 90-12952
ISBN 978-0-442-00450-7

Library of Congress Cataloging-in-Publication Data

How, R. Brian (Richard Brian)
Marketing fresh fruits and vegetables / R. Brian How.
p. cm.
Includes bibliographical references and index.
ISBN 978-0-442-00450-7 ISBN 978-1-4615-2031-3 (eBook)
DOI 10.1007/978-1-4615-2031-3
1. Fruits—Marketing. 2. Vegetables—Marketing. I. Title.
HD9240.5.H68 1990
635'.068'8—dc20 90-12952
CIP

To Janet,
and to our children
Sarah, Katie, and George

Contents

Preface / ix
Acknowledgments / xiii

Part 1 Markets, Sources, and the Marketing System / 1

1. The United States Market for Food / 3
2. The United States Market for Fresh Fruits and Vegetables / 23
3. Sources of Fresh Fruits and Vegetables / 37
4. Major Sources of Supply: California, Florida, and Mexico / 53
5. The Marketing System and Firms Involved: An Overview / 74
6. Marketing Systems for Three Major Fruits and Vegetables: Oranges, Apples, and Tomatoes / 92

Part 2 The Marketing Environment / 115

7. Market Information: Agricultural Statistics, Grading and Inspection, Market News, and Other Information Sources / 117
8. Market Prices and Price Analysis / 129
9. Trade Practices, Credit Ratings, and Regulation of Trading (Perishable Agricultural Commodities Act) / 151
10. Cooperative Marketing / 163
11. Marketing Orders / 171
12. Pesticide Use and Food Safety / 179
13. Nutritional Quality and Nutrition Marketing / 204
14. Generic, Brand, and Private Label Advertising and Promotion / 217

Part 3 Marketing Operations and Firms / 229

15. International Trade / 231
16. Shipping Point Operations and Firms / 253

17. Long Distance Transportation / 266
18. Wholesaling at Destination and Terminal Market Facilities / 284
19. Food Retailers and Retailing / 301
20. The Foodservice Industry / 319
21. Direct Marketing by Farmers to Consumers / 328

Part 4 Epilogue / 341

22. Future Prospects / 343

Index / 349

Preface

This book has evolved out of experience gained during 15 years of teaching a course on fruit and vegetable marketing to Cornell University undergraduates. Initially it was difficult to assemble written material that would introduce the students to the industry and provide examples to illustrate marketing principles. Apart from a few major studies like the U.S. Department of Agriculture's survey of wholesale markets that came out in 1964 or the report of the National Commission on Food Marketing published in 1966 there was little research to turn to in the early 1970s. Trade association meetings, trade papers, and personal contacts with members of the industry were the major sources of information. It became necessary to collect information from many different sources to fill the need for a descriptive base. Now there are many good research reports and articles being published on various phases of the industry. There still remains a pressing need, however, to consolidate and interpret this information so that it provides an understanding of the total system and its various parts.

Fresh fruit and vegetable marketing is different in many respects from the marketing of other agricultural and nonagricultural products. Hundreds of individual commodities comprise the total group. Each product has its own special requirements for growing and handling, with its own quality attributes, merchandising methods, and standards of consumer acceptance. However, many do share the general characteristics of being highly perishable and produced far from the market. Fruits and vegetables are marketed through many firms, some of which perform only a single function such as storage or transportation on a single commodity. Other firms, such as service wholesalers or retailers, handle a full line of fresh products. In an effort to preserve quality throughout the marketing system, these companies are increasingly using the latest technology in grading, storage, transportation, and merchandising.

Recent advances in production technology are further influencing the manner in which fresh products are packaged, sold, shipped, and displayed at the retail level. Transactions between sellers at shipping points and buyers close to the market are generally made verbally with the product sight unseen. The informality with which these verbal agreements are made requires a high degree of confidence and trust. Trade terms and rules governing transactions are complex and specific to the produce industry.

The need for a text on this subject stems not only from the unusual nature of the business but also from the recent expansion in economic importance that has created new opportunities for existing firms and attracted the interest of other firms. The changes which are occurring in marketing practices, technology, and business organization point to the need for an even finer understanding of the industry and its peculiarities. American consumers' heightened interest in, and recent obsession with physical fitness, weight control, and the nutritional value of the foods we eat has greatly increased the total market for fresh fruits and vegetables, especially those that are unusual or exotic in nature. Over the past 20 years the volume of many individual commodities marketed in this country has increased several times, and total marketing charges for fresh fruits and vegetables has risen faster than for any other group of farm products. This added demand, along with improvements in transportation and the reduction in trade barriers has expanded the geographic area from which supplies are drawn. Large companies, some with nationally and internationally known brands of other foods, have been entering the business of marketing fresh products in increasing numbers. Existing firms have been fighting hard to maintain, if not increase, their share of the market.

The underlying approach taken in this book is that in order to evaluate the issues and alternatives, as well as the strengths and weaknesses, of the system it is first necessary to understand how each phase operates and how it has evolved. Only then is it possible to determine what factors will influence future directions and what the impact of various alternatives might be.

Logically this book is divided into four unequal parts. Part 1 provides an overview of the system by first considering the ultimate market for food and for fresh fruits and vegetables, the sources of supply in general and three major sources in particular, the total system that links sources and markets, and specific examples of the system for three products. After this introduction Part 2 describes the environment in which marketers must operate. This includes the study of information sources and communication networks, market prices, trade practices, farmer cooperatives, marketing orders, pesticide use, nutritional attributes, and advertising and promotion programs. Part 3 then looks at the marketing functions and firms involved in the actual process of moving these commodities through the system from international trade through shipping point, long distance transportation, destination wholesaling, retailing, foodservice, and direct marketing. Part 4 considers prospects for the future and the needs of the industry.

This book will provide members of the trade who are already involved in specific segments of the industry with a general background covering all phases of marketing from growing operations to retail merchandising, including past experiences, current developments, and emerging trends and issues that must be addressed by the private and public sectors. Those with experience in marketing other products who seek to enter the fresh fruit and vegetable arena will find this material a valuable resource in their attempts at better understanding produce practices and the forces that shape the business. The management of foreign businesses which are considering or are already marketing fresh fruits and vegetables in the United States will be able to gain greater knowledge of the fresh produce marketing system in this country. Many professionally trained workers at educational institutions who are employed on research or public educational programs to improve the growing, harvesting, and post-harvest handling of fresh fruits and vegetables will be guided by the additional knowledge of the economic aspects of the total marketing system. And the many public employees who estimate crop production, administer government regulations, report prices and product movement, and aid in policy development will more fully understand their role and their contribution to the total effort through the material contained in this book.

The objective of producing a book such as this goes far beyond simply providing a supplementary text for an undergraduate course or a reference for members of the industry. Ultimately it is hoped that a better understanding of the fresh produce marketing system and some knowledge of the structure and performance of the individual parts will contribute to better communication and coordination, improved service to the consumer, and greater distribution efficiency. It is further anticipated that the book will offer appropriate rewards to those members of the industry who wish to comprehend the issues, alternatives, and emerging trends and manage their operations accordingly.

Acknowledgments

The attempt to describe and analyze such a complex and diverse industry as produce marketing is an ambitious project, and can only be accomplished with help from many different quarters. In a single chapter so many topics have to be covered, each one of which really requires specialized knowledge and deserves much fuller treatment. Any success in this task is due in large part to the help and encouragement provided from many willing sources.

I owe a great debt to the students who took the course over the years. Preparing to face a group of eager undergraduates two or three times a week for several months each year can be a real learning experience. Some had no previous knowledge of the subject, while for others produce marketing was born and bred in the bone, yet all contributed. Several have gone on to positions of responsibility in the industry, in government, or in academia, and my thanks go to them all.

The guest lecturers and field trip hosts were a great help in instructing me and my students, and in keeping me in touch with the real world. We were fortunate to visit, or have come to class, representatives from every phase of the industry and supporting public and private agencies. From them I learned about many aspects of the business that would otherwise have escaped my attention. I am reluctant to single out any for fear of offending others. However, I would like to mention two who came faithfully every year. Barney Mayrsohn kept us up to date on international trading, and Mel Nass reflected on his marketing experience including selling New York grapes to California buyers.

I have relied heavily on information from the U.S. Department of Agriculture. The Agricultural Marketing Service, the Foreign Agricultural Service, and the National Agricultural Statistical Service have provided a wealth of data. The Outlook and Situation reports and the food marketing analyses prepared by members of the Commodity Economics Division of the Economic Research Service have been invaluable sources, as citations in the text will testify. Their help is much appreciated.

At various stages in the project several of my colleagues at Cornell have reviewed chapters and made suggestions. I would particularly like to thank Bruce Anderson, David Blandford, Enrique Figueroa, Gene German, Bill Lesser, Ed McLaughlin, and Bill Tomek for their help. The manuscript has benefited greatly from their input, but they should not be held responsible for any remaining deficiencies.

Others who have helped include Wes Kriebel, Hal Linstrom, John Love, and Nancy Tucker. I owe a special debt to David Marguleas who, as a student, saw the potential for such a book, and more recently read a complete draft. I hope I have come close to living up to his expectations.

And finally, all members of my family have been remarkably supportive. Long ago when this endeavor was just a gleam in my eye they offered enthusiastic support. As it developed and came closer to fruition they continued to extend their encouragement, and have aided me in many ways. My wife, Janet, typed the first draft into the computer some time ago, and has recently provided editorial comment on the whole manuscript. In between she contributed in many ways, from relieving me of household chores to offering suggestions on the general presentation. Each of the children has also helped in their own way, as has my son-in-law, Jim Alexander, and this has meant a great deal to me.

Marketing Fresh Fruits and Vegetables

Part I

Markets, Sources, and the Marketing System

Chapter 1
The United States Market for Food

INTRODUCTION

The markets for all goods and services are interrelated, some more closely than others. Directly or indirectly the changes in consumer preferences or in supplies for one food product affect many others. We therefore begin the study of fresh fruit and vegetable marketing in the United States by examining the total market for all foods and related services in this country.

This nation constitutes a tremendous market for food, but food purchases represent a relatively minor expenditure for many people. We eat many different foods and obtain them through several different marketing channels. Food costs and consumption patterns have changed in recent years. We need to be aware of recent trends in life-styles and eating habits, and to recognize the reasons for these changes. People differ in their diets and eating habits, and this is related somewhat to their social, economic, and demographic characteristics. Knowing how consumers differ with respect to background, personal characteristics, living arrangements, employment, and income, and how these differences are related to current and prospective purchases, is what makes food marketing such an exciting challenge. Current trends in economic and demographic characteristics enable us to use this information to project future consumption patterns.

THE UNITED STATES MARKET FOR CONSUMER GOODS AND SERVICES

The United States today is a gigantic market for consumer goods and services. In 1988 our population of about 246 million had an aggregate personal disposable income of almost $3.5 trillion, or an average of $13,114 per person (Table 1.1). Personal consumption expenditures amounted to $3.2 trillion. Expenditures for nondurables, of which food was the major component but which also included clothing and shoes and many other

Table 1.1. U.S. Allocation of Disposable Personal Income, 1988.

	AMOUNT	PERCENT
	billions	
Disposable personal income	$3,472.9	100.0
Total personal consumption expenditures	3,226.0	93.2
Nondurables	1,047.2	30.2
Food, clothing, fuel and gas, etc.		
Durables	449.9	13.0
Motor vehicles, furniture etc.		
Services	1,728.9	49.8
Housing, medical care, transportation, etc.		
Savings	147.0	4.2
Other*	99.9	2.9

*Includes interest paid by consumers to businesses and personal transfer payments to foreigners.
Source: USDA ERS, 1989.

items, amounted to $1,047 billion in 1988. Food expenditures that year were $409 billion, and represented 12.7 percent of total disposable personal income. Expenditures for durables, mainly motor vehicles and parts and furniture and household equipment, came to $450 billion. Expenditures for services such as housing, medical care, and household operation totaled $1,729 billion. The remainder of disposable personal income went to savings and other miscellaneous items.

There have been major changes in how we have spent our personal disposable income over the past 25 years. In 1965 almost 40 percent of personal disposable income went for nondurables compared to 30 percent today, while expenditures for services rose from under 40 percent to almost 50 percent of income. The proportion going for durables is about the same, while the savings rate is lower and the percentage going for other items is higher.

Most people in this country live in households, and the household represents the primary decision-making unit with respect to many types of purchases, including food. Food purchases have been found to be closely related to the size, composition, and total income of the household. In 1986, according to the Bureau of Labor Statistics, a typical urban household consisted of 2.6 people with an income after taxes of $23,119 or $8,917 per person. About two-thirds of the average annual expenditures of this household went for housing, transportation, and food, with the remainder divided among many other different types of expense (Figure 1.1). Housing, consisting of shelter, fuel, and utilities, household operation, and furnishings and equipment was by far the most important and accounted for 30.3 percent. Transportation, both public and private, took 21.1 percent, and

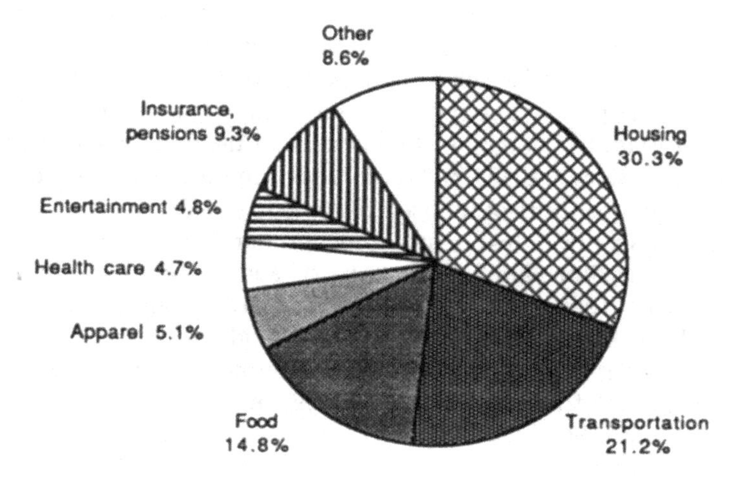

Figure 1.1. Average Annual Expenditures, by Major Categories, All U.S. Consumer Units, 1986. *Source:* Adapted from Bureau of Labor Statistics, 1989.

food 14.8 percent of expenditures. Other important expense categories were entertainment, apparel, and services, and health care. Health care paid for directly by the household unit represents only a small proportion of the total medical costs since much is provided by employers or the government.

As a nation we allocate a smaller proportion of our total expenditures to food purchases than do people of any other country. The 13.6 percent of personal disposable income that we spent for food in 1987 (Putnam, 1989) was several percentage points lower than in other developed countries such as the Netherlands, Canada, and the United Kingdom, while people in less developed countries such as India and China spent more than half of their total income for food.

U.S. Food Expenditures

Personal consumption expenditures for food in 1988 were estimated to amount to $409 billion. The total value of food consumed in this country is larger than that, however, since it also includes the value of additional food obtained from home production, government donations, meals supplied in military and prison mess halls, and expense account meals. In 1988 these additional food costs were estimated to amount to $63 billion, bringing the total value of food consumed in the United States that year to $473 billion (USDA ERS, 1989).

An important distinction is made in food marketing between food purchased for use at home, largely bought in retail stores, and food eaten away from home, largely in restaurants, fast food outlets, and institutions. In 1988 the value of food consumed at home came to $262 billion or 55.5 percent of the total, of which $254 billion consisted of purchases and $8 billion of home production and donations. Away-from-home meals were valued that year at $210 billion, 44.5 percent of the total, and consisted of $185 billion in sales and $25 billion in foods supplied and donated.

Tremendous quantities of food are required to satisfy the American appetite. The food supply in this country comes primarily from U.S. production, which is enhanced or diminished by exports and imports and additions to or withdrawals from stocks. In 1987 we used 41.2 billion pounds of meat (beef, veal, pork, and lamb), 19.0 billion pounds of chicken and turkey, 145.9 billion pounds of all dairy products in terms of milk equivalent, 25.0 billion pounds of fresh fruits, 22.1 billion pounds of commercially produced fresh vegetables, and 11.5 billion pounds of fresh potatoes (Putnam, 1989).

The many different foods consumed in this country can be grouped according to the type of commodity. Commodity groups differ in economic importance. Official estimates of food expenditures by commodity group are only published for food products originating on U.S. farms. This excludes food from other U.S. sources such as seafood, and does not include imports. Consumer expenditures for foods originating on U.S. farms amount to about 90 percent of total food costs. Meat products constitute 29 percent of the total expenditures for U.S. farm produced foods, outlays for fruits and vegetables, both fresh and processed, amount to 23 percent of the total, while dairy products come to 14 percent (Figure 1.2). Together

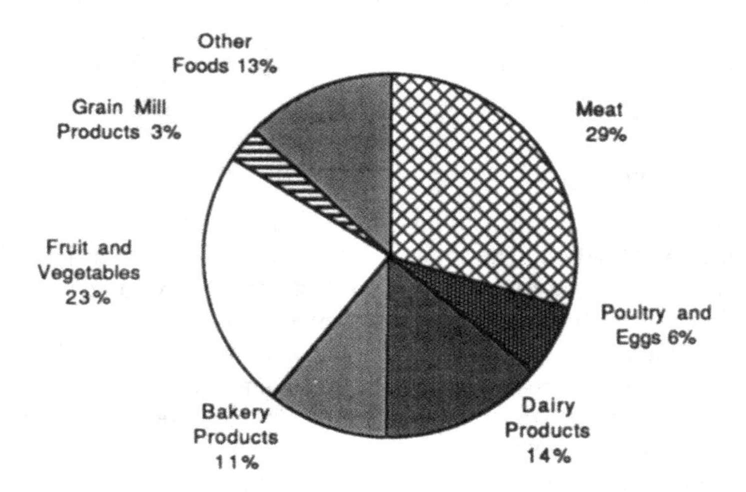

Figure 1.2. U.S. Consumer's Allocation of Expenditures for U.S. Farm Produced Foods, 1987. *Source:* Adapted from Putnam, 1989.

these three major commodity groups account for almost two-thirds of consumer expenditures for all U.S. farm foods.

Changes in Food Expenditures

From 1960 to 1987 the U.S. population increased from 180 to 242 million or by one-third, and per capita disposable income in real terms went up by about two-thirds. The combination of the larger population and greater per capita income resulted in a gain in constant dollars of total disposable personal income of close to 125 percent over this period. Expenditures for food and beverages increased but only by about 60 percent in real terms. The proportion of income spent for food declined by 40 percent from 20 percent in 1960 to 12 percent in 1987 (Figure 1.3).

This decline in the proportion of total income spent for food has not been consistent from year to year, nor have changes in the proportions spent for food at home and away from home exhibited a similar pattern. The proportion of income spent for food at home today is only about 60 percent of what it was following World War II, while the proportion of income spent for food away from home is only a little larger.

The Marketing Bill for Food

In 1988, consumers in the United States spent $395 billion at eating places and food stores for foods originating on U.S. farms. U.S. farmers received $97.3 billion for foods provided U.S. consumers in 1988, or 25 percent of consumer expenditures for these foods. The marketing bill, or the value added by marketing firms, came to $297.6 billion.

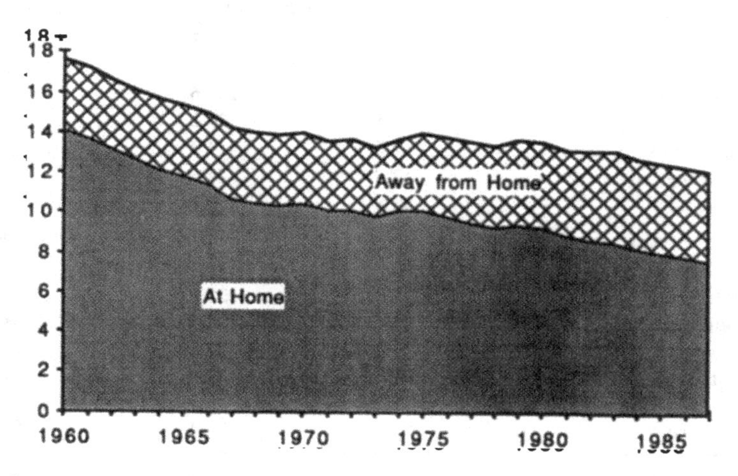

Figure 1.3. Percentage of Disposable Personal Income Spent By U.S. Consumers for Food Eaten At Home and Away From Home, 1960–87. *Source:* Adapted from Putnam, 1989.

The marketing bill for U.S. farm foods provided to U.S. consumers can be further subdivided according to major functional categories. Processing costs in 1988 were estimated to amount to $97.4 billion, intercity transportation costs to $18.2 billion, wholesaling costs to $33.3 billion, retailing costs to $55.1 billion, and food service costs to $88.4 billion that year (Table 1.2).

Between 1981 and 1988 the marketing bill for U.S. farm foods rose from $204.5 billion to $297.6 billion, an increase of $93.1 billion or 45.5 percent. This increase came about in part through the general increase in costs but also through changes in marketing channels and marketing services. In 1981 about 80 percent of U.S. farm foods were sold through retail stores and 20 percent through food service establishments. By 1988 almost 26 percent of

Table 1.2. Marketing Function Components of Consumer Expenditures for U.S. Farm Foods, 1981 and 1987.

EXPENDITURES AND COMPONENTS	1981	1987	PERCENT INCREASE
	billion dollars		percent
		Expenditures for all foods	
Total expenditure	287.7	394.9	37.2
Farm value	83.2	97.3	16.9
Marketing bill	204.5	297.6	45.5
Processing cost	73.0	97.4	33.4
Intercity transportation	14.3	18.2	27.3
Wholesaling cost	22.8	33.3	46.1
Retail/Foodservice cost	94.4	143.5	52.0
		Expenditures for food at foodstores	
Total	194.0	239.8	23.6
Farm value	66.6	72.2	8.4
Marketing bill	127.4	167.6	31.6
Processing cost	59.5	74.2	24.7
Intercity transportation	11.6	14.3	23.3
Wholesaling cost	17.5	24.0	37.1
Retailing cost	38.8	55.1	42.0
		Expenditures for food eaten away from home	
Total	93.7	155.1	65.5
Farm value	16.6	25.1	51.2
Marketing bill	77.1	130.0	68.6
Processing cost	13.5	23.2	71.9
Intercity transportation	2.7	3.9	44.4
Wholesaling cost	5.3	9.3	75.5
Foodservice cost	55.6	88.4	59.0

Source: Adapted from Dunham, 1989.

U.S. farm foods moved through food service establishments and less than 75 percent through retail stores. Retailing costs rose $16.3 billion from $38.8 billion to $55.1 billion or 42 percent, but food service marketing costs rose $32.8 billion from $55.6 billion to $88.4 billion, or 59 percent.

Marketing costs differed for major U.S. produced farm foods. Marketing costs for meats in 1988 were estimated to amount to $81.8 billion, for fruits and vegetables (fresh and processed) to $773.1 billion, for dairy products to $38.8, for bakery products to $36.3, for poultry to $15.0, and for grain mill products, eggs, and other foods to $53.6 billion in 1988.

Labor was the major cost component in 1988. Direct labor costs in marketing came to $136.5 billion or 46 percent of the total marketing bill and 35 percent of the total consumer expenditures for domestically produced farm foods. Packaging materials cost $32.3 billion, intercity rail and truck transportation $17.8 billion, fuel and electricity $14.2 billion, corporate profits before taxes $11.4 billion, and many other different expenses came to $85.4 billion.

PER CAPITA FOOD DISAPPEARANCE

Currently there is available for consumption in this country about 1,375 pounds of food per person per year (Table 1.3). How much is actually eaten and how much thrown away we do not know, so we tend to use the terms

Table 1.3. Selected Food Items, Annual Per Capita Availability, 1987.

	AVAILABILITY PER CAPITA	PERCENT OF TOTAL
	pounds	percent
Meat, Poultry, Fish	228.6	16.5
Eggs	31.6	2.3
Dairy Products, Excluding Butter	299.3	21.6
Fats and Oils, Including Butter	65.7	4.7
Fresh Fruit	98.6	7.1
Processed Fruit, excluding Juices	15.9	1.1
Citrus Juice	46.4	3.4
Selected Fresh Vegetables	78.6	5.7
Processed Vegetables, Farm Weight	104.2	7.5
Fresh Potatoes, Sweet Potatoes	49.5	3.6
Frozen Potatoes	23.2	1.7
Dry Beans, Peas, Nuts	16.6	1.2
Flour and Cereals	173.7	12.6
Total Sweeteners, Dry Weight	151.6	11.0
Totals	1,383.5	100.0

Source: Adapted from Putnam, 1989.

disappearance or *utilization* rather than consumption. Our food supply contains more food energy than is needed, and sufficient amounts of all essential nutrients.

The foods we eat differ greatly in food energy and nutrients. Dairy products are the most important group in terms of pounds of food in our diet, but in contribution to food energy the meat, poultry, and fish group, and the fats and oils, flour and cereal products, and sugar and sweeteners groups are all more important. About 42 percent by weight of the food we consume each year is of animal origin, including red meats, poultry, fish, eggs, and dairy products. The other 58 percent consists of plant products such as vegetables, fruits and melons, flour and cereal products, sugar and sweeteners, potatoes in various forms, fats and oils of vegetable origin, and miscellaneous other products including beverages. The foods of animal origin contribute about 36 percent of our energy supply, the foods of plant origin the remaining 64 percent.

Fresh and processed fruits, vegetables, potatoes, and sweet potatoes make up about 30 percent of the quantity of food we consume each year but contribute less than 10 percent to the food energy or calories in our diet. This is no doubt why people seeking to lose weight often turn to fruits and vegetables. This group of products is also an important source of many necessary nutrients, and as we shall see, also serves in many other ways to enhance our diet.

RECENT CHANGES IN OUR NATIONAL DIET

There have been many significant changes in our national diet over the past 30 or 40 years (Learn et al 1987, Levenstein 1988). Currently we are eating more chicken and fewer eggs, drinking more frozen orange juice and eating fewer fresh oranges, eating more lettuce and buying less fresh potatoes, drinking more lowfat milk and soft drinks and less whole milk and coffee (Putnam, 1989). The trends have not been entirely consistent during this period. Beef consumption, for example, rose rapidly during the 1960s and early 1970s then dropped sharply, and has since leveled off. Fresh fruit and vegetable consumption, which had been declining in the 1950s and 1960s, changed direction and began to increase in the early 1970s. These developments can be largely attributed to changes in life-style and socioeconomic conditions on the part of the consuming public, and improvements in the technology of production and distribution of these perishable products.

In the United States the period immediately following World War II was characterized by an increase in family formation, a shift toward suburban living, and a rapid rise in the birthrate that together brought about a whole new way of living. The proliferation of home appliances such as freezers and dishwashers coupled with the demands of raising a young family in the home placed an emphasis on convenience foods and casual living. The homemakers' day was busy enough caring for two or three children, and a major form of entertainment was a barbeque in the back yard. Canned and frozen foods enjoyed increased sales and TV dinners were introduced at this time. The baby boom lasted only from the late 1940s into the early 1960s, but this generation was destined to have a tremendous impact on the economy for many years to come.

The 1960s again gave rise to a profound change in eating habits and food consumption. The combination of the onset of the Vietnam War, the concern with civil rights, and the graduation of baby boomers into high school and college brought about significant economic and social change. Guns and butter could only be obtained with an increase in the workforce made possible by the greater participation of women, and the proportion of married women working outside the home shot up during this period. Young women flocked to professions or types of work that few women had entered before. Eating out became more attractive and economically possible, at fast food restaurants for families on the go and at full service restaurants for young couples and business people. Households where all members assembled for three meals a day, or even two, became increasingly scarce. Breakfast was strictly on the run, and lunch for many was eaten at school or at work.

By the early 1970s a new factor became a major influence on the food market. Added to the decline in the birth rate and the consequently smaller household size, the entry of baby boomers into the work force, the increased employment in service industries requiring less physical labor than manufacturing, the growing proportion of women working outside the home, and the consequent expansion in discretionary income that could be spent on new food products and new foodservices, now came a greater concern with health and physical fitness.

What sparked this interest in diet and health at this time is hard to say. Certainly the relationship between food consumption and exercise on general health had been known for many years. But the Surgeon-General's report on smoking and lung cancer in the late 1960s and the accumulation of medical evidence on cholesterol and heart disease, undoubtedly had an effect. Along with jogging and other forms of exercise many people, especially young adults, began to seek out what they considered a more healthful diet and avoided fattening or high cholesterol foods. This was to be a major influence on food consumption for the future.

In recent years the influx of people from other countries with different food preferences and eating habits and the interest of native born citizens in sampling the fare from other countries and cultures has greatly diversified our national diet. The growing number of people of Hispanic or Asian background and the proliferation of restaurants offering foods from countries all around the world is now a major consideration in food marketing. Food safety and quality in terms of chemical constituents, additives, and pesticides also has become a major concern.

These changes in life-style, food preferences, and socioeconomic conditions have been accompanied by an infusion of new technology in food production and marketing and a massive restructuring of the distribution system. Changes in consumer demand have been able to be realized largely because of improved methods of growing, harvesting, storing, and shipping perishable products. The response has also been made possible by better communications, reduced trade barriers, and more efficient management practices on the part of food marketing firms. Growth in the business of food distribution, especially of perishables, has retained and attracted many of the most progressive firms and individuals to this line of work.

Changes in Consumption of Some Individual Foods and Beverages

The steady and significant increase in the consumption of beef that occurred in the 1960s and early 1970s largely reflected the increasing affluence of the American consumer, the trend to casual living, and the greater frequency of eating out especially in fast food restaurants (Figure 1.4). But by the mid-1970s the increased concern with weight control and physical fitness, and for cholesterol and its impact on cardiovascular problems, is believed to have affected the market for beef and brought a significant reduction in beef supplies and consumption. By contrast the steady increase in poultry consumption over this period has been both supply and demand driven. Chicken, especially broilers, has benefited from major reductions in production costs, greater availability, and greater nutritional acceptance on the part of consumers.

Two major changes on the part of consumers are considered to have caused the consistent decline in the consumption of eggs in recent years. One, of course, has been the growing awareness of their high cholesterol content. The other, and possibly more important, has been the change in life-style that for many has relegated breakfast from the status of a formal meal of bacon and eggs to a bowl of cold cereal or a doughnut or bagel on the run.

Three other products whose consumption has also been affected by changes in consumer tastes and preferences are lettuce, frozen potatoes, and American cheese (Figure 1.5). Lettuce marketings, largely of the iceberg or crisphead type, benefitted from the expanding interest in low calorie foods and salads at least through the 1970s. The leveling off in consumption of

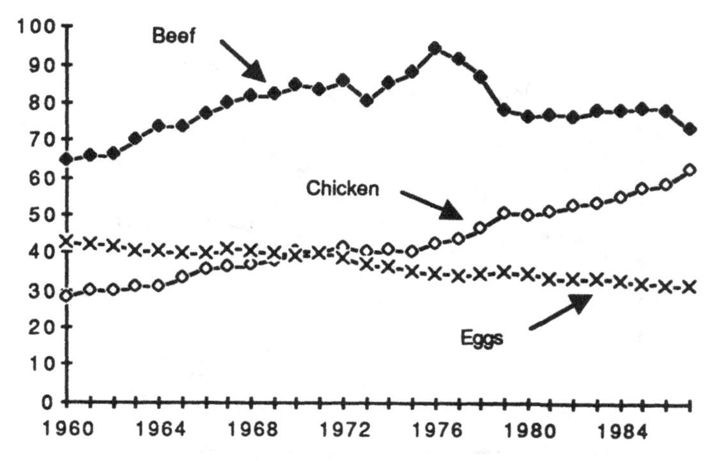

Figure 1.4. Per Capita Disappearance of Beef, Chicken, and Eggs, Pounds per Year, 1960–87. *Source:* Adapted from Putnam, 1989.

this type of lettuce in the 1980s, however, has apparently been due to consumers shifting demand to other types of salad greens and other more unusual fresh vegetables. Frozen potatoes have experienced a spectacular increase in consumption in recent years that could hardly be attributed to nutritional awareness but rather to their availability at fast food restaurants in acceptable quality at reasonable cost. The steady increase in consumption

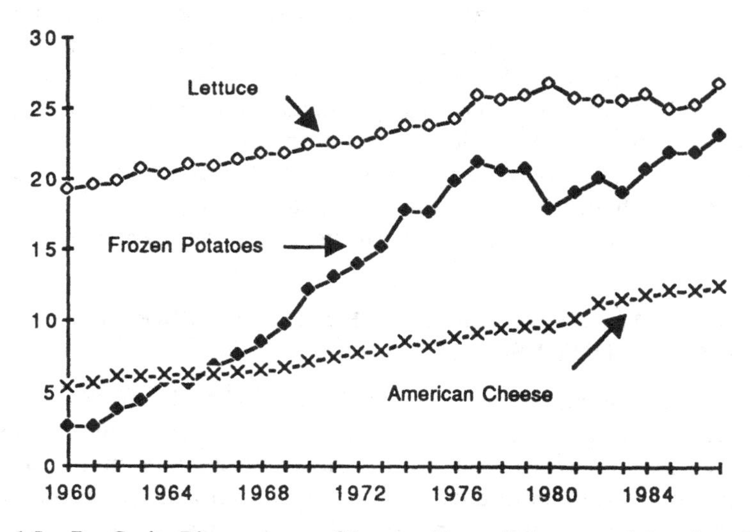

Figure 1.5. Per Capita Disappearance of Lettuce, Frozen Potatoes, and American Cheese, Pounds per Year, 1960–87. *Source:* Adapted from Putnam, 1989.

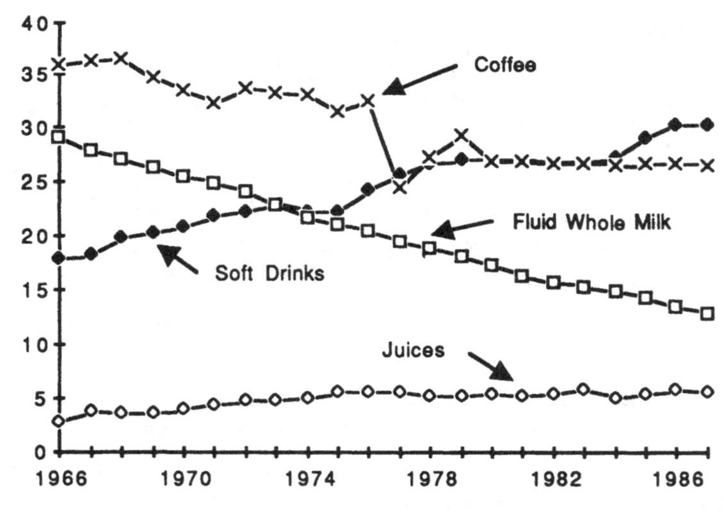

Figure 1.6. Per Capita Disappearance of Coffee, Fluid Whole Milk, Soft Drinks, and Juices, Gallons per Year 1966–87. *Source:* Adapted from Putnam, 1989.

of American cheese, although not as dramatic as frozen french fries, can also be attributed in part to fast food popularity.

The human digestive system is somewhat more flexible in the quantities of beverages it can handle than of solid foods. While the past few years have seen relative stability in the total quantity of food consumed, there have been some major changes in both total beverage consumption and in individual beverages (Figure 1.6). Dairymen are concerned with the continual decline in fluid whole milk consumption as consumers turned to lowfat milk drinks. Consumption of fruit juices has exhibited a consistent increase as supplies became available not only from this country but from around the world to fill the growing consumer demand. Among beverages, the most spectacular change has been the increased consumption of soft drinks, a growing portion of which are low calorie, that reflects a changing life-style.

INDIVIDUAL DIFFERENCES IN FOOD CONSUMPTION AND EXPENDITURES

Changes in annual average per capita availability fail to reveal the great differences in individual food consumption and expenditure patterns in this country. Although the amount of food available contains more than enough energy and essential nutrients for all of us, hunger is still a serious problem. Many young pregnant women also fail to consume sufficient minerals and vitamins. Ignorance as well as poverty still contributes to nutritional deficiencies. Excessive food consumption leading to obesity is considered by some to be the most serious nutritional problem in this country today. Excess weight is thought to contribute to diseases of the heart and circulatory system as well as to certain forms of cancer.

Within the realm of what are considered nutritionally acceptable diets, however, there is still a great diversity of food habits. Strict vegetarians forego animal products entirely and rely on fresh fruits, vegetables, grains, beans, and related products to obtain an adequate diet. Other people may subsist on very little fresh fruit or vegetables. While different eating habits are based on individual preferences and choices they are often linked to socioeconomic or demographic characteristics such as ethnic origin, income, education, type of employment, region of residence, urban–rural location, age, or household size. Some foods are eaten by almost everybody, while consumption of others is concentrated in a relatively small proportion. There are probably many reasons for these differences and they may be difficult to determine, but the more we know about them the more effective we can be in marketing.

A wealth of information on the variation in consumer food expenditures according to various socioeconomic characteristics is now provided annually by the Continuing Consumer Expenditure Survey (CCES) conducted by the Bureau of Labor Statistics. Average weekly per person expenditures of urban households, and the percentage of households purchasing food items, are available for many individual foods and food groups according to household type and size, region and city size, season, housing tenure, income quintile and class, race, householder's age, and number of earners.

The 1986 CCES found average weekly food expenditures per person amounted to $23.92, of which 37.7 percent was spent for food away from home and the remainder for food at home (Table 1.4). Even among the broad groups delineated by the Bureau, total weekly food expenditures varied considerably. Variation was greatest according to household type and size, income, and race, and considerable variation also existed according to housing tenure, age of householder, and region. Expenditures tended to be highest in households consisting of a married couple only, in smaller households, in households with higher income, in those headed by a person of white race, and in households headed by older householders. The percentage of weekly food expenditure for food away from home also varied widely among household groups within these categories. In households of only one member food expenditures amounted to $32.94 per week, of which 50.8 percent was spent for food to eat away from home. In households of six or more members average weekly food expenditures came to $15.60, of which only 23.1 percent was spent away from home.

Many of these household characteristics are, of course, related. There are

Table 1.4. Income and Average Weekly Food Expenditures Per Person of Urban Households According to Selected Household Characteristics, 1986.

SELECTED CHARACTERISTIC	ANNUAL INCOME PER PERSON	WEEKLY FOOD EXPENDITURES			PERCENT AWAY FROM HOME
		FOOD AT HOME	FOOD AWAY FROM HOME	TOTAL	
		dollars			percent
All Households	10,338	14.90	9.02	23.92	37.7
Household type					
Married couple, couple only	16,004	19.15	12.74	31.99	39.8
Single parent, female head, own children only	4,328	10.92	4.05	14.97	27.1
Household size					
One member	14,384	16.20	16.74	32.94	50.8
Six or more members	4,750	12.00	3.60	15.60	23.1
Region					
Northeast	11,242	16.68	10.17	26.85	37.9
South	10,473	13.68	8.76	22.44	39.0
Season					
Spring	10,084	15.04	9.28	24.32	38.2
Fall	10,468	15.60	9.13	24.73	36.9
Housing tenure					
Homeowner without mortgage	10,172	17.39	9.11	26.50	34.4
Renter	8,251	13.43	8.45	21.88	38.6
Income quintile					
Lowest fifth	2,583	13.44	5.44	18.88	28.8
Highest fifth	19,300	17.69	13.59	31.28	43.0
Race					
White	11,007	15.39	9.71	25.10	38.7
Black	5,581	11.71	4.57	16.28	28.1
Age of Householder					
Under 25 (non student)	7,840	11.47	10.34	21.81	47.4
Age 55-64	12,938	18.15	9.37	27.52	34.0
Number of earners					
One	10,210	14.93	9.98	24.91	40.1
Three	10,640	15.54	8.56	24.10	35.5

Source: Adapted from Smallwood, 1990.

wide differences in income per person between households of different type and size, as well as between households headed by persons of different race or of different ages. For example, households headed by white householders were smaller (2.47 compared to 2.8 persons) and had a greater degree of home ownership (62 compared to 41 percent), as well as a substantially higher annual income per person ($11,007 compared to $5,581) compared to households headed by black persons (Smallwood, 1990).

Individual demand for food is the result of the complex interactions of many factors that differ for each individual. This makes it difficult to generalize about food consumption patterns and impossible to predict what any one person will actually eat. Each of us possesses a set of food prefernces that we express in different ways, depending on our particular social and economic environment. Persons of similar ethnic background or demographic group tend to have somewhat similar food habits but still often exhibit many differences. Foods also can be thought of as having a set of attributes that contribute to satisfaction or utility differently for different people.

Individual food preferences are hard to identify and explain (Price, Broan, 1984). They may be the result of habits formed in childhood, or simply the wish to gratify desires based on current fads. There may be complex underlying psychological or physiological reasons for some food preferences. Concern for health and physical fitness coupled with nutritional knowledge can be important for some, as well as the desire to avoid injurious pesticides or additives. The type of job and the amount of physical labor involved can have a bearing on the amount and kind of foods consumed. Convenience is becoming more and more important in many households, and this influences not only in what form and where foods are purchased but also the kinds of foods that are eaten. Cost is still a major consideration for many, even though there may be some at upper income levels that can ignore the expense.

In order to target merchandising and promotion efforts more effectively, many food marketers seek to identify groups of individuals that may have particular food preferences and buying habits, on the basis of their socioeconomic characteristics (Mitchell, 1983). Combinations of age, employment, household composition, racial or ethnic background, and urbanization are often the basis for targeting particular foods or marketing services. Groups, for example, that have been recognized as innovators or trend setters in food buying are the young upwardly mobile professionals (yuppies) or couples with double income and no kids (dinks). Market researchers have even considered values and life-styles as a basis for differentiating potential customers into groups such as belongers, achievers, inner-directed, and need-driven, each differing in food buying behavior.

THE SEPARATE EFFECT OF INDIVIDUAL CHARACTERISTICS ON FOOD CONSUMPTION

Although it is clear that differences in income, household size and composition, race, age, and other factors are clearly associated with differences in food buying practices, the relationship of each individual factor is not readily apparent since many factors are closely associated. Income per person tends to be higher in smaller households headed by older persons. Blacks tend to live in larger households and have lower incomes than whites. The separate relationship of individual characteristics is difficult to determine but can be estimated using rigorous research methods.

Income followed by household size and composition are two of the most important demand factors that help explain food consumption variation among households (Blaylock and Smallwood, 1986). Using the 1980–81 Continuing Consumer Expenditure Survey Blaylock and Smallwood identified the separate association of income and household size and also included geographic region of household residence, season of the year, and race to improve the analysis. They found that income is a significant determinant of consumer expenditures for all food groups analyzed, except for eggs, milk and cream, and margarine. Higher income households spend more than their lower income counterparts on all food groups analyzed, all else held constant. In general, higher income households prefer beef and fish to pork and poultry, cheese to other dairy products, and butter to margarine. The product groups most responsive to a change in income are total food, food away from home, beef, fish, cheese, vegetables, butter, and miscellaneous foods.

Increased expenditure for individual foods and food groups comes about through both a higher proportion of households using the foods and by households buying more of these foods. Market entry, as compared to increased use, is more important for some foods than others, and more important for individual foods than for larger groups. For example, 37 percent of the demand response for meat, poultry, fish, and eggs is due to changes in the proportion of household consuming these foods, while 68 percent of the total demand response for fish is attributed to this factor. The response of increased expenditures to changing incomes tends to diminish as income rises partly because fewer nonusers are available to become potential market participants.

Household characteristics and factors other than income that were thought to influence consumer demand for food included household age composition, region of household residence, race, and season. The separate effect of each of these characteristics holding the others constant was analyzed. Older people spent less on foods eaten away from home but more on eggs, cereals and bakery products, fruits, vegetables (especially fresh), sugar and sweeteners, fats and oils, and prepared foods, and less on beef than do younger people.

Nonblacks spent about 11 percent more than blacks on total food but significantly more on dairy products, especially cheese, after accounting for differences in income, age of householder, region of residence, and season of the year. Blacks spent more on meat, poultry, fish, and eggs, and especially fish and poultry products. Regional and seasonal differences were identified that may be partly attributed to differences in prices that were not taken into account.

FUTURE TRENDS IN FOOD CONSUMPTION

Food buying and eating habits are bound to change in the future. To some extent these changes will be due to changes in socioeconomic and demographic characteristics that can be predicted with some degree of certainty. But they will also be influenced by improvements in technology, in production and distribution, international trade policy, and many other factors bringing changing values and life-styles.

Changes in living and eating habits are difficult to predict. The change in consumer demand of the early 1970s attributed largely to increased concern with health and physical fitness was not foreseen, and not even recognized by many until several years later. The current trend toward eating prepared food at home may be the wave of the future. Certainly there has been increased interest in carryout and home delivery. This may be partly due to the advent of the videocassette player and the availability of cassettes, illustrating once again how technology can shape our lives.

But there is no question that the size and composition of the population and the level and distribution of real income will still be important determinants of the total demand for food. Many people who will be part of the population 10 and 20 years from now are already here, so the size and composition of the population in the year 2000 can be forecast fairly accurately, given some assumptions regarding birth and death rates and emigration and immigration flows. Past trends provide a guide to general changes in real income but forecasts for specific years are risky because of the business cycles of recession and high economic activity.

Census projections indicate that our population will continue to grow in the next few years, but at a declining rate, growing from 250 million in 1990 to 268 million by 2000, and 283 million by the year 2010 (Table 1.5). The biggest changes in age groups will be in the youngest and the oldest categories. The percentage of people under 25 will decline from 36.0 percent in 1990 to 33.9 in 2000 and to 31.8 percent by the year 2010. Those over age

Table 1.5. Projections of Population by Age Group, Sex, and Race for 1990, 2000, and 2010.

	1990	2000	2010
Total Population (1,000s)	250,410	268,266	282,575
Percentage of Total			
Under 5	7.4	6.3	6.0
5–17	18.2	18.2	16.2
18–24	10.4	9.4	9.6
25–34	17.5	13.8	13.3
35–44	15.1	16.4	13.2
45–54	10.2	13.9	15.8
55–64	8.5	9.0	12.5
65–74	7.3	6.8	7.4
75 and over	5.3	6.2	6.5
Male	48.4	48.9	49.0
Female	51.2	51.1	51.0
White	84.1	82.6	81.0
Black	12.4	13.1	13.7
Other Races	3.5	4.3	5.3

Source: Bureau of the Census, 1989.

65 will increase from 12.6 in 1990 to 13.9 percent in the year 2010. In 1990 the baby boomers born from the late 1940s to the early 1960s were in the 25 to 44 age group and constituted 32.6 percent of our population. Twenty years later they will be in the 45 to 64 age group and still account for 28.3 percent of the total. Women will still outnumber men, but the margin is expected to narrow.

The racial mix is projected to change also. The percentage of whites will decline from 84.1 percent in 1990 to 81.0 percent in 2010 if present trends continue. Blacks will increase from 12.4 to 13.7 percent, and those of other races will increase from 3.5 percent to 5.3 percent. The numbers of those of Hispanic origin will continue to increase faster than any other ethnic or racial group. From under 20 million in 1990 the number is expected to exceed 30 million by the year 2010, a growth of more than 50 percent.

Regional shifts in the population come about through shifts in economic activity and the changing age distribution. High oil prices and many retirements brought migration to the South and West a few years ago, but now a stronger automobile business keeps people in the North Central region. The South and West, however, are likely to continue to attract a larger share of the population.

The U.S. Department of Agriculture, using Census Bureau projections

Table 1.6 Projections of National Food Spending Based on Expected Changes in Demographics and a 2 Percent Per Year Increase in Real Income, 1980 = 100.

	1990	2000	2010	2020
		index 1980 = 100		
Total Food	119.0	138.9	160.0	181.8
Food Away From Home	123.6	148.9	177.8	212.2
Food at Home	115.7	131.7	147.7	161.7
Meat, Poultry, Fish, Eggs	116.2	133.2	150.2	164.1
Cereals and Bakery Products	113.9	127.9	140.5	150.7
Dairy Products	113.5	126.5	138.6	149.1
Fruits	116.4	134.0	154.3	176.9
Fresh	116.3	134.9	157.1	183.6
Processed	116.8	133.1	149.8	165.3
Vegetables	117.6	135.4	153.4	169.0
Fresh	118.0	136.4	155.2	172.2
Processed	116.5	133.1	148.9	161.6
Sugars and Sweeteners	113.9	127.8	139.4	149.4
Nonalcoholic Beverages	113.7	127.3	139.7	149.5
Fats and Oils	115.5	130.8	145.6	158.1
Miscellaneous	116.1	131.4	146.7	160.9

Source: Blaylock, Smallwood, 1986.

of population and a 2 percent rate of increase in real income per year has estimated that total spending for food in 1990 will be 19 percent above 1980, by 2000 will go up 39 percent, and by the year 2020 will be 82 percent over 1980 (Table 1.6).

The biggest growth in spending will be for food away from home which is expected to increase by 112 percent by 2020 compared to food for home use which will increase only 62 percent. Among the groups of foods for home consumption the greatest increase will be in fresh fruits at 84 percent and fresh vegetables with a 72 percent increase, followed by the meat, poultry, fish, and eggs group with 64 percent. Dairy products will be among the slower growing groups, with a projected increase of only 49 percent by 2020 over 1980.

Many factors may modify these projections. These would include the introduction of new technology that makes certain products more attractive to consumers, more readily available, or more affordable. Merchandising and promotion efforts either of private brand products or of industry generic programs will undoubtedly influence the acceptance of specific foods in the future.

REFERENCES

Blaylock, James R., David M. Smallwood, 1986. *U.S. Demand for Food: Household Expenditures, Demographics, and Projections.* U.S. Department of Agriculture, Economic Research Service, Technical Bulletin No. 1713.

Bureau of Labor Statistics, 1989. *Handbook of Labor Statistics: 1989.* U.S. Department of Labor, August.

Bureau of the Census, 1989. *Statistical Abstract of the United States: 1989.* (109th edition). U.S. Department of Commerce, January.

Dunham, Denis, 1989. *Food Cost Review, 1988.* U.S. Department of Agriculture Economic Research Service, Agricultural Economic Report No. 615.

Learn, Elmer, Gordon A. King, Robert Sommer, Desmond A. Jolly, Shaun E. Lakin, 1987. Demographic Shifts and Other Factors Affecting Demand. In: *Marketing California Specialty Crops: Worldwide Competition and Constraints.* University of California, Agricultural Issues Center.

Levenstein, Harvey A., 1988. *Revolution At the Table: The Transformation of the American Diet.* New York: Oxford University Press.

Mitchell, Arnold, 1983. *The Nine American Lifestyles: Who We Are and Where We Are Going.* New York: Macmillan.

Price, Charlene C., Judy Broan, 1984. *Growth in the Health and Natural Foods Industry.* U.S. Department of Agriculture, Economic Research Service, Staff Report No. AGES840501.

Putnam, Judith Jones, 1989. *Food Consumption, Prices, and Expenditures 1966–87.* U.S. Department of Agriculture, Economic Research Service, National Economics Division, Statistical Bulletin No. 773.

Smallwood, David M., 1990. *Food Spending in American Households, 1980–86.* U.S. Department of Agriculture, Economic Research Service, Statistical Bulletin No. 791.

U.S. Department of Agriculture, Economic Research Service, 1989. *Food Marketing Review, 1988.* Agriculture Economic Report No. 590.

——— 1989. *National Food Review,* Volume 12 Issue 2 April–June.

Chapter 2

The United States Market for Fresh Fruits and Vegetables

INTRODUCTION

The United States market for fresh fruits and vegetables has been radically transformed over the past 40 years. Changes in population, in economic conditions, in life-styles, and eating habits have had their impact on the quantity of these products produced and consumed, where and in what form they were purchased, and the prices consumers paid. In recent years there has been a great expansion in sales of fresh fruits and vegetables, especially of the unusual and formerly unfamiliar ones. While the market for most products has expanded, it has declined for others. Fresh fruits and vegetables today enjoy great popularity among some segments of the population. Prospects are for increased total consumption of a wide variety of fresh produce items, especially unusual and exotic ones, in the future.

CHANGING FRESH FRUIT AND VEGETABLE CONSUMPTION

The period following World War II saw some major changes in food habits, especially in fruit and vegetable consumption (Weimar, Stevens, 1974). For the first 25 years or so the market for fresh fruits and vegetables generally declined. Consumers shifted to canned and frozen fruits and vegetables, prompted by the introduction of new technology in manufacturing and distribution and the wider availability of home freezing and refrigeration units. For many products the declining consumption was the continuation of trends that had existed for many years. Year to year changes in consumption can often be blamed on the vagaries of the weather, but persistent changes usually reflect adjustments in underlying economic forces.

The decline in fresh utilization was more severe for some products than others. Sales of fresh oranges and grapefruit dropped in part due to changing breakfast habits but also to the introduction of frozen concentrated orange juice. Fresh apples suffered in competition with canned applesauce and apple slices. Fresh potatoes were hit on two fronts—the time and effort required to have them boiled or mashed, and the availability and acceptability of frozen french fries and potato chips. But some fresh fruits and vegetables that were minor in terms of per capita consumption enjoyed increased popularity during this period. These included strawberries, cherries, pineapples, broccoli, cucumbers, and peppers.

Somehow during the early 1970s, as discussed in the previous chapter, several factors combined to bring about major changes in consumption patterns that amounted in many cases to the reversal of long-time trends. On the demand side these factors included increased interest in nutrition and physical fitness, the shift to a service economy and less heavy physical labor, and rising real income in part due to the greater participation of women in the work force. On the supply side many improvements in growing, harvesting, and handling fresh products, especially those of a more perishable nature, were coupled with improved presentation in retail food stores and public eating places. What triggered the change in direction at this particular time is difficult to determine.

Whatever the cause, the per capita utilization of fruit, both fresh and processed, has increased substantially since the early 1970s (Figure 2.1). Sales of fruit in fresh form have risen from about 75 pounds per person to almost 100 today. Frozen citrus continued to gain in popularity. The quan-

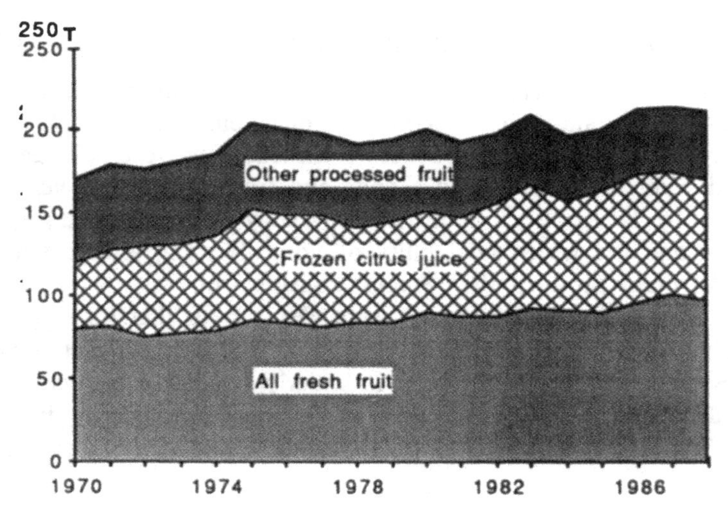

Figure 2.1. Per Capita Utilization of Fruit in Pounds, Fresh Weight Equivalent, 1970–88. *Source:* Adapted from USDA ERS, 1989b.

tity of citrus fruit used for frozen juice has escalated from less than 50 pounds 20 years ago to almost 80 pounds in a recent year. Fruit used for other processed products has declined from about 50 pounds to 40 pounds over this period.

Again the individual fruits have fared differently in the fresh market in the last few years (Table 2.1). Among the citrus fruits, oranges and grapefruit have continued to decline in fresh use while lemons and limes, and tangerines and tangelos have increased.

Utilization of most of the noncitrus fruits has increased. Bananas, the leader in per capita use, has increased its lead with a substantial gain. Use of apples, as with many other fruit, varies with the crop size, and the apple

Table 2.1. Changes in Per Capita Utilization of Fresh Fruits and Melons, Retail Weight, 1970–72 to 1985–87.

	PER CAPITA UTILIZATION 1970–72 TO 1985–87		PERCENT CHANGE 1970–72 TO 1985–87
	pounds		percent
Citrus Fruits			
Oranges	15.2	13.3	−12.5
Grapefruit	8.2	6.1	−25.6
Lemons and Limes	2.1	3.1	+47.6
Tangerines and Tangelos	2.2	3.0	+36.4
Total Citrus	27.5	25.5	−7.3
Noncitrus Fruits			
Bananas	17.9	24.7	+38.0
Apples	15.8	18.1	+14.6
Peaches	4.8	4.2	−12.5
Grapes	2.2	6.7	+204.5
Pears	2.1	2.9	+38.1
Plums and Prunes	1.2	1.6	+33.3
Strawberries	1.6	2.9	+81.2
Avacadoes	0.7	1.6	+128.6
Pineapples	0.7	1.6	+128.6
Nectarines	0.7	1.4	+100.0
Cherries	0.5	0.5	0
Other Noncitrus*	0.5	1.1	+120.0
Total Noncitrus	48.3	59.6	+23.4
Total Fruit	75.9	92.7	+22.1
Melons			
Watermelons	12.5	na	na
Cantaloupes	6.9	na	na
Honeydews	0.9	2.2	+144.4

*Apricot, figs, cranberries, papayas, kiwifruits.
Source: Adapted from Putnam, 1989.

crop tends to vary greatly from year to year. But average fresh use has been higher in recent years than in the earlier period. Peaches are a major crop recording a decline in use, possibly due to the difficulty of providing consumers with the texture and flavor they expect. Use of many other of the minor noncitrus fruits has more than doubled over the period from the early 1970s.

Among the melons official estimates of use are only available for honeydews, which have shown a sharp increase. The market for watermelons and cantaloupes is still large, but how large is not officially measured.

Total per capita utilization of commercially produced vegetables, not including potatoes, rose from just over 175 pounds in 1970–72 to more than 200 pounds in 1986–88 (Figure 2.2). There are many different vegetables marketed in fresh and processed form, and production and use is only recorded for a few of these in selected locations. Of the 10 currently tracked, fresh use went from 70 pounds to 98 pounds, or from 40 to 50 percent of the total. The quantity per person going for canning dropped from 95 to 85 pounds in this period, while that directed to freezing increased from 13.4 to 16.7 pounds.

Changes in fresh vegetable use also differed between vegetables (Table

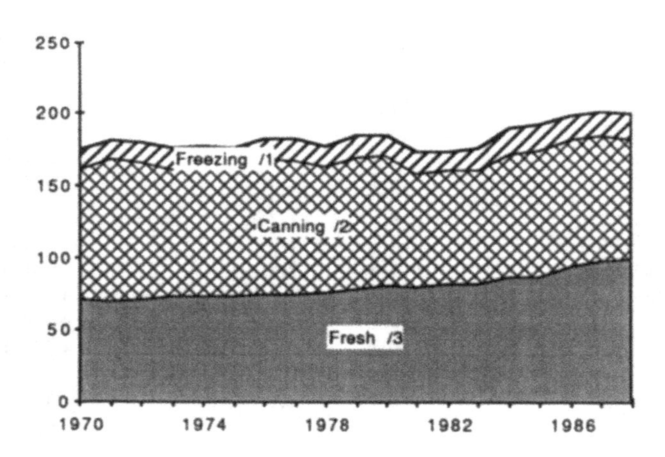

[1] Includes asparagus, snapbeans, carrots, sweet corn, green peas, pickles and tomatoes
[2] Includes asparagus, broccoli, carrots, cauliflower, celery, sweet corn, lettuce, onions, tomatoes, and honeydews
[3] Includes asparagus, snapbeans, broccoli, carrots, cauliflower, sweet corn, and green peas

Figure 2.2. Per Capita Utilization of Commercially Produced Vegetables, Fresh Weight Equivalent, 1970–88/1. *Source:* Adapted from USDA ERS, 1989a.

2.2). The greatest gains in pounds were recorded for lettuce, tomatoes, onions and shallots, and carrots with use of each increasing between 6 and 7 pounds per person. Percentagewise the greatest gains were for broccoli and cauliflower. Honeydews were reported with melons in Table 2.1 as well as Table 2.2. Absence of official data makes it impossible to determine what happened for major vegetables such as cabbage, cucumbers, green peppers, green beans, and spinach.

Between 1970 and 1987 the quantity of potatoes used per person on a farm equivalent basis has remained at about 120 pounds per year, varying up or down depending on the size of the crop (Figure 2.3). How these potatoes were used, however, has changed in a fairly consistent fashion. Fresh or table stock use has declined from an average of 58 pounds in 1970–72 to 48 pounds in 1985–87. Most of this decline took place in the early 1970s. Use now fluctuates from year to year without showing much upward or

Table 2.2. Changes in Per Capita Utilization of Fresh Vegetables and Potatoes, 1970–72 to 1985–87.

	PER CAPITA UTILIZATION		PERCENT CHANGE
	1970–72	1985–87	1970–72 TO 1985–87
	pounds		percent
Fresh Vegetables			
Lettuce	20.8	26.6	+27.9
Tomatoes	10.2	17.3	+69.6
Onions and Shallots	11.9	17.6	+47.9
Carrots	6.0	12.5	+108.3
Cabbage	8.3	na	na
Celery	6.7	7.3	+9.0
Sweet Corn	7.1	7.1	0
Cucumbers	2.9	na	na
Green Peppers	2.3	na	na
Honeydews	0.9	2.2	+144.4
Broccoli	0.6	3.8	+533.3
Green Beans	1.5	na	na
Cauliflower	0.6	2.8	+366.7
Spinach	0.4	na	na
Asparagus	0.6	0.6	0
Total 10 Fresh Vegetables	70.9	98.0	+38.2
Potatoes			
Irish Potatoes	58.8	47.7	−18.9
Sweet Potatoes	5.0	4.8	−4.0

na Not available
Source: Adapted from Putnam, 1989.

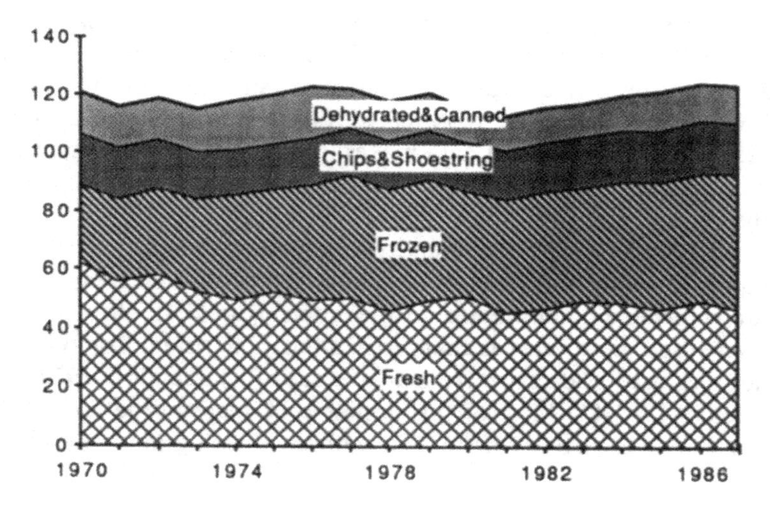

Figure 2.3. Per Capita Utilization of Potatoes, Pounds per Year Farm Weight, 1970–87. *Source:* Adapted from Putnam, 1989.

downward trend. The quantity going for frozen products, mainly french fries, has increased fairly steadily from 28 to 45 pounds. Use for chips and shoestrings has increased from about 17 pounds to 18 pounds over this period, with considerable year to year fluctuation. Dehydration and canning use took about 14 pounds in 1970–72, declining to 12 in 1985–87. Use of sweet potatoes has changed very little, remaining close to 5 pounds per person annually.

CHANGES IN EXPENDITURES

The expansion in sales of fresh fruits and vegetables that took place since the early 1970s was not the result of relatively low prices for this group of commodities. In fact prices of a fixed group of fresh fruit and vegetable items in what is called the market basket of food products originating on U.S. farms rose higher than did fixed groups of other major commodities (Figure 2.4). On an index based on 1982–84 = 100 the retail cost of the market basket group of fresh fruit and vegetables had risen by 1987 to 126.8 compared to 108.9 for processed fruits and vegetables, 109.7 for meat products, and 91.5 for eggs, all of which had experienced declining consumption. Prices for another commodity group, fish, which also had enjoyed increased market utilization had also risen substantially to 129.9 by 1987.

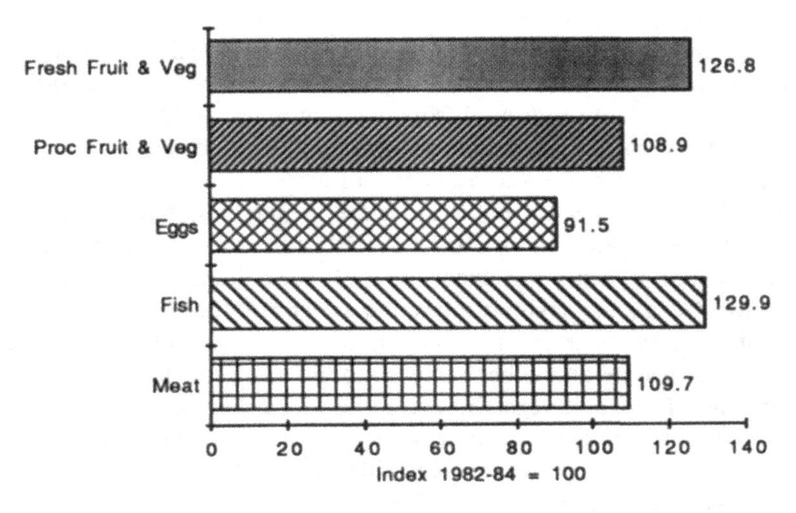

Figure 2.4. Consumer Price Index (1982–84 = 100) for Selected Food Groups, 1987. *Source:* Adapted from Putnam, 1989.

INDIVIDUAL VARIATION IN FRESH FRUIT AND VEGETABLE CONSUMPTION

Individuals differ considerably in their consumption of fresh fruits and vegetables, as they do with other foods. There are great differences in what items and in what quantities people buy, where and when they buy them, and how they use them in the home. Being able to identify the purchase characteristics of different segments of the population can, as we have already seen, be important in targeting merchandising and promotion efforts. Public and private agencies continue to gather information that can enhance our knowledge of group differences in purchasing patterns.

The 1977–78 National Food Consumption Survey

The most comprehensive data on household food consumption and nutrition come from the U.S. Department of Agriculture's Nationwide Food Consumption Surveys (NFCS) which are conducted about every 10 years. The latest published NFCS at time of writing was the survey conducted in 1977–78. This sample consisted of approximately 15,000 households chosen to represent the people of the 48 coterminous states. Households were surveyed between April 1977 and March 1978 with approximately equal numbers reporting in each of the four seasons. Information on household characteristics and food use was obtained through personal interview. Two kinds of information were obtained: (1) 1-week recall of kinds, quantities, values, and sources of foods used from home supplies, and (2) an individual intake record of each household member listing the kinds and quantities of foods eaten both at home and away from home.

Expenditures per person on fresh fruits, vegetables, and potatoes by households participating in the 1977–78 NFCS amounted to $1.30 per week. Of this amount 60 cents was spent on fresh vegetables, 56 cents on fresh fruit, and 14 cents on fresh potatoes and sweet potatoes (Table 2.3). Purchases of fruits and vegetables by individuals and households tended to vary according to income and other socioeconomic characteristics. Consumers also obtained fruits and vegetables from home gardens and as gifts.

Consumers in the highest income quintile spent more than those at lower income levels for fresh fruits and vegetables but the least for fresh potatoes to eat at home. Expenditures for fresh fruits and vegetables per person were highest in the Northeast, but expenditures for fresh potatoes were greatest in the South. Consumers in the central cities faced higher prices, as might be expected, and spent the most for fresh produce. Nonwhite/nonblack people spent more than those of other races for fresh fruits and vegetables, and blacks spent the most for fresh potatoes. Expenditures for fresh vegetables were greatest in the spring, and for fresh fruits greatest in the summer. The smaller the household the greater the expenditure per capita for fresh fruits, vegetables, and potatoes.

Many of these factors are interrelated. The average income of white people was higher than for black and nonblack/nonwhite people. Income per person tends to be higher in smaller households than in larger ones. Those living in suburbia tend also to have higher incomes than those in the central city or nonmetropolitan areas.

Rigorous statistical analysis is necessary to isolate and measure the relationships between food consumption and expenditures and separate individual characteristics, apart from the effect of other variables. The impact of income and other household characteristics on per person expenditures for fruits, vegetables, and potatoes was analyzed using data from the 1977–78 NFCS (Smallwood, Blaylock, 1984). Information was obtained on the relationship of income and other household characteristics to (1) changes in the proportion of consumers using the product, and (2) changes in the level of expenditures by those already using the item. The household characteristics analyzed included income, household size and age composition, region and urban location of the household, race, season of the year, and participation in the federal food stamp program.

For purposes of analysis, vegetables, fruits, and potatoes were grouped

Table 2.3. Weekly Expenditures Per Person for Fresh Fruits, Vegetables, and Potatoes for At-Home Use by Selected Demographic Groups 1977–78.

DEMOGRAPHIC GROUP	FRESH FRUIT	FRESH VEGETABLES	FRESH POTATOES	TOTAL
			dollars	
All	.60	.56	.14	1.30
Income Quintile				
I—lowest	.64	.54	.17	1.35
II	.60	.54	.16	1.30
III	.54	.51	.13	1.18
IV	.56	.55	.12	1.23
V—highest	.68	.65	.12	1.45
Region				
Northeast	.73	.67	.14	1.54
North Central	.94	.55	.14	1.63
South	.54	.43	.16	1.13
West	.71	.68	.11	1.50
Urbanization				
Central City	.74	.67	.15	1.56
Suburban	.62	.57	.13	1.32
Nonmetropolitan	.46	.46	.14	1.06
Race				
White	.60	.57	.13	1.30
Black	.61	.50	.17	1.28
Nonwhite/Nonblack	.75	.69	.13	1.57
Season				
Spring	.66	.56	.14	1.36
Summer	.61	.69	.13	1.43
Fall	.55	.49	.14	1.18
Winter	.60	.51	.14	1.25
Household Size				
1 person	.99	.90	.16	2.05
2	.82	.73	.16	1.71
3	.64	.57	.14	1.35
4	.53	.50	.13	1.16
5	.46	.47	.12	1.05
6 or more persons	.40	.41	.12	.93

Source: Smallwood, Blaylock, 1984.

into 32 categories. Vegetables were subdivided into fresh, canned, frozen, juice, and dried, and fresh vegetables were further subdivided into dark green, deep yellow, light green, tomatoes, and others. Fruits were divided into fresh, canned, frozen, juice, and dried. Fresh fruits were subdivided into citrus, other vitamin C, and other. Fruit juice was subdivided into fresh, canned, and frozen. Potatoes, including sweet potatoes, were divided into fresh, canned, frozen, dehydrated, and chips, sticks, and salads. The following are some of the highlights of this study.

Analysis found that a 10 percent increase in income, apart from other socioeconomic and demographic factors, was associated with a 1.5 percent increase in expenditures for fresh vegetables, a 1.9 percent increase in expenditures for fresh fruits, but a 0.6 percent decline in expenditures for fresh potatoes and sweet potatoes (Table 2.4). About half of this response was due either to people buying fresh fruits or vegetables that had not bought them during the survey week, or people not buying potatoes or sweet potatoes that had bought them during the survey week.

The expected response to changes in income was about the same for each individual group of fresh fruits and vegetables except for the group consisting of fresh fruits high in vitamin C other than citrus. A 10 percent increase

Table 2.4. Fresh Fruit, Vegetable, and Potato and Sweetpotato Expenditure Response Associated wtih a 10 Percent Increase in Income, 1977–78.

	WEEKLY EXPENDITURE PER PERSON	EXPENDITURE RESPONSE			SHARE OF TOTAL RESPONSE DUE TO MARKET ENTRY
		TOTAL	MARKET ENTRY	EXPEND. LEVEL	
	cents		percent		percent
Vegetables, Fresh	60	1.51	.67	.84	45
Dark green[1]	7	1.57	1.20	.36	76
Deep yellow[2]	4	1.93	1.41	.52	73
Light green[3]	20	1.64	.90	.74	55
Tomatoes	10	1.72	1.22	.50	71
Other[4]	20	1.90	1.10	.80	58
Fruits, Fresh	56	1.90	.93	.97	49
Citrus[5]	12	2.19	1.56	.63	71
Other vitamin C[6]	5	5.64	4.78	.89	85
Other[7]	39	1.70	.91	.79	54
Potatoes, Fresh, including sweet	14	−.62	−.34	−.28	55

[1] Includes collards, spinach and related greens, broccoli, and peppers.
[2] Includes carrots, carrots and peas, pumpkin, and winter squash.
[3] Includes asparagus, lima beans, snap beans, wax beans, kidney beans, cabbage, lettuce, okra, peas, artichokes, soybeans, bean curd, and brussel sprouts.
[4] Includes celery, cucumbers, onions, garlic, leeks, beets, cauliflower, corn, turnips, eggplant, mushrooms, radishes, summer squash, and mixed vegetables.
[5] Includes grapefruit, lemons, limes, oranges, and other citrus fruit.
[6] Includes canteloupes and papayas, muskmelon, strawberries, mangoes, guava, currants, and persimmons.
[7] Includes apples, bananas, berries, cherries, melons other than canteloupes, peaches, pears, pineapple, and other fruit.
Source: Adapted from Smallwood, Blaylock, 1984.

in income was associated with a 5.6 percent increase in expenditures for this group which consisted of canteloupes and papayas, muskmelons, strawberries, mangoes, guava, currants, and persimmons. For this group 85 percent of the expenditure response was considered due to an increase in the number of people buying the item, and only 15 percent to greater expenditures by those already purchasing the item. The study specified that response would decline as income and probability of use increased.

The separate effect of income, region of the country, urbanization, race, season of the year, age group, and food stamp status was also determined. The study concluded that people with annual per capita incomes of $8,000 would spend almost 30 percent more on fresh fruit, about 23 percent more on fresh vegetables, but almost 8 percent less on fresh potatoes than those with annual incomes of only $2,000 (Table 2.5).

People in the Northeast, other things equal, spent more for fresh fruits and vegetables than people in the North Central and South, but about the same as those in the West. There were greater differences in purchases of subgroups. People in the North Central region spent 49 percent less for dark green vegetables and 45 percent less for tomatoes than did those in the Northeast, while people in the South spent 37 percent less for dark green and deep yellow vegetables. The biggest regional differences in fresh fruit consumption were between people in the South who as a group spent 44 percent less for citrus than did those in the Northeast, while people in the West spent 62 percent more than those in the Northeast for fruits high in vitamin C other than citrus.

Fresh fruit and vegetable expenditures also varied according to urbanization, but not fresh potato expenditures. People living in central cities spent more for fresh fruits and vegetables than those in suburban and nonmetropolitan areas. People living in nonmetropolitan areas, who may have had better access to gardens and to less expensive produce, spent 49 percent less for dark green vegetables, 41 percent less for high vitamin C fruit other than citrus, and 36 percent less for tomatoes.

After adjusting for income and other factors the study found that white and black people spent about the same amount per week for fresh fruits, vegetables, and potatoes but allocated their money differently. Blacks spent 143 percent more for the dark green vegetables but 48 percent less for the deep yellow group. They also spent 26 percent more for fresh citrus, but 56 percent less for the noncitrus fruits high in vitamin C.

There was a seasonal pattern in expenditures that differed for fresh vegetables, fresh fruits, and potatoes which was probably due to seasonal variation in prices and availability. Expenditures for fresh vegetables tended to be highest in the Spring and moderately lower in the Fall, with Summer and Winter expenditures in between. Spending for tomatoes in the Fall was 31

Table 2.5. Simulated Weekly Expenditures Per Person for Fresh Fruits, Vegetables, and Potatoes According to The Separate Effect Of Various Household Characteristics, 1977-78.

HOUSEHOLD CHARACTERISTIC	FRESH FRUIT		FRESH VEGETABLES		FRESH POTATOES	
	EXPEND.	% OF BASE	EXPEND.	% OF BASE	EXPEND.	% OF BASE
	cents	percent	cents	percent	cents	percent
Income/capita						
$2,000	57.9	100.0	67.4	100.0	15.8	100.0
4,000	66.3	114.5	75.0	111.3	15.1	95.8
6,000	71.4	123.4	79.6	118.1	14.8	93.4
8,000	75.2	129.9	83.0	123.1	14.5	91.7
Region						
Northeast	75.0	100.0	85.0	100.0	15.7	100.0
North Central	64.6	86.1	62.2	73.2	14.7	93.8
South	54.3	72.4	69.4	81.7	16.8	106.7
West	74.4	99.2	83.0	97.7	12.5	79.6
Urbanization						
Central City	73.2	100.0	86.0	100.0	16.0	100.0
Suburban	63.5	86.8	73.2	85.1	15.2	95.0
Nonmetropolitan	56.9	77.8	53.6	72.5	14.7	91.7
Race						
White	63.8	100.0	72.2	100.0	15.0	100.0
Black	63.3	99.2	78.0	106.5	17.0	113.3
Nonwhite/ nonblack	80.6	126.4	90.3	125.1	17.0	113.4
Season						
Spring	64.1	100.0	78.6	100.0	14.9	100.0
Summer	78.7	122.8	75.6	96.2	13.8	92.8
Fall	58.1	90.6	67.1	85.4	14.2	95.1
Winter	58.6	91.4	72.9	92.8	16.5	110.9
Age Group						
0-2	50.0	72.1	40.4	49.3	11.7	64.4
3-12	70.1	101.1	65.1	79.5	12.7	70.3
13-19	61.6	88.9	66.3	81.0	11.5	63.6
20-39	53.2	76.8	70.4	85.9	12.9	71.0
40-64	69.3	100.0	81.9	100.0	18.1	100.0
65 and over	75.3	108.7	80.5	98.3	19.9	110.0
Food Stamp Status						
Nonrecipients	64.4	100.0	72.8	100.0	15.0	100.0
Recipients	62.1	96.4	79.4	109.1	18.5	123.1

Source: Adapted form Smallwood, Blaylock, 1984.

percent below the Spring outlay, probably due to the availability of home grown supplies. Fresh fruit expenditures were highest in the Summer and lowest in the Fall and Winter. Spending for fresh citrus in the Summer was 51 percent below the Spring, but 29 percent above the Spring level in the Winter.

Persons in the age 40–64 group spent more per person for fresh vegetables on an adult equivalent basis than did people in other age groups. Spending for infants up to 2 years old on fresh vegetables was lower than for others even taking into account relative food needs. But patterns varied: Persons 65 and over spent about 25 percent more for deep yellow vegetables than did the 40–64 age group. Older folk also spent more for fresh fruits, especially for the noncitrus high vitamin C group for which they spent 85 percent more than the 40–64-year-old group.

Food stamp recipients, after taking other personal and household characteristics into account, spent a little less for fresh fruits and a little more for fresh vegetables than did nonrecipients but substantially more for fresh potatoes. Food stamp recipients especially favored dark green vegetables and fresh citrus compared to nonrecipients.

Some of these differences may not seem large. The comparison, however, is between groups that differ only in a single major socioeconomic or demographic characteristic. The differences magnify, of course, when several characteristics are involved. For example, high income black people over 65 who live in the Northeast are likely, according to this study, to eat a lot more fresh citrus than do low income white people aged 20–39 who live in nonmetropolitan areas in the South.

FUTURE PROSPECTS

Prospects for the future have to be based largely on the projection of past trends. Per capita utilization of fresh fruits and vegetables in the United States has been increasing steadily in recent years, and all indications are that this will continue. The prospects are especially favorable for specialty items such as bamboo shoots, snow peas, bok choy, jicama, and papayas to name only a few (Beck, 1984).

In an effort to serve the market more effectively many marketing firms try to identify particular consumer groups that are heavy buyers of their products, or are good potential customers. The results of such studies are seldom released to the public. But in addition to the U.S. Department of Agriculture and the land-grant colleges several commercial organizations and trade associations conduct studies and make their findings available for a fee. The Vance Publishing Company, for example, has funded major research on consumer buying practices with respect to fresh fruits and vegetables that identifies frequency of purchase and attitudes toward fresh fruits and vegetables. The fourth survey, conducted by a Chicago based research firm in October 1989, yielded 1,260 responses from a nationally representative sample of U.S. consumers (The Packer, 1990). Full statistical data and reports are contained in *Fresh Trends 1990, A Profile of Fresh Produce Consumers.* Separate reports have been prepared on Fresh Fruits, Fresh Vegetables, Shopping for Fresh Produce, and Fresh Produce Nutrition and Industry Issues. The percentage of people in the 1989 study reporting increased consumption of fresh fruits and vegetables was up substantially over the previous year.

A New York City based research firm, FIND/SVP, The Information Clearinghouse Inc., completed a second fresh produce study in May 1989 (The Packer, 1990). The study was called *The Fresh Produce Market, A Competitive Intelligence Report,* and focused largely on emerging industry trends and issues, such as branded and organic produce. As reported in *The Packer Focus* (1990) the FIND/SVP study predicts an annual growth rate in the industry of 1 to 3 percent, and the challenge to growers of an oversupply of fresh produce in the United States that may lead to depressed prices.

The industry has expanded rapidly during a period of highly favorable economic conditions. It appears likely that growth will continue, but economic conditions may not be as favorable and successful operation may require even greater skill and ability than in the past.

REFERENCES

Beck, Bruce, 1984. *Produce: A Fruit and Vegetable Lover's Guide.* Photographs by Andrew Unangst. New York: Friendly Press.

Nutrition Marketing Task Force, 1988. PMA Nutrition Marketing Resource Produce Marketing Association, Newark, Delaware.

Putnam, Judith Jones, 1989. *Food Consumption, Prices, and Expenditures, 1966–87* U.S. Department of Agriculture, Economic Research Service, Statistical Bulletin No. 773.

Smallwood, David M., James R. Blaylock, 1984. *Household Expenditures for Fruits, Vegetables, and Potatoes.* U.S. Department of Agriculture, Economic Research Service, Technical Bulletin No. 1690.

The Packer, 1990. *The Packer: Focus.* Lincolnshire IL: Vance Publishing.

U.S. Department of Agriculture, Economic Research Service, 1989a. *Vegetables and Specialties Situation and Outlook Report,* TVS–248 August.

——— 1989b. *Fruit and Tree Nuts Situation and Outlook Yearbook,* TFS–250 August.

Weimer, Jon, Patricia Stevens, 1974. *Consumers' Preferences, Uses, and Buying Practices for Selected Vegetables: A Nationwide Survey.* U.S. Department of Agriculture Marketing Research Report No. 1019.

Chapter 3

Sources of Fresh Fruits and Vegetables

INTRODUCTION

Our fresh fruits and vegetables come from many different sources. The seasonal availability, quality, and cost of product from each source differs, and these differences are continually changing. Growers and shippers, wherever they are, need to be aware of their competitive position relative to other sources, while wholesalers and retailers must keep abreast of the changing sources of supply. Sourcing, as it is called, is one of the most important functions in marketing fresh produce.

United States consumers today have access to an amazing variety of fresh fruits and vegetables all year round. Most still comes from domestic production, but there has been a marked increase in recent years in the kinds and quantities of imported commodities. Different regions supply different markets, depending on the market location with respect to the growing areas. Sources also change during the year, as shifts occur in seasonal production. Over the years in this country the output of many crops has become increasingly concentrated in geographical locations where climate, topography, soils, and local economic conditions provide a competitive advantage. At the same time there has been an upsurge of interest in growing specialty crops close to major markets.

HISTORICAL CHANGES IN PRODUCTION LOCATION AND MARKET SOURCES

The regional production of fruits and vegetables is constantly changing, often gradually but sometimes rapidly. The sources of supply that particular markets draw on to fill their needs are never stable for very long.

When our country was first settled most people lived on farms or in rural areas and many grew their own fruits and vegetables. Those that did not depended largely on local supplies. With the development of urban centers a ring of market garden farms came into being around our towns and cities, especially to supply the more perishable items during the local growing season. In the northern states only the storable commodities like root crops, cabbage, and apples were available in fresh form in the winter.

As methods of domestic transportation progressed from horse drawn wagons to canal barges to railroads the growing of less perishable items expanded in more favorable areas of production. About 100 years ago the development of railway refrigerator cars and a network of icing stations combined to enable perishable crops to be shipped long distances. In more recent years the combination of mechanically refrigerated tractor-trailers and the interstate highway system has made the location of U.S. production much less dependent on transportation considerations. Irrigation development has opened up many highly productive growing areas. Meanwhile competition from urban growth as well as from distant growing areas has all but eliminated the traditional market garden farms around our major cities.

The changes are continuing, both with respect to regions and crops. Specialized growing areas are responsible for a larger and larger share of total production. In the last 25 years California's share of our domestic production of fresh vegetables and deciduous fruits increased from one-third to about one-half. Florida would have had a much larger share of total citrus output today had it not been for several severe freezes in recent years. Changes in the location of some crops such as citrus have been greater within states than between them, due largely to urban pressure. In California citrus has moved from the southern counties over the mountains into the San Joaquin Valley. In Florida orange production has shifted from the coast to the central region.

Some crops like dates or artichokes require very specific growing conditions and are restricted to very limited areas that have the necessary climatic and soil conditions. Their production does not shift readily. But even crops that can be widely grown, like apples and potatoes, are tending to become concentrated in specialized areas. In the last 25 years the proportion of our apple crop grown in Washington State has gone from one-fifth to one-third, while the proportion of our potato production grown in Idaho and Washington has increased from 20 to 40 percent.

Information on sources of fresh fruits and vegetables can be obtained from data on farm production, on commercial shipments, and on arrivals at major markets. Because of the many different commodities, production areas, and uses, all these three sources have limitations with respect to coverage in any year and consistency from year to year. The shipment and arrival information, although incomplete, seems to provide the most useful data on sources of supply of fresh produce.

U.S. SHIPMENTS

In 1988 the Market News Service of the U.S. Department of Agriculture reported the shipment of 572.7 million hundredweight of fresh fruits and vegetables from United States vineyards, farms, and orchards. This does not include all domestic production but mainly shipments from major growing areas destined for the domestic and export fresh market. The coverage is not complete but the information is consistent from year to year and especially useful to identify patterns and trends.

Six states accounted for three-quarters of the reported quantity of fresh shipments led by California with 39.3 percent, Florida with 14.2 percent, Washington with 7.4 percent, and Idaho, Arizona, Texas accounting for the remaining 13.7 percent (Table 3.1). California ships a large volume of major crops such as lettuce and oranges but also smaller quantities of many different commodities. Florida is heavily into tomatoes, grapefruit, oranges, and many other fruits and vegetables. Washington is a major supplier of fresh apples and potatoes, but also ships pears and onions in volume. Idaho is, of course, noted for potatoes but supplies dry onions, apples, apricots, cherries, and plums–fresh prunes. Arizona and Texas each contribute significant volumes of many fruits and vegetables especially during the winter season.

The six leading commodities in terms of reported shipments accounted for 53.2 percent of the total (Table 3.2). These were table potatoes with 19.0 percent, lettuce with 10.2 percent, oranges with 7.6 percent, and apples, onions, and tomatoes sharing the remaining 16.4 percent.

Shipments of table potatoes are reported for 23 states, but six states or regions account for 70 percent led by Idaho with 27 percent. The other

Table 3.1. U.S. Shipments from Major States of Fresh Fruits and Vegetables for Domestic and Export Markets, 1988.

	TOTAL SHIPMENTS	PERCENT OF TOTAL
	1,000 cwt	percent
California	225,285	39.3
Florida	81,350	14.2
Washington	42,236	7.4
Idaho	32,913	5.7
Arizona	25,843	4.5
Texas	25,795	4.5
Other States	139,315	24.3
Total	572,737	100.0

Source: USDA AMS, 1989d.

Table 3.2. U.S. Shipments of Major Fresh Fruits and Vegetables, 1988.

	TOTAL SHIPMENTS	PERCENT OF TOTAL
	1,000 cwt	percent
Potatoes, Table	109,004	19.0
Lettuce	58,350	10.2
Oranges	43,358	7.6
Apples	37,129	6.5
Onions, Dry	30,607	5.3
Tomatoes	26,072	4.6
Other	268,017	46.8
Total	572,737	100.0

Source: USDA AMS, 1989d.

five are Colorado, California, Washington, Maine, and the North Dakota-Minnesota region mainly along the Red River Valley. Two states account for 94 percent of reported lettuce shipments, California with 74 percent, and Arizona with 20 percent. Although Florida leads the nation in the production of oranges a large proportion of the crop goes for processing, and California consequently leads in fresh orange shipments, contributing over three-quarters of the total. Florida had 17.7 percent in 1986, and Arizona and Texas shared the remaining 5 percent.

Washington leads by a wide margin in shipments of fresh apples, with 71 percent of the total. Other important states or regions are Michigan, New York, and Appalachia (the Shenandoah Valley section of Virginia, West Virginia, Maryland, and Pennsylvania). Many states ship dry onions. Among the leaders are Texas, Colorado, Oregon, California, New York, and Michigan. Florida is the leading shipper of fresh tomatoes with just over 58 percent in 1988. California followed with 32 percent. The remaining 10 percent of reported U.S. shipments was shared by South Carolina, Virginia, North Carolina, Florida, Texas, and Michigan.

IMPORTS

The U.S. Department of Agriculture (USDA) reported imports of 129 million hundredweight of fresh fruits and vegetables into this country in 1988 (Table 3.3). Bananas accounted for almost half the total, but there were 70 other individual commodities or groups of commodities reported. These commodity groups consisted of miscellaneous oriental vegetables, miscellaneous tropical fruits and vegetables, herbs, and mixed-miscellaneous melons. The five other leading fruits and vegetables imported in addition to bananas were tomatoes, table grapes, cucumbers, table potatoes, and

Table 3.3. U.S. Imports of Major Fresh Fruits and Vegetables, 1988.

	TOTAL IMPORTS	PERCENT OF TOTAL
	1,000 cwt	percent
Bananas	63,157	48.9
Tomatoes	7,917	6.1
Grapes, Table	6,873	5.3
Cucumbers	4,278	3.3
Potatoes, Table	4,019	3.1
Cantaloupes	3,492	2.7
Other	39,450	30.5
Total	129,186	100.0

Source: USDA AMS, 1989d.

cantaloupes. These amounted to 20.5 percent of the total, with all the rest adding up to 30.5 percent.

Bananas are imported from nine Central and South American countries, the three leaders being Ecuador, Honduras, and Costa Rica. Tomatoes come primarily from Mexico with small quantities from seven other countries, some as far away as Israel and Spain. Almost all the grapes we import come from Chile during the winter, but significant supplies were reported from Mexico and Canada and even a small shipment from Brazil. Cucumbers and cantaloupes also come primarily from Mexico, but other Central and South American countries are more important suppliers of cantaloupes than cucumbers. Major shipments of cantaloupes came from Honduras, Guatemala, and the Dominican Republic. Canada is the only country from which we import potatoes. Other countries are excluded because of phytosanitary restrictions.

With large shipments of many products, Mexico easily ranks as the most important foreign supplier of fresh fruits and vegetables with 30 percent of the total (Table 3.4). The next four countries—Ecuador, Honduras, Costa Rica, and Colombia—rank high mainly because of bananas but they also ship other crops. Ecuador is a major shipper of plantains and other fruits, Honduras ships cantaloupes, pineapples, and plantains in volume. Costa Rica is also significant in pineapples and miscellaneous tropical fruits and vegetables. This category includes such exotic items as apio, arum, batatas, breadfruit, calabaza, chayote, dasheen, gandules, ginger root, honeyberry, malanga, quenapas, tamarind, taro, yams, yautiá, and yucca. Colombia is a major supplier of plantains as well as bananas. Canada ships a large number of different fruits and vegetables to our markets, chief among them being potatoes, apples, carrots, turnips-rutabagas, table grapes, cabbage, blueberries, and raspberries. Chile has become a major contraseasonal sup-

Table 3.4. U.S. Imports of Fresh Fruits and Vegetables from Leading Countries, 1988.

	TOTAL IMPORTS	PERCENT OF TOTAL
	1,000 cwt	percent
Mexico	38,761	30.0
Ecuador	17,412	13.5
Honduras	14,594	11.3
Costa Rica	14,498	11.2
Colombia	11,305	8.8
Canada	9,295	7.2
Chile	8,639	6.7
Other	14,682	11.4
Total	129,186	100.0

Source: USDA AMS, 1989d.

plier of fresh fruits and vegetables. In addition to grapes Chile also supplies peaches, grapes, pears, plums–fresh prunes in volume.

ARRIVALS AT MAJOR MARKETS

Valuable information on sources of supply at specific markets is provided by statistics on arrivals of fresh fruits and vegetables. The Fruit and Vegetable Division of the U.S. Department of Agriculture collects arrival data for 23 U.S. and five Canadian cities. Jobbers, carlot receivers, chain store buyers, and other members of the produce industry report truck arrivals, and railroad, air, and steamship agents report rail, air, and boat arrivals. The coverage of arrivals is also not considered complete, but is reasonably consistent from year to year.

Arrival data is believed to cover more than 90 percent of the volume of shipped in produce received at these major markets. The coverage differs from market to market. Arrival data does not include local production sold directly by growers to retail stores, to food service operations, or to consumers. These nearby sources can be very important for some commodities in some markets during the local growing season, but reliable data on direct sales of locally grown produce is not available.

The 23 U.S. cities for which arrival data is gathered constitute 18 major markets for fresh fruits and vegetables in this country. In some cases two cities are combined into a single market area. Markets range in size from very large ones like New York/Newark and Los Angeles to smaller ones like Columbia, South Carolina, and Buffalo, New York. During the period 1986–88 a significant quantity of fresh fruits and vegetables was received at the 23 cities from every state except Alaska, and from the territory of Puerto Rico and 33 foreign countries.

Arrivals are reported separately for major fresh fruits and vegetables and in groups for minor commodities. More than 160 different fruits, vegetables, and herbs are identified in the arrival data. This does not begin to count the different varieties and types of each fruit or vegetable or the different grades, sizes, and packages that are available on each market.

SOURCES OF SUPPLY FOR MAJOR MARKETS

During 1986–88 the arrivals of fresh fruits and vegetables at 23 major cities in the United States averaged 232.9 million hundredweight per year (Table 3.5). A total of 83.7 percent of the total originated within the country, and 16.3 percent was imported. California farms and ranches supplied 92.6 million hundredweight, 39.8 percent of the total or 47.5 percent of the domestic supply. Florida was next with 10.2 percent of the total, then Washington with 6.6 percent. Along with Idaho and Texas these five leading states ac-

Table 3.5. Yearly Average Quantity of Fresh Fruit and Vegetable Arrivals in 23 U.S. Cities by State or Country of Origin, 1986–88.

ORIGIN	QUANTITY	PERCENT OF DOMESTIC/IMPORT	PERCENT OF TOTAL
	1,000 cwt	percent	percent
Domestic			
California	92,578	47.5	39.8
Florida	23,744	12.2	10.2
Washington	15,267	7.8	6.6
Idaho	9,960	5.1	4.3
Texas	7,458	3.8	3.2
Other States	45,833	23.5	19.8
Total Domestic	194,840	100.0	83.7
Imports			
Unspecified (Bananas)	21,930	57.6	9.4
Mexico	10,418	27.4	4.5
Chile	2,028	5.3	0.9
Canada	1,543	4.1	0.7
Other Countries	2,133	5.6	0.9
Total Imports	38,052	100.0	16.3
Total Arrivals	232,892		100.0

Source: USDA AMS, 1989a.

Table 3.6. Arrivals of 7 Leading Fresh Fruit and Vegetable Commodities at 23 U.S. Cities, 1986–88 Average.

COMMODITY	QUANTITY	PERCENT OF TOTAL
	1,000 cwt	percent
Potatoes, Table	33,837	14.5
Bananas	21,877	9.4
Lettuce, Iceberg	19,996	8.6
Tomatoes	12,397	5.3
Onions	12,159	5.2
Oranges	12,096	5.2
Apples	11,842	5.1
Other Fruits and Vegetables	108,688	46.7
Total Arrivals	232,892	100.0

Source: USDA AMS, 1989a.

counted for almost two-thirds of the total quantity of fresh fruits and vegetables arriving at these 23 cities during this period.

Arrivals from foreign countries in 1986–88 amounted to 38.1 million hundredweight, or 16.3 percent of the total. Almost 60 percent of the imports consisted of bananas, unspecified as to country or origin in this report in order not to reveal individual company activities.

A greater quantity of fruits and vegetables came from Mexico than from any other foreign country to these 23 cities during 1986–88. The 10.4 million hundredweight arriving from that country represented just over one quarter of total arrivals from foreign sources. Canada and many other countries accounted for the remainder.

Seven items, three fruits, three vegetables, and potatoes, accounted for over half of the total volume of fruits and vegetables reported to have arrived at the 23 markets (Table 3.6). These seven, with the percentage of total volume of unloads represented by each, were potatoes 14.5 percent, bananas 9.4 percent, iceberg lettuce 8.6 percent, tomatoes 5.3 percent, onions 5.2 percent, oranges 5.2 percent, and apples 5.1 percent.

REGIONAL AND SEASONAL DIFFERENCES IN SOURCES OF SUPPLY

At any given time of the year most fruits and vegetables come from a few major production areas that provide a large proportion of our total needs, but a number of secondary areas around the country also contribute smaller quantities to total U.S. supply. Several foreign countries are often also potential sources of product especially during the winter season. Individual markets consequently often draw on many different sources, depending somewhat on geographical location. Information on sources of supply by states and countries as well as the names and addresses of specific firms shipping various commodities can be obtained from publications prepared by trade associations and publishing houses.

The United Fresh Fruit and Vegetable Association (1986) publishes a Supply Guide showing the total annual available supply of 70 fresh fruits and vegetables from major producing states, and their monthly availability as a percentage of total annual supply. In addition there is a supply guide for 65 produce specialities for which statistical information is not availale.

A trade paper *The Packer* (1988) publishes a produce availability and merchandising guide that contains a chart showing the availability and shipping seasons for 75 major fruits and vegetables. A listing of firms by state and town that ship each item is also included. The publication also contains useful information on varieties, care, grades, merchandising hints, and nutritional qualities for the major items, and some information on about 120 specialty items.

The arrivals of tomatoes at Saint Louis illustrate a typical pattern of sources of supply. In 1988 tomato arrivals on the Saint Louis market were reported to amount to 208,000 hundredweight. Florida was the major source, accounting for 110,000 hundredweight, followed by California with 47,000 and Mexico with 24,000 hundredweight (Figure 3.1). Arkansas with 15,000 hundredweight was also a significant source. Several other states including Missouri, Illinois, Michigan, South Carolina, and Kentucky shipped

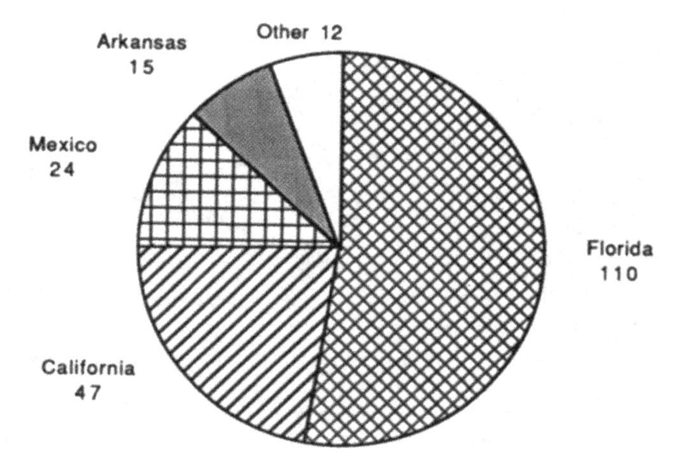

Figure 3.1. Arrivals of Tomatoes in St. Louis By State or Country of Origin, 1,000 cwt, 1988. *Source:* Adapted from USDA AMS, 1989c.

the remaining 12,000 hundredweight. Tomatoes were also undoubtedly available in quantity from home gardens and nearby growers during the summer and early fall that were not included in the arrival data.

Commercial arrivals of tomatoes on the Saint Louis market in 1988 peaked during the spring and early summer, then declined during the fall and winter (Figure 3.2) Mexico and Florida were the major sources during the winter and early spring. Arkansas and other nearby states took over in June and July, augmented by California. California and Michigan provided the bulk of tomato supplies during fall. Small quantities came from Mexico during the fall, probably from the Baja California region. Arrivals from Florida began again in November. Such a pattern of tomato arrivals at Saint Louis is generally consistent from year to year, varying somewhat depending on growing conditions and markets.

The way markets differ in their sources of supply can be illustrated by comparing the arrivals of tomatoes at Boston and San Francisco (Figure 3.3). During the 3 year period 1986–88 Boston reported the arrival of an average of 1,193,000 hundredweight of tomatoes and San Francisco 906,000 hundredweight. Mexico supplied 45.7 percent of the tomatoes arriving on the western market, California 41.7 percent, and Florida the remaining 12.5 percent. At Boston Florida supplied 60.8 percent of the tomatoes, and California another 13.7 percent. The remaining 25 percent came from many different states such as South Carolina and Virginia as well as Mexico.

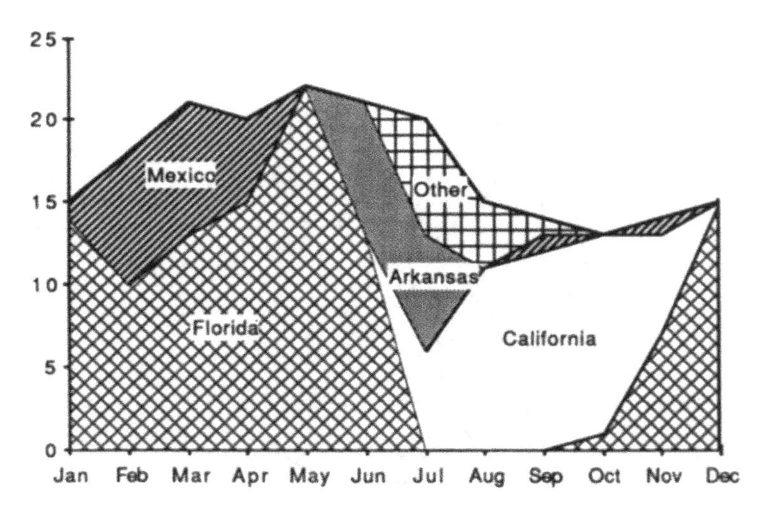

Figure 3.2. Sources of Tomato Arrivals at St. Louis By Months, 1,000 cwt, 1988. *Source:* Adapted from USDA AMS, 1989c.

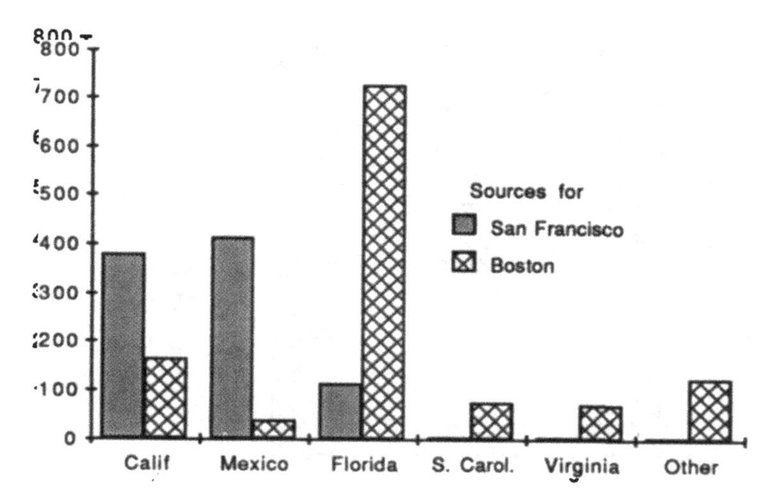

Figure 3.3. Sources of Tomato Arrivals at San Francisco and Boston By State or Country of Origin, 1,000 cwt, 1986–88. *Source:* Adapted from USDA AMS, 1989b, USDA AMS, 1989c.

Sources are continually changing. From the early 1970s to the late 1980s there have been some major shifts in the sources of tomatoes coming on the Boston market (Figure 3.4). Florida has increased in importance as have South Carolina and Virginia and several other states. The percentage of total arrivals from California, Massachusetts, and Mexico has declined. In the case of Massachusetts it is possible that the decline in reported arrivals

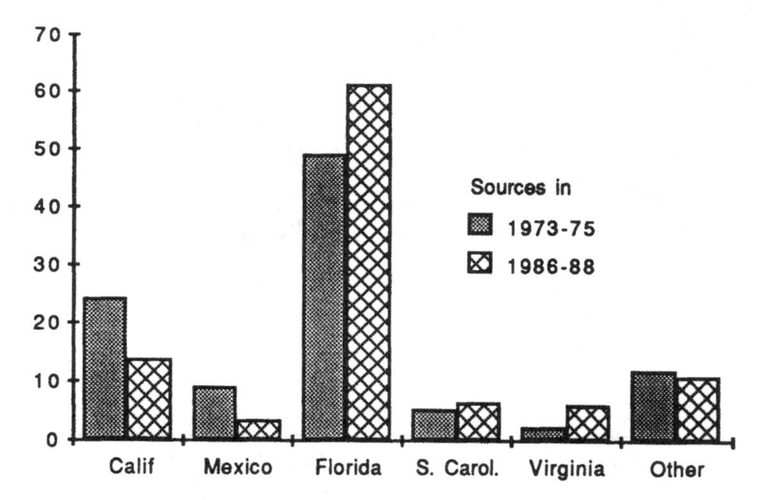

Figure 3.4. Changes in the Source of Commercial Tomato Arrivals in Boston, 1973–75 to 1986–88. *Source:* Adapted from USDA AMS, 1989d.

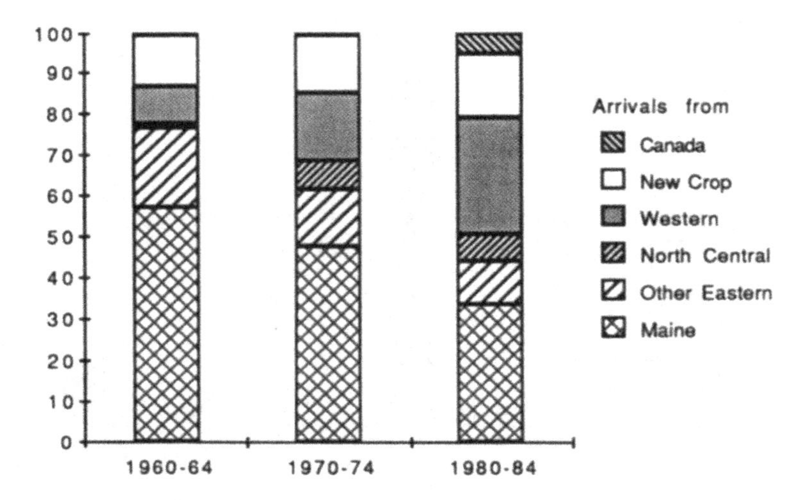

Figure 3.5. Change in Table Potato Market Percentage Shares in 11 Eastern Cities by Production Region, 1960–64, 1970–74, 1980–84. *Source:* Adapted from Buckley, Mai, 1986.

from that state reflects an increase in direct sales by growers rather than a decline in production.

The substantial shift over the past 30 years in the source of fresh potatoes on eastern markets illustrates what can occur (Figure 3.5). In the early 1960s Maine and other Eastern states accounted for almost 80 percent of the fresh

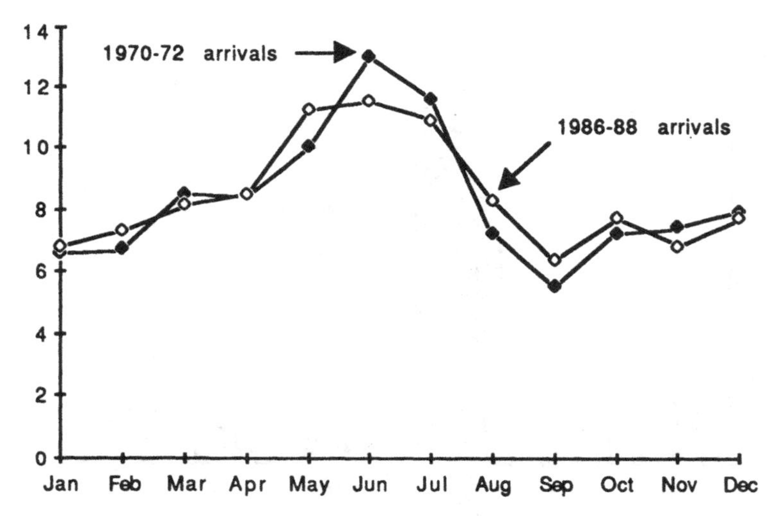

Figure 3.6. Changes in Percentage of Total Arrivals of Tomatoes By Months, Boston, 1970–72 to 1986–88. *Source:* Adapted from USDA AMS, 1989b.

potato arrivals on eastern markets, but by the early 1980s this had dropped to about 45 percent. Western and North Central states had captured a larger share of the market, and imports from Canada constituted a small but significant amount. New crop sources had remained about the same.

The pattern of arrivals throughout the year can also change. For most commodities the change is very gradual, but arrivals are tending to become more uniform throughout the year. In Boston the percentage of total tomato arrivals during the peak months of June and July in 1970–72 has declined and May became the peak month by 1986–88 (Figure 3.6). The percentage of total arrivals also increased between these two periods during the low months of August and September so tomato arrivals now are a little less variable during the year.

REGIONAL LOCATION OF PRODUCTION

Shifts in sources of supply from one region to another are not easy to predict, yet the ability to anticipate such changes could be very useful. We know the general principles involved in production location and can explain past history, but determining prospects for future production may involve predicting changes in technology and in consumer demand. Yet some idea of future economic opportunities is essential for those planning regional developments or starting new enterprises (Greig, Blakeslee, 1978).

The principle that applies to the location of production is known as that of comparative advantage. In general terms this states that producers (individuals, regions, or nations) will tend to specialize in the production of the commodities in which they have the greatest comparative advantage or the least comparative disadvantage, and to obtain by trade the commodities in which they have the least comparative advantage or greatest comparative disadvantage (Bressler, King, 1978). The advantage need not be absolute, for that would rule out economic activity entirely in many situations, but only comparative or relative.

In practice this is translated in many cases into a Darwinian economic condition of survival of the fittest. Businesses prosper and grow where they have a clear competitive advantage, and fail where the opposite condition exists. A region is said to have a competitive advantage if: (1) the region maintains lower production and marketing costs relative to a competing region; (2) the region receives a higher average price for its product relative to that received by a competing region; or (3) both (Buckley, 1986). Farming operations are particularly sensitive to economic conditions that may affect competitive advantage, since they depend so heavily on natural resources which cannot be shifted or easily modified.

For the individual considering production of a commodity in such small quantity that it would not affect market prices the determination of whether the business is likely to succeed is relatively straightforward. Given current costs and prices is the enterprise likely to make a profit? There may be some difficulty in estimating costs and returns for crops not currently grown in the area. Yields per acre and market prices are difficult to predict. Consumers are sensitive to many different product attributes such as color, flavor, and texture and may not necessarily favor locally grown products. Wholesale buyers may require a certain volume and consistency or continuity of supply that may be difficult to meet.

When growers and marketers in different regions are competing for the same market or markets both differences in costs and prices must be taken into account in determining which is likely to succeed. Prices of the same fruits and vegetables coming from different areas will often differ because of differences in the conditions under which the items were grown or harvested and packed. Average prices for the season may differ because of fluctuations in prices and differences in shipments during the season. Costs may differ because of different production practices and yields, and also because of different opportunity costs for land, financial capital, and management. Because land is not of uniform quality and prices of production inputs often change with changes in volume of production the costs of production may change with changing levels of output, as may market prices. This often explains why regional shifts in production do not eliminate output entirely, but simply move it toward a new lower equilibrium.

In considering prospective changes in competitive advantage it is dangerous to focus too much attention on any single cost element since other factors may be changing also (Beilock, Dunn, 1982). Rising costs of water in irrigated areas that might be considered to introduce a considerable disadvantage may be largely offset by the adoption of more efficient methods of application. Rising fuel costs affecting transportation rates may be a relatively minor part of the total delivered cost and more than offset by changes in other costs. Only a total systems approach can provide some indication of the competitive advantage of one region versus another.

FUTURE PROSPECTS

Based on the current situation what shifts appear likely to occur in production in the future? We have observed the production of many crops becoming concentrated in major producing areas, the decline of market garden farms, and the increase in output of specialty items grown close to market. At the same time output in other countries has expanded bringing increased competition in export markets and growing competition here at home from imports.

There may not be any major reversal in the trends toward increased concentration of production in favored growing areas nor expanded output in other countries, but the system does seem to have become more fragile and thus likely to be more easily disrupted by unexpected events. Within this country urban development, water charges and availability, labor regulations, pesticide labeling, and fuel costs are some factors that will have a differential effect on our growing areas. In other countries their economic conditions, farm policies, foreign trade agreements, and technological development will certainly have an impact on our export markets and import competition.

The principle of comparative advantage based on natural resources and market prices and costs may not be allowed to operate in international trade and be replaced by what has been termed competitive advantage. Under these conditions trade is fostered or retarded by government policies such as export subsidies or import quotas that modify the effect of market forces.

There have been many changes in the primary sources of supply for fresh fruits and vegetables in this country in the past. Now we are part of an international economy. The primary sources for many of our fresh produce items has shifted to foreign countries, and we may expect such competition to intensify in the future. Until recently most imports of fresh fruits and vegetables came in during our winter season when our domestic production was in short supply. We may see more competition from imports during our main growing seasons in the future.

During the 1960s and 1970s production of fruits and vegetables in California expanded somewhat at the expense of other regions of the country. Now that state, as we shall see in the next chapter, is experiencing pressure on production from urban development, and competition in traditional markets from other countries. What this implies for the future is a major concern.

REFERENCES

Beilock, Richard P., James W. Dunn, 1982. *An Econometric Model of the U.S. Potato Industry Emphasizing the Effect of Changes in Energy Costs.* Pennsylvania State University College of Agriculture Bulletin 839.

Bressler, Raymond G., Jr., Richard A. King, 1978. *Markets, Prices, and Interregional Trade.* Raleigh, NC: Norman-Weathers Printing Co.

Buckley, Katherine C., 1986. Competitive Advantage in Producing Winter Fresh Vegetables in Florida and West Mexico. *Vegetable Outlook and Situation.* U.S. Department of Agriculture, Economic Research Service, TVS 238.

Dunn, James W., Stanley M. Beard, Jr., 1982. *The Effect of Higher Energy Prices on Interregional Competition in Peaches.* Pennsylvania State University College of Agriculture Bulletin 841.

Greig, W., Smith, LeRoy Blakeslee, 1978. *Potatoes: Optimum Use and Distribution with Comparative Costs by Major Regions of the U.S.* Washington State University, College of Agriculture Research Center, Bulletin 865.

The Packer, 1988. *Produce Availability and Merchandising Guide.* Vance Publishing, Shawnee Mission, Kansas.

United Fresh Fruit and Vegetable Association, 1986. *Supply Guide.* United Fresh Fruit and Vegetable Association, Alexandria, Virginia.

U.S. Department of Agriculture, Agricultural Marketing Service, 1989a. *Fresh Fruit and Vegetable Arrivals in Eastern Cities by Commodities, States, and Months 1988,* FVAS-1 Calendar Year, 1988.

—— 1989b. *Fresh Fruit and Vegetable Arrivals in Western Cities by Commodities, States, and Months 1988,* FVAS–2 Calendar Year 1988.

—— 1989c. *Fresh Fruit and Vegetable Arrival Totals for 23 Cities 1988,* FVAS-3 Calendar 1988.

—— 1989a. *Fresh Fruit and Vegetable Shipments By Commodities, States, and Months 1988,* FVAS-4 Calendar Year, 1988.

Chapter 4

Major Sources of Supply: California, Florida, and Mexico

As we have seen, a large proportion of fresh fruits and vegetables from domestic sources comes from California and Florida, and Mexico is the most important source of imports. This chapter provides a brief description of some of the characteristics of these regions, the reasons why they compete so effectively in our markets, and the challenges they may face in maintaining a major share of the market in the future.

CALIFORNIA

Commercial shipments of California fresh fruits and vegetables for domestic and export use, not counting production for local consumption, amounted in 1988 to over 225 million hundredweight per year and about 40 percent of the total commercial shipments reported in the United States (Table 4.1). The leading commodity in terms of volume was iceberg lettuce, with almost 20 percent of the California total and almost three-quarters of the national supply, followed by oranges with 15 percent of the California volume and close to 77 percent of the U.S. total. Next in line were a group of commodities with from 3 to 6 percent of total California volume and 70 to 97 percent of the nationally reported shipments. These are grapes, celery, cantaloupes, broccoli, lemons, and strawberries. Shipments of table potatoes constituted almost 5 percent of the California total, but less than 10 percent of the national total. California tomatoes also only amounted to about one-third of the domestic shipments.

In addition to these major commodities California accounts for all or almost all the commercial shipments of many specialty crops. California shipments constitute over 90 percent of the total reported commercial shipments of broccoli, nectarines, juice grapes, artichokes, melons, and cherry

Table 4.1. California Major Fresh Fruit and Vegetable Shipments for Domestic and Export Use, 1988.

	QUANTITY	PERCENT OF CALIF. SHIPMENTS	PERCENT OF U.S. SHIPMENTS
	1,000 cwt	percent	percent
Lettuce, Iceberg	43,603	19.4	74.8
Oranges	33,699	15.0	77.4
Grapes, Table	13,101	5.8	97.2
Celery	11,888	5.3	74.9
Carrots	11,325	5.0	78.8
Potatoes, Table	10,441	4.6	9.6
Cantaloupes	10,385	4.6	71.2
Tomatoes	8,356	3.7	32.3
Broccoli	7,884	3.5	92.1
Lemons	7,029	3.1	72.3
Strawberries	6,721	3.0	93.5
Other	60,853	27.0	24.1
Total	225,285	100.0	39.3

Source: USDA AMS, 1989.

tomatoes, and over 80 percent of the avocados, plums–fresh prunes, asparagus, miscellaneous oriental vegetables, and Chinese cabbage.

California shipments probably rank higher nationally in terms of value than in weight since so many high value specialty crops are grown there. Other states, however, also grow some crops that are not included in national agricultural statistics. Broccoli shipments for example, of which California accounts for 92 percent, are only reported for Arizona, California, Florida, and Texas. In several other states this crop is also grown commercially in volume.

This large and diverse volume of fruit and vegetable production is due to the extensive land areas in the state with favorable weather, productive soils, and a sufficient supply of water. It is said that California has some of the hottest, coldest, wettest, and dryest areas of the United States. The state extends some 800 miles along the Pacific coast from about 32 N latitude to 42 N latitude (Figure 4.1). Within approximately 200 miles in width lie the coastal ranges on the west stretching almost from the Oregon border to Los Angeles, and the Sierra Nevada mountains to the east. Between these two mountain ranges lies the Central Valley, an area of major agricultural importance. The coastal range is interspersed with several smaller valleys and plains also of significant agricultural value.

California weather along the coast is influenced by Pacific Ocean currents

Figure 4.1. Map of California.

that moderate extremes in temperature and rainfall. Moist ocean breezes deposit rainfall on the western slopes of the mountain ranges, with the resulting rivers and streams providing water for irrigation. Farming areas range in distance and climate from Tulelake–Butte Valley along the Oregon border where potatoes and onions are grown in the summer, to the Imperial Valley on the Mexican border where crops such as lettuce, tomatoes, and watermelons are grown in winter. Major irrigation developments have brought water to the west side of the San Joaquin Valley and the desert areas to the south.

The Central Valley is the dominant feature of California agriculture, stretching from Kern County in the south to Tehana County in the north. The valley consists of two parts, the Sacramento Valley in the north and the San Joaquin Valley in the south, each drained by rivers that gave their names to the valleys. The Sacramento River flowing south joins the San Joaquin flowing north to form the Delta region before the combined flow empties into San Francisco Bay. The Tehachapi Mountains mark the southern boundary of the San Joaquin Valley, joining the Sierra Nevada and the coastal range at their southern extremities.

The Sacramento Valley is noted for fruits such as apricots, grapes, pears, and plums, for honeydew and other melons, for tomatoes, and for watermelons (Nuckton and Johnston, 1985). The Delta region has significant production of asparagus, sweet corn, potatoes, and tomatoes. The San Joaquin Valley from San Joaquin County to Kern County produces a tremendous quantity of many different fruits and vegetables. This valley is the principal source of many fruits and nuts, especially melons, grapes, and soft fruits in the summer and onions, potatoes, and carrots in late winter and spring.

The North Coast that includes the Napa and Sonoma Valleys is an important area for wine grapes, apples, pears, prunes, and walnuts. The coastal valleys and plains still produce large volumes of many different crops in spite of increased urban pressure. The Salinas Valley stretching southeast from Monterey Bay is the major source of lettuce and other cool season crops. Irrigation development has enabled desert areas in the south to become significant sources of winter warm season vegetables.

The production of horticultural crops in California has expanded tremendously in the past 30 years. Irrigation development has brought thousands of acres under cultivation. Profitable markets have been found both in the United States and abroad for this increased output. Some of this additional supply has replaced production in other parts of the country, and some has served to provide a net addition to the total sales.

California natural resources are well suited to the production of large volumes of horticulture products having widely different physical requirements, but the state may be hard pressed to maintain production in the future. Limitations in the supply and quality of water for irrigation, possible shortages of seasonal labor at competitive wage rates, pressures from urban development, and problems in controlling plant pests may raise production costs, while competition from other regions may narrow potential marketing margins.

Water

California agriculture depends heavily on irrigation water. Rain and snow fall mainly in the wrong places at the wrong times, and water must be stored and transported to where it is needed. According to the 1987 Census of Agriculture, all the commercial vegetable acreage and over 98 percent of the orchard and vineyard acreage was irrigated. About 95 percent of all California harvested cropland is under some form of irrigation. Of the water put to use in California about 85 percent goes initially to agriculture; the remainder is used by municipalities and industries.

Currently about 40 percent of the irrigated acreage is supplied by water that must be pumped from underground aquifers. Some wells are shallow, others deep and costly to drill and use. The other 60 percent of the acreage is supplied by surface sources such as local streams or rivers or by the systems of dams, canals, and ditches that transport water long distances.

California farmers obtain water from their own wells, directly from nearby rivers and streams, from locally owned irrigation systems, or from municipal, state or federal projects (Highstreet, Nuckton, Horner, 1980). Major municipalities have developed water storage and distribution systems to serve their needs. The State Water Project has constructed a series of dams, reservoirs, power plants, and canals to provide water for farming, industry, and home use. As the demand for water by these three groups has increased so has the controversy surrounding the allocation of water between them and also regionally, since most of the precipitation is in the north and most of the people live in the south. At issue currently is the state plan to divert water around the Delta region and provide additional storage and distribution facilities in the south.

The federal government, with the passage of the Reclamation Act of 1902, has undertaken major irrigation development in 18 western states. In the 1930s the Bureau began the Central Valley Project that currently includes a network of dams and canals to supply agricultural and domestic water to much of the northern two-thirds of the state. The west side of the San Joaquin Valley has been the latest region to benefit from this development. Farmers on the west side had obtained water in the 1930s through drilling deep wells, but these were showing signs of drying up.

Federal reclamation projects generally have had multiple objectives, and charges for irrigation water generally have been kept relatively low through the arbitrarily high allocation of costs to nonfarm uses and the artificially low charges for capital. Federal policy consequently has been to try to distribute the benefits of public irrigation development as widely as possible. The 1902 Reclamation Act stipulated (1) that no water from a reclamation project could be delivered to any landowner who had title to more than 160 acres, and (2) that no water could be delivered to any landowner who did not reside on the land or in the neighborhood. A means of redistributing land at the "prewater" price was instituted. The Act has since been amended many times, tested in the courts, and original provisions relaxed somewhat. New legislation was passed in 1982 that increased to 960 acres the amount of land that could be owned, and changed other provisions. The legislation appears, however, to have had little effect on farm size in federal reclamation projects.

Rising water costs and farm wages are changing the way water is applied and the cropping patterns in many areas (Highstreet, Nuckton, and Horner, 1980). Cheap surface water had encouraged liberal application by furrow or flooding. The use of sprinkler and drip application methods that require substantial initial capital investment but use less labor and less water is on the increase.

Water requirements vary by crop and location. Alfalfa in the Imperial Valley takes about 8 acre feet compared to 3.2 feet in the central coastal region. Fruits and vegetables in the San Joaquin Valley generally require 2 to 4 acre feet each season.

Water costs have represented a small proportion of total production costs for some crops. In 1975 the total cost of applying 2.8 acre feet of water in growing lettuce in the central coast region amounted to less than 2 percent of the estimated $2,000 per acre total production costs (Highstreet, Nuckton, Horner, 1980). In other areas and for other crops irrigation expense is a more important component. On grapes in the San Joaquin region, for example, where water is obtained from both surface and ground sources and applied by furrow, border, sprinkler, and drip irrigation methods the irrigation costs in 1975 ranged all the way from 3 to 10 percent of total production costs, averaging about 4 percent. In the westside San Joaquin region where water costs run up to $55 per acre foot total irrigation costs ranged up to 20 percent of production costs and averaged close to 10 percent. This level of water costs has encouraged the installation of solid set sprinkler systems to cover 25 percent of the grape acreage and drip systems another 10 percent. For crops like alfalfa hay or irrigated pasture the irrigation costs constitute an even larger percentage of total production costs.

As the demand for irrigation water has increased, costs of water have risen, and efforts have been made to economize in water use; the problems of salinity and drainage have become even more critical. The westwide of the San Joaquin Valley has experienced serious drainage problems. Salinity problems have come in two forms, high water tables of salty water and downstream degradation of the San Joaquin River. The Bureau of Reclamation in 1968 began construction of the San Luis drain to carry water northward to the Delta region but the project has not been completed and the brackish outflow in the Kesterson Wildlife Refuge has aroused the ire of ecologists.

Even more serious have been the salt and soil saturation problems of the Imperial Valley that developed shortly after Colorado River water became available there in 1901. Beginning in the 1940s expensive surface and subsurface drainage systems and improved water management by farmers have enabled a favorable salt balance to be maintained, but the salt buildup in the Salton Sea may threaten this balance in the future. Of critical concern also is the availability of Colorado River water in the future. Competition for Colorado water has been intense between several western states and Mexico.

In 1968 the U.S. Supreme Court established Colorado River water allocations for seven western states situated along its banks. In the past Arizona has not taken its full 2.8 million acre feet allotment, allowing California to exceed the 4.4 million acre feet alloted to that state. The rapidly growing population in Arizona, however, is expected to require the full allotment soon, and California agriculture will have to compete with industry and urban users for limited supplies. The Imperial District has been entitled to 3.85 million acre feet of the California allotment, with the rest shared by the Metropolitan District and several small commercial and industrial users. Major conservation projects are planned to meet this emergency.

Federal long-term water contracts which obligate the Government to supply water to about 300 districts in California's Central Valley were first signed 40 years ago and are beginning to come up for renewal. Under announced Department of Interior policy the new water contracts will be subject to changes in price, conservation, time, manner, and efficiency of delivery but not in quantity supplied.

California has had successive years of drought in 1976–77 and 1987–88 as measured by the flow of the Sacramento River (Rausser, 1988). During the drought of 1976–77 there was no significant reduction in vegetable acreage. Rausser (1988) has suggested that the reason is that most vegetable acreage lies in central and southern California, which receives most of its water from the Colorado River, and has extensive reservoir diversions from the north to fill its water needs. Diversion from the Colorado River may be limited in the future as needs in Arizona increase.

Seasonal Labor

California growers of fruits and vegetables are heavily dependent on seasonal agricultural workers, employing an estimated 300,000 annually. Employment is for a short time, the work is hard, and many seasonal workers have been illegal immigrants. Immigration policy, unionization, and alternative employment opportunities can have a significant impact on the California seasonal labor supply.

Prior to World War II much of California's farm labor was supplied by unemployed workers from other states, such as the Okies of the 1930s. After the war legislation permitted guest workers from Mexico to come in temporarily under what was known as the Bracero program. At the termination of the program in 1964 major adjustments were made. The entry of seasonal workers was greatly restricted. Harvest mechanization took a major step forward with the introduction of the mechanical harvester for processing tomatoes, but there was also additional incentive for illegal immigration.

A sample of California farm workers interviewed in 1983 found that 73 percent were born in Mexico, 7 percent in other foreign countries, and 20 percent in the U.S. (Mines, Martin, 1986). Two-thirds of the immigrant workers had green cards that enabled them to live and work legally in the U.S., but the validity of the cards was not checked. Twenty-five percent were clearly illegal or undocumented, and 5 percent had their legal status pending. Illegal workers were not distributed uniformly in the state. The coastal vegetable areas had fewer illegals than the Central Valley, where 30 to 50 percent of the young men involved in harvesting citrus, grapes, and tree fruit were illegal.

The California farm labor profile distinguishes between casual, seasonal, and year-round workers. Casual workers did less than 12 weeks of farm work, accounted for 34 percent of the sample, but did only 12 percent of the work. Seasonal workers worked 13 to 37 weeks, constituted 50 percent of the sample, and contributed 56 percent of the work. Year-round workers working 38 weeks or more made up 16 percent of the sample and contributed 32 percent of the farm work. The average married male in the survey did farm work for 26 weeks, did nonfarm work for 3 weeks, was unemployed 21 weeks, and was out of the U.S. for 2 weeks.

The California survey found that farmworkers earned an average of $5.10 per hour in 1983 but there was great variation. Few workers were on salary. Most were paid on a piecework basis and the remainder by the hour. Pieceworkers generally earned more per hour than hourly paid workers. Lettuce harvest crews in the Salinas Valley earned $15 to $18 per hour. The wages were not apparently high enough to hold career workers, and many move on to other work as soon as they can. The turnover poses problems for the future.

Mines and Martin (1986) reported that total demand for in-field labor in

California in the late 1970s was approximately 128,000 workers on a year-round basis, but that in 1978 there were 298,000 farmworkers who earned more than $800. This surplus labor phenomenon of more than twice as many workers as there were jobs has given rise to the crew leader recruitment and supervisory system. This recruitment style is used by farm labor contractors, by small growers in tree fruits and vineyards, and by larger growers for hoeing and thinning operations. In contrast some employers, particularly larger ones in regions where farm labor unions have been active, use more formal labor management practices.

Federal legislation makes no provision for the process of unionization for farm workers, but the California Agricultural Labor Relations Act of 1975 provides for farm worker unionization. Jurisdictional disputes between the United Farm Workers and the Teamsters as well as between the unions and farm operators have disrupted the industry from time to time. The secondary boycott has also been used as, for example, when food stores that sell lettuce harvested and packed by the firm being struck by the farm union have been picketed. Whatever the cause, ending the Bracero Program, farm unionization, or other reasons, farm wages have risen relative to nonfarm wages over the past 20 years but still are relatively low.

Illegal immigration continued in volume during the 1970s and early 1980s. In an attempt to control this flow the Immigration Reform and Control Act (IRCA) of 1986 was passed. This legislation has three elements that are expected to affect the farm labor market (Martin, 1988). One is that the Act imposes sanctions on employers who knowingly hire illegal alien workers. A second is the Special Agricultural Worker (SAW) program that permitted illegal aliens who did at least 90 days of qualifying farmwork in the 12 months ending May 1, 1986 to apply for temporary and eventually permanent U.S. legal status. Qualifying farmwork is defined as doing or supervising fieldwork on seasonal perishable commodities. The third element of the IRCA consists of the provisions to provide supplemental farmworkers as needed under the H-2A and the Replenishment Agricultural Worker (RAW) programs. This provides for bringing in foreign workers either temporarily under the H-2A program or with the prospect of permanent residence under the RAW program.

Time alone will tell how effective the IRCA will be in controlling illegal immigration and providing a stable and productive farm work force. Martin (1988) has reported that under the SAW program many more illegal aliens applied for and were granted temporary legal status than were expected. Many of these have also continued in farm work rather than migrating to cities. The impact of the legislation will probably depend to a large extent on just how it is administered.

Urbanization

The proportion of California land devoted to homes, roads, parking lots, and public buildings is relatively small, but the competition with farming for the use of prime real estate and water, and the effect of urbanization on air quality, has had a profound impact on California agriculture. Urbanization and smog have reduced production of some crops and forced the relocation of many others.

Perhaps the best known recent effect of urbanization has been the conversion in the 1950s and 1960s of part of Santa Clara County south of San Francisco, known as Silicon Valley, from an area of intense production of specialty crops such as prunes, apricots, cherries, and walnuts to an outpost of the computer industry. Of possibly greater significance was the forced exodus of the navel orange groves from the Los Angeles basin area to the southern San Joaquin Valley region following the influx of people and the deterioration in air quality after World War II. Crops such as artichokes in the central coastal region and dates in the southern desert area, that are highly sensitive to climate and soil requirements, have experienced reductions in production due to urban pressure rather than simply shifting to a higher cost situation. The shift of specialty crops to the Central Valley has intensified pressures on natural resources, especially irrigation water.

California agriculture is currently feeling the effects of increased competition in domestic markets from foreign countries such as Mexico and Chile, and not only heated competition abroad from these countries and others but also restrictions on sales due to protectionist policies on the part of major potential markets such as Japan (Moulton, 1988).

Pest Control

The conditions under which most crops have been grown in California have not generally been conducive to the development of insect pests and plant diseases. The combination, however, of the high concentration of production of certain crops and the natural selection of some pests has led to problems in some areas. Through the judicious application of pesticides most of these insects and diseases have been kept under control, under the watchful eye of the California Department of Food and Agriculture. Others have been prevented from establishing a foothold in the state through a strict quarantine. But it has become increasingly difficult to control certain pests, and to keep others out.

In recent years the white fly, which carries a virus that attacks lettuce, has periodically devastated the winter lettuce crop. Supplies have been disrupted, and prices have soared.

The Mediterranean fruit fly (Medfly) has penetrated the quarantine several times in recent years (USDA ERS, 1989). Medflies are apparently brought by travelers into urban areas which makes them difficult to eradicate. If the Medfly became established in California it is estimated that the state's citrus industry alone would suffer a $150 million annual loss. In the summer of 1989, 40 adult Medflies and 40 larvae were discovered in a 3-square-mile of Los Angeles County.

In an attempt to eradicate this pest the U.S. Department of Agriculture released sterile male flies, the California Department of Food and Agriculture sprayed malathion over a 14-square-mile area, and a quarantine was imposed on the movement of fruits and vegetables out of a 70-square-mile area. At the same time the USDA also quarantined a 65-square-mile area of Los Angeles following the discovery of 15 Oriental fruit flies during routine trapping surveys. Some householders in the quarantine area have raised objections to the pesticide spraying.

FLORIDA

Florida shipped just over 81 million hundredweight of fresh fruits and vegetables to domestic and export markets in 1988, a little over one-third the California volume (Table 4.2). Grapefruit led the list of commodities with 17.6 million hundredweight or close to 22 percent of the Florida total, and 86 percent of all U.S. grapefruit shipments. Not far behind was tomatoes with 18.5 percent of the state's shipments and 52 percent of the reported

Table 4.2. Florida Shipments of Fresh Fruits and Vegetables for Domestic and Export Use, 1988.

	QUANTITY	PERCENT OF FLORIDA QUANTITY	PERCENT OF U.S. QUANTITY
	1,000 cwt	percent	percent
Grapefruit	17,637	21.7	86.2
Tomatoes	15,078	18.5	52.3
Watermelons	7,366	9.1	35.0
Oranges	7,691	9.5	17.7
Sweet Corn	3,463	4.3	56.9
Cucumbers	2,920	3.6	53.1
Celery	2,406	3.0	15.2
Peppers	3,118	3.8	50.3
Potatoes, Table	1,759	2.2	1.6
Other	19,918	24.5	6.2
Total	81,356	100.0	14.2

Source: USDA AMS, 1989.

national total. Watermelons and oranges each accounted for just under 10 percent of Florida shipments. Florida also ships about half the country's sweet corn, cucumbers, and peppers. As in the case of California, Florida potatoes do not constitute a large share of the national market annually, but are important during the late winter. Many different fruits and vegetables, including some of tropical origin, make up the remaining 25 percent.

The state of Florida consists largely of a peninsula at the southeastern tip of the United States (Figure 4.2). The land area extends southward from 31 N. latitude almost 450 miles to 25 N latitude, much of it lying between the Atlantic Ocean and the Gulf of Mexico. Florida has little variation in climate and topography but can be divided geographically into three main regions. These are the Atlantic coastal plain, the East Gulf coastal plain, and the uplands lying in between. The uplands are higher than the sur-

Figure 4.2. Map of Florida.

rounding regions, but the average elevation is only 200 to 300 feet above sea level, and only 345 feet at the highest point.

Much of Florida enjoys a long frost-free period. Many years no killing frosts occur in southern Florida, but every few years low temperatures devastate the citrus and winter vegetable crops. Winters are mild and summers are warm and humid. The state has an average annual rainfall of 53 inches, two-thirds falling in the 6 months from April through September.

Florida soils vary from organic material to coral limestone. There are many shallow lakes, the largest being Lake Okeechobee which covers some 730 square miles in the southern part of the state. Most of the land south of Lake Okeechobee is naturally swamp and marsh, but much has been developed for farming.

Florida has about 850,000 acres of citrus trees, of which about 750,000 acres are in bearing. Citrus is grown commercially in 36 of Florida's 67 counties. The industry was once located primarily in the central ridge area and a narrow strip along the east coast. A shortage of desirable available land, freezes in 1957–58, 1962, and the 1980s, and competition from urban development have pushed the industry to expand inland along the east coast and in the southwest portion of the state. The largest acreages are located in the ridge counties of Polk and Lake, but the second- and third-place counties, St. Lucie and Indian River, are on the east coast. Rainfall on the east coast, where the Indian River flows, is the highest in the state, and the region is known for the highest quality citrus.

There are about 3,000 acres of tropical fruits grown in Florida, mostly in Dade County. The mango is the most important fruit of this group. The acreage of tropical fruits is increasing slowly.

Vegetable crops are harvested in Florida from early fall through June, with spring and fall crops being grown in the northern counties and crops for winter harvest in the south. There are many different vegetable growing areas in the state. Florida is an important source of fresh tomatoes, especially in the late fall and early spring. Risk of frost damage limits shipments during midwinter, and that niche is largely left to Mexico.

Florida agriculture faces many challenges in the future. Many of these are common to other farming regions as well, such as labor, urbanization, and markets. Of particular importance in Florida are the risks from unusual frost damage, and the impact of plant insect and disease pests.

Unusual Frost Damage

A major problem facing Florida citrus growers, and to a lesser extent vegetable growers, in recent years has been the frequency of freezing temperatures that have damaged the citrus crop, killed trees, and severely injured winter vegetables. Frosts in 1939/40, 1957/58, and 1962/63 had significantly reduced the citrus crops, but the frequency of freezing temperatures seems to be on the increase, with low temperatures inflicting serious damage in 1980/81, 1982/83, 1983/84, 1984/85, and 1989/90. This damage occurred in spite of the shift in plantings to more southern counties that has been going on ever since the complete devastation that followed unusually cold conditions back in 1885/1886.

The development of frozen concentrated orange juice following World War II brought increased tree plantings, some on soils less well suited to citrus production or in areas with greater susceptibility to frost. Production of citrus rose fairly steadily, apart from the brief period following the frost of 1962/63, and reached a peak of 285 million boxes in 1979/80. Since that time there has been a substantial drop in production almost every year. Were it not for the importation of frozen concentrate from Brazil, severe shortages and high prices would certainly have resulted.

One of the worst freezes in recent years occurred in January 1985, when the weather stayed cold for several days. Many of the oranges were able to be salvaged for juice processing, but growers reported that about half the living citrus acreage was killed. Much of what was lost will either be moved into other crops or used for development.

Cold weather also adversely affected the vegetable crops. Yields are reduced, crops are delayed, fields have to be replanted, and the flow of product to market is interrupted. The replanting of fields that have been damaged by frost tends to result in excessive supplies later by adding to production from the regularly scheduled plantings in other areas. Many Florida growers have therefore begun to concentrate more on the late fall and early spring crops and leave the winter fresh vegetable sales to the Mexican growers.

Disease and Insect Pests

The heavy concentration on citrus in Florida makes that state's agricultural economy particularly vulnerable to insects and diseases that might attack this crop. Citrus canker is a virulent disease that exists in other parts of the world and has appeared from time to time in Florida. Continued effort is necessary to eradicate infestations of this pest when they appear, and to operate strict quarantines to prevent entry.

MEXICO

Mexico is our leading source of imports of fresh fruits and vegetables, shipping 38 million hundredweight to this country in 1988 (Table 4.3). Tomatoes

Table 4.3. U.S. Imports of Mexican Fresh Fruits and Vegetables, 1988.

	QUANTITY	PERCENT OF TOTAL FROM MEXICO	PERCENT OF TOTAL U.S. IMPORTS
	1,000 cwt	percent	percent
Tomatoes	7,752	20.0	97.9
Cucumbers	4,078	10.5	95.3
Watermelons	2,970	7.7	97.6
Cantaloupes	2,511	6.5	71.9
Onions, Dry	2,290	5.9	81.3
Squash	2,177	5.6	97.7
Bananas	1,982	5.1	3.1
Onions, Green	1,513	3.9	100.0
Limes	1,203	3.1	95.7
Tomatoes, Cherry	835	2.2	100.0
Mangoes	651	1.7	80.0
Other	10,799	27.9	28.5
Total	38,761	100.0	30.0

Source: USDA AMS, 1989.

are by far the most important commodity, accounting for 20 percent of Mexican shipments and close to 98 percent of all our tomato imports. Mexico is also an important source of cucumbers, watermelons, cantaloupes, squash, green onions, cherry tomatoes, limes, and mangoes. These vegetables are shipped mainly during the winter season, although an increasing quantity is being shipped from the northern Baja Peninsula during the late summer and fall. In all, Mexican fresh fruit and vegetable shipments to this country accounted for 30 percent of our total imports in 1988. Excluding bananas Mexico supplied 55.7 percent of our fresh fruit and vegetable imports that year.

Fresh fruits and vegetables have been shipped from Mexico and the Carribbean countries to the United States during the winter at least since the early 1900s. Producers in this country did not, however, begin to consider these imports a significant threat until the late 1950s (Cook, 1956). Much of the imported produce, then as now, was grown under the auspices of U.S. firms. Florida growers were attracted to operations in Cuba and west coast producers to operations in Mexico. By the end of the 1950s, however, U.S. producers began to feel alarm over Mexico's growing share of the U.S. winter vegetable market.

The development of irrigation and transportation facilities in western Mexico, the embargo on the importation of products from Cuba into the U.S. in 1962, the termination of the Bracero program in 1964 under which special provisions were in effect covering the entry of foreign seasonal agricultural workers, all contributed to the expansion of exports of winter vegetables from Mexico to this country and Canada during the early 1960s (Heertford, 1971). At the same time the acreage of tomatoes and other warm season crops for winter sales declined in California and Texas where weather conditions were not as favorable. In Florida though, especially in the southeast where much of the winter vegetables are grown, the weather in winter is usually favorable for warm season crops.

A report by Emerson (1980) describes the economic aspects of Mexico's winter vegetable growing and marketing, and the factors underlying its impact on the U.S. vegetable industry. Mexico's vegetable industry expanded dramatically in the 1970s, with about a fourth of the total output destined for export. Mexican policymakers continually raised crop support prices so that the competitive position of grains, oilseeds, sugarcane, and cotton improved more than that of horticultural crops. But despite the rising prices of competing crops, vegetable farming in 1980 was considered one of the most lucrative farming activities in Mexico because of improved farm productivity and excellent financial returns from vegetable exports.

In Mexico five states are important in the growing of vegetables and fresh fruit for export to this country. The most important state is Sinaloa which is on the west coast opposite the southern tip of the Baja California peninsula. The Sinaloa region has accounted for two-thirds of the vegetables and melon production of the country (Figure 4.3). The city of Culiacan is the capital of Sinaloa. Michoacan and Guanajuato, west of Mexico City, are important in the production of cantaloupes and strawberries. Sonora, across the border from Arizona, and Baja California also are sources of vegetables and strawberries for export. Yucatan, on the tip of the east coast peninsula also produces significant quantities of winter vegetables.

Shipments of produce from Mexico enter the United States mainly through Nogales, Arizona; Laredo and McAllen, Texas; and Key West, Florida. Most of these exports are destined for United States markets. A significant portion, however, eventually goes to Canada and lesser amounts go to Europe and to other Latin American countries. Tomatoes are by far the most important crop, and Sinaloa Province around Culiacan the most important producing area.

Mexican agricultural production increased dramatically during the period 1940–65, primarily as a result of major land reforms and irrigation programs begun earlier (Heertford, 1971). During 1963–65 about 15 percent of total public investment was applied to water projects. In the state of Sinaloa alone almost 800,000 hectares (2 million acres) were scheduled to be under irrigation when all projects were completed.

A major government program has been undertaken to broaden the distribution of farm operation. Between 1917 and 1965 over 46.5 million hectares

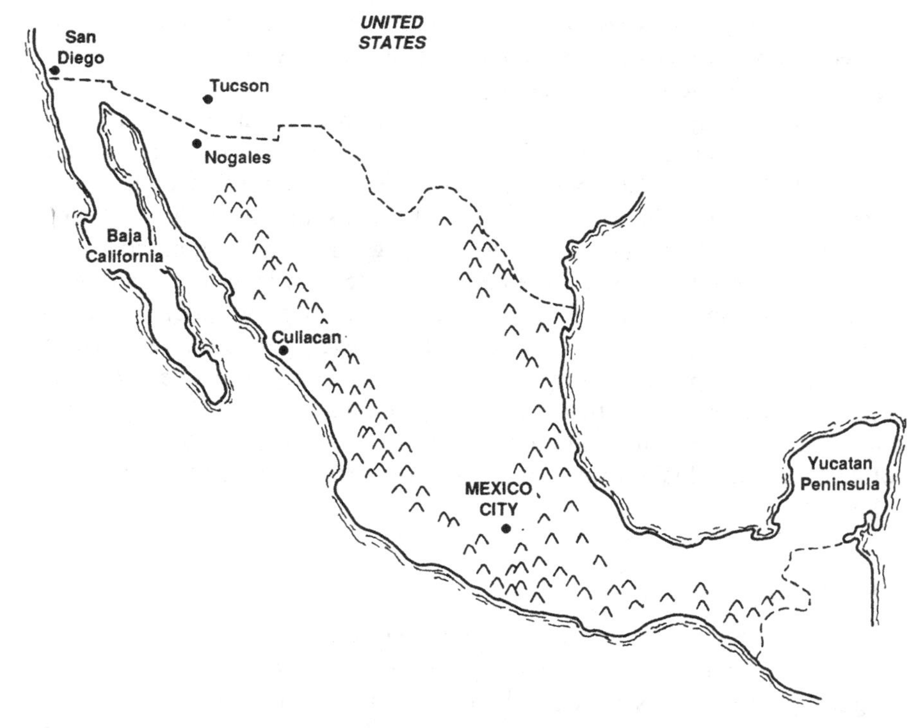

Figure 4.3. Map of Mexico.

were expropriated from large estates and distributed to 2.3 million previously landless peasants, or *ejidatores,* who now represent a substantial proportion of Mexico's farmers. Returns to family labor on the *ejido* are low, partly because the Mexican Agrarian Code prohibits the rental or sale of ejido farms. But even though productivity on these farms is low, net increases in output are believed to have resulted from the transfer of resources from the private to the ejido sector.

The Mexican vegetable industry initially developed largely as a result of investment from U.S. sources. Many growers obtained credit and market outlets by associating with a broker or produce handler from the United States. Some U.S. growers have also been directly involved in Mexican production. Most of these arrangements take the form of partnerships since Mexico does not permit foreign nationals to own land within 50 kilometers (32 miles) of a coastline or 100 kilometers (64 miles) of an international boundary without special arrangements. Mexican producers are now becoming less dependent on U.S. firms for financing. Growers have developed a high degree of crop technology and use modern machinery in growing and packing fresh vegetables. Large numbers of hand laborers are still needed, however, especially for harvesting. Wage rates, of course, are substantially below those in this country.

Government supported grower associations play an important role in Mexican vegetable production. The national association of horticultural producers (UNPH) is composed of affiliated associations in each major state, and local associations within each of these states. Sinaloa has one of the strongest state associations, CAADES. Legal authority is granted by the Mexican government through the national, state, and local associations to conduct programs of acreage control, export quantity and quality control, and other industry efforts. The associations have formulated regulations for packing, shipping, crossing the border, and selling vegetables. Recommended acreages for planting are based on an analysis of market demand and what can be produced in an area. The local associations apportion allotments to individual growers, based on past performance and size of operations.

Sinaloa Region

There is a heavy concentration of winter vegetable production in the State of Sinaloa near the city of Culiacan, between 500 and 600 miles south of Nogales, Arizona. This is the southern extremity of the Sonoran desert that stretches all the way from southern Arizona. Dams impound the rain that falls on the western slopes of the Sierra Madre mountains during the summer and would otherwise flow down the many rivers that run westward across the coastal plain. An extensive system of irrigation ditches delivers the water to growers during the dry winter season.

This region is capable of producing many different warm season crops such as oilseeds, cotton, melons, sugarcane, and citrus as well as vegetables. The winter vegetables are mainly grown on large private farms that have the latest equipment for grading, packing, and shipping. The vegetables for export to the United States and Canada are shipped north by truck to the CAADES compound just south of the U.S. border. Here the shipments are checked by USDA inspectors to make sure they meet the minimum CAADES specifications for export before moving across the border to the U.S. Customs station. At Customs the shipment is checked against the manifest and duties determined, and random shipments are inspected by the Animal and Plant Health Inspection Service (APHIS) for violative pesticide residues. Import brokers arrange for the payment of duty before the tractor–trailer load is permitted to move to one of the many Nogales, Arizona, wholesaler warehouses.

The tariff on fresh tomatoes from Mexico during the early 1980s was 1.5

cents per pound from September 1 through November 14, 2.1 cents per pound from November 15 to the end of February, and 1.5 cents per pound from March 1 through July 14. The tariff is a significant but not a major expense. A recent study (Buckley et al., 1986) found that export costs for a 25-pound carton of tomatoes from Sinaloa to Nogales, Arizona, consisting of transport costs, export fees, and duties amounts to $2.28. This was equivalent to 37 percent of total production and marketing costs for Mexican tomatoes. By comparison, export costs for peppers amounted to 47 percent of production and marketing costs, for cucumbers 54 percent, for green beans 33 percent, for eggplant 45 percent, and for squash 31 percent. Florida, on the other hand, has a competitive advantage primarily because of its location relative to the U.S. market as VanSickle has reported (1989). Mexico, however, has maintained a substantial share of the U.S. winter vegetable market.

Northern Baja California

The region of Baja consists of a peninsula that extends 800 miles south of the southern California border. The farming areas are separated by the central mountains running the length of the peninsula which create numerous coastal valleys. The three leading farming regions are Mexicali, located in northeastern Baja, bordering California's Imperial Valley and Arizona's Yuma Valley; San Quintin, on the Pacific coast 150 miles south of the U.S. border; and Santa Domingo, also on the west coast and 600 miles south of the border. Nearby Mexicali is the San Luis Valley of Sonora, bordering Arizona.

A few years ago a USDA study (Zepp and Simmons, 1980) concluded that high production costs, limited availability of good water, disease problems, weather extremes, and uncertain market prices made future substantial expansion of tomato exports from the Baja area unlikely. These challenges have apparently been overcome, at least in part.

Fresh market tomato export production has been expanding in Baja California. Tomato production in the San Quintin Valley has competed directly with that of southern California in the spring and fall. According to a California study quoted in Moulton (1988) the rapid growth in vine ripened tomato exports was due to a significant cost advantage in 1987. The majority of this advantage was due to lower labor costs, $3 per day compared to $40 in San Diego, and rent for unimproved land of $10 per acre compared to $210. Major sources of investment have been mainland (Sinaloan) Mexican growers, and growers from the U.S. Baja production enables Sinaloan growers to extend their season and maintain a more stable labor supply. Strawberry production has also expanded in recent years in Baja California.

The California report concludes that the future development of vegetable production in Baja and San Luis depends on several factors, including: (1) the availability of good water and suitable land; (2) the availability of sufficient labor to support expanded production; (3) policies of the Mexican government regarding U.S. investment in agriculture and partnerships of both U.S. and Mexican growers with ejidos; (4) the development of the necessary infrastructure, and (5) the U.S. climate for fresh produce imports (e.g., country of origin labeling and consumer attitudes on food safety).

REFERENCES

Buckley, Katherine C., John J. VanSickle, Maury E. Bredahl, Emil Belibasis, Nicholas Guterrez, 1986. *Florida and Mexico Competition for the Winter Fresh Vegetable Market.* U.S. Department of Agriculture, Economic Research Service Agricultural Economic Report No. 556, June.

Cook, A. Clinton, 1956. *Survey of Mexican Winter Vegetable Production.* U.S. Department of Agriculture, Foreign Agriculture Service, FAS-M-3.

Ely, George, Jim Gray, Marc Leinwald, 1977. *Small Scale Farming in the Westlands: A Rural Land Appraisal.* Institute of Government Affairs, University of California, Davis.

Emerson, Leonidas P. Bill, Jr., 1980. *Preview of Mexico's Vegetable Production for Export.* U.S. Department of Agriculture, Foreign Agriculture Service FAS M-297.

Heertford, Reed, 1971. *Sources of Change in Mexican Agricultural Production 1940–65.* U.S. Department of Agriculture, Foreign Agriculture Economic Report No. 73.

Highstreet, Allan, Carole Frank Nuckton, Gerald L. Horner, 1980. *Agricultural Water Use and Costs in California.* University of California, Giannini Information Series 80-2.

Martin, Philip L., 1988. *Harvest of Confusion.* Boulder, Colorado, Westview Press.

Mines, Richard, Philip L. Martin, 1986. *A Profile of California Farmworkers.* University of California, Division of Agriculture and Natural Resources, Giannini Information Series No 86-2, July.

Moulton, Kirby, Study Group Leader, 1988. *Marketing California Specialty Crops: Worldwide Competition and Constraints—Competitiveness at Home and Abroad.* University of California, Agricultural Issues Center.

Nuckton, Carole Frank, Warren E. Johnston, 1985. *California Tree Fruits, Grapes, and Nuts: Location of Production and Trends in Acreage, Yields, and Production, 1946–1983.* University of California, Giannini Foundation Information Series, No. 85-1.

Rausser, Paige D., 1988. California Vegetables: Water Needs in 1989. *Vegetables and Specialties.* U.S. Department of Agriculture, TVS-245.

U.S. Department of Agriculture, Agricultural Marketing Service, 1989. *Fresh Fruits and Vegetable Shipments by Commodities, States, and Months,* FVAS-4 Calendar Year 1988.

——— Economic Research Service, 1989. *National Food Review,* Volume 12, Issue 4, October–December.

VanSickle, John J., 1989. Import Competition in the U.S. Winter Fresh Vegetable Industry. *Vegetables and Specialties Situation and Outlook Report.* U.S. Department of Agriculture, Economic Research Service, TVS-247, March.

Zepp, G. A., R. L. Simmons, 1980. *Producing Fresh Tomatoes in California and Baja California: Costs and Competition.* U.S. Department of Agriculture, Economics, Statistics, and Cooperative Service, ESCS-78.

Chapter 5

The Marketing System and Firms Involved: An Overview

The United States marketing system for fresh fruits and vegetables is unusual in many respects. Many of the commodities handled are highly perishable, and move long distances to market. This has given rise to specialized storage and transportation methods, and to specialized firms that have developed their own methods of operation. This special character of the marketing system exists all the way from farm production through to retail and food service operations. This chapter will provide a broad overview of the system and the firms involved.

THE MARKETING SYSTEM

This country annually produces, on a farm weight basis, about 68 billion pounds of fruits, vegetables, and potatoes for fresh consumption (Table 5.1). This total includes an estimate of the production of several items not reported officially such as watermelons and cantaloupes, cabbage, and many minor fruits and vegetables. Almost 12 billion pounds of fresh produce is imported, more than offsetting the 5 billion pounds exported. Almost all the bananas we eat are imported, and they account for about half our total fruit imports. Very little fresh citrus or potatoes are imported. We do now bring in a substantial quantity of tropical and minor deciduous fruit. About one-quarter of the citrus produced for fresh sale is exported. We currently have available per person per year about 304 pounds of fresh fruits and vegetables of which fresh fruit constitutes 103 pounds, fresh vegetables 121 pounds, melons of all kinds 22 pounds, and potatoes both white and sweet 58 pounds.

U.S. grown fresh produce moves to domestic markets through many different channels. There are four major stages in the marketing process.

Table 5.1. U.S. Supply and Disposition of Fresh Fruits, Vegetables, and Potatoes, Late 1980s.

				DOMESTIC UTILIZATION	
	PRODUCTION	IMPORTS	EXPORTS	TOTAL	PER CAPITA
	million pounds				pounds
Fresh Fruit					
Bananas	12	5,938	—	5,950	24.1
Citrus	8,256	672	2,059	6,869	28.2
Apples	5,632	264	667	5,229	21.3
Other Fruit	6,470	1,245	645	7,070	28.7
Total	20,370	8,119	3,371	25,118	103.3
Major Vegetables	23,764	2,559	1,139	25,184	101.5
Other Vegetables[1]	4,710	100	50	4,760	19.3
Melons, all kinds[1]	4,800	744	174	5,370	21.8
Fresh Potatoes					
Irish Potatoes	13,267	399	126	13,540	54.6
Sweet Potatoes	900	—	—	900	3.8
Total	67,811	11,921	4,860	74,872	304.3

[1] Author's estimate.

Source: Adapted from Putnam, 1989; USDA ERS, 1989a; USDA ERS, 1989b; USDA ERS, 1989c; USDA ERS, 1989e.

These consist of shipping point operations close to the area of production, long distance domestic transportation, wholesale operations at terminal or destination markets, and retailing or food service to the final consumer. Almost all produce moves through all four stages. Small quantities, however, go directly from farms to consumers or directly to local retail stores and food service operations, without involving long distance transportation or destination wholesaling.

Exports usually move directly from shipping point through border crossing stations or ports of embarkation to their final destination. The exporter may be located anywhere along the line. Imports from Canada and Mexico enter at many different points; imports from overseas generally arrive through the major ports.

Complete information is not available on the volume of fresh fruits and vegetables moving through each major stage of the marketing system nor the value added at each stage. Based on available data, however, it has been possible to estimate the quantity of produce moving through each stage and the value added at each stage (Table 5.2 and Figure 5.1). These data should only be considered approximate, but do reflect the importance of the var-

Table 5.2. Value Added in Marketing Fresh Fruits and Vegetables in the United States, Late 1980s.

		OPERATIONS AT SHIPPING POINT		DOMESTIC LONG DIST. TRANSPORTATION		DESTINATION WHOLESALING		RETAILING AND FOOD SERVICE	
	BEGINNING VALUE	VALUE ADDED	ENDING VALUE	VALUE ADDED	ENDING VALUE	VALUE ADDED	ENDING VALUE	VALUE ADDED	ENDING VALUE
	billion dollars								
	Disposition of U.S. Farm Production and Imports								
U.S. Farm Sales for									
Direct Mkt.	$0.40							$0.20	$0.60
Home Use	6.40	$5.60	$12.00	$2.00	$14.00	$1.00	$15.00	7.50	22.50
Food Service	1.80	1.80	3.60	0.60	4.20	0.60	4.80	7.20	12.00
Export	0.80	0.60	1.40	0.25	1.65				
Total	$9.40	$8.00	$17.00	$2.85	$19.85	$1.60	$19.80	$14.90	$35.10
Imports for									
Home Use	$1.60			$0.32	$1.92	$0.30	$2.22	$1.11	$3.33
Food Service	1.20			0.22	1.42	0.18	1.60	2.40	4.00
Total	2.80			0.54	3.34	0.48	3.82	3.51	7.33
Grand Total	$12.20	$8.00	$17.00	$3.39	$23.19	$2.08	$23.62	$18.41	$42.43

	Source of Produce for Home Use, Food Service and Export								
For Home Use									
Direct Sales	$0.40							$0.20	$0.60
U.S. Shipmts.	6.40	$5.60	$12.00	$2.00	$14.00	$1.00	$15.00	7.50	22.50
Imports	1.60			0.32	1.92	0.30	2.22	1.11	3.33
Total	$8.20	$5.60	$12.00	$2.32	$15.92	$1.30	$17.22	$8.81	$26.43
For Food Service Use									
U.S. Farms	$1.80	$1.80	$3.60	$0.60	$4.20	$0.60	$4.80	$7.20	$12.00
Imports	1.20			0.22	1.42	0.18	1.60	2.40	4.00
Total	$3.00	$1.80	$3.60	$0.82	$5.62	$0.78	$6.40	$9.60	$16.00
Export Sales from U.S. Farms									
Total	$0.80	$0.60	$1.40	$0.25	$1.65				
Grand Total	$12.00	$8.00	$17.00	$3.39	$23.19	$2.08	$23.62	$18.41	$42.43

Source: Adapted from USDA ERS, 1989a; USDA ERS, 1989c; USDA ERS, 1989e; USDA NASS, 1989; and Author's estimates.

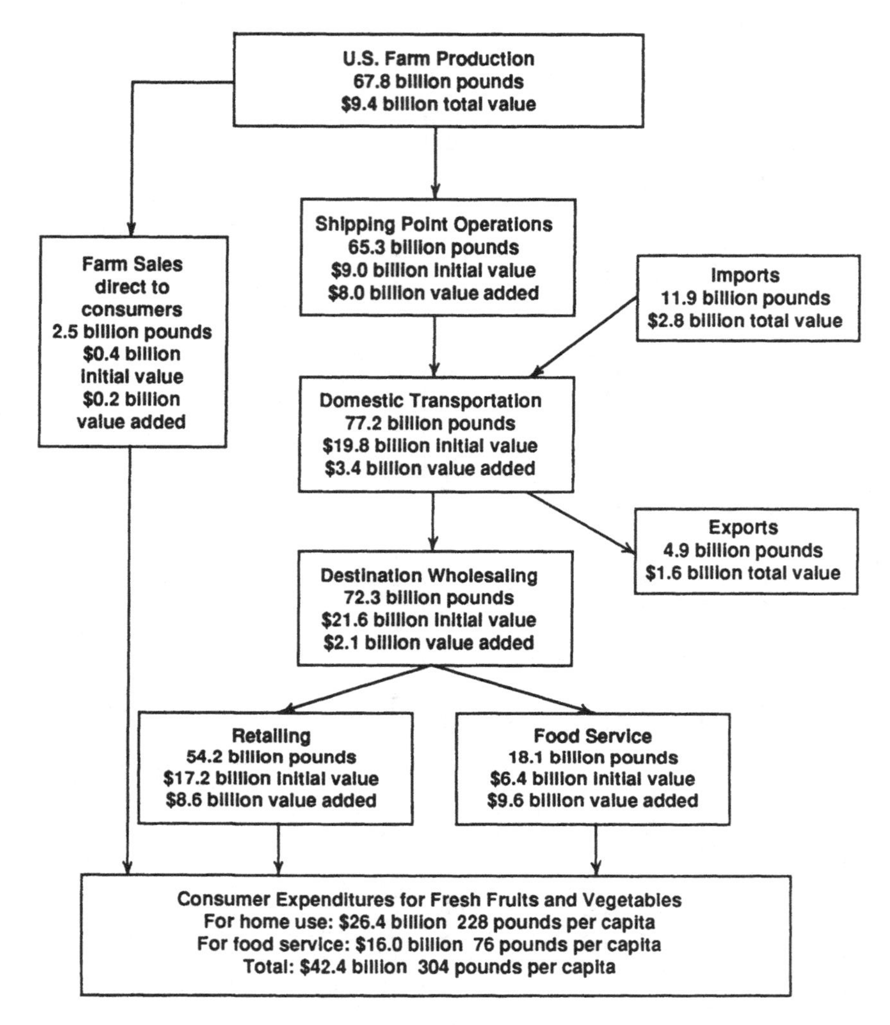

Figure 5.1. U.S. Marketing System for Fresh Fruits and Vegetables, late 1980s. *Source:* Adapted from Table 5.2.

ious channels and marketing stages. The value added represents the additional value of the commodity resulting from all the operations and services performed at each stage and not just the added value of labor, management, and capital as the term usually is defined. The value added in this case corresponds to what is sometimes called the marketing margin or charge.

According to this analysis U.S. production of fresh fruits and vegetables destined for the domestic market had a farm value of about $8.6 billion in the late 1980s, and a retail value of $35.1 billion. Fresh produce destined for export has a value of $0.8 billion at the farm that grew to $1.65 billion when shipped. Imports valued at $2.8 billion on arrival in this country increase in value to $7.3 billion by the time they reach the point of consumption. Thus the total consumer bill for U.S. produce as well as imported fresh fruits and vegetables amounts to $42.4 billion annually.

Fresh fruits and vegetables for home use have an initial value at the farm or on arrival in this country of about $8.2 billion. This increases to $26.4 billion by the time they reach the consumer. Fruits and vegetables destined for food service use have an initial value of $3.0 billion that expands to $16 billion by the time of final consumption. Of the total final value of $42.4 billion, the proportion provided through food service is valued at $16 billion or almost 38 percent of the total compared to $26.4 billion or 62 percent sold for home use.

The estimate of value added at each stage provides an indication of the relative importance of the marketing functions at these stages. Farm production is valued at $9.4 billion and imports at $2.8 billion, $8.0 billion is added in value at shipping point, $3.39 billion is charged for long distance domestic transportation, wholesaling at destination adds $2.08 billion, and retailing and food service operations together contribute $18.41 billion to bring the total value to $42.43 billion. This analysis shows that the largest single major sector is that of food service, which adds $9.6 billion to the total final bill for U.S. consumed fresh produce, exceeding even retailing at $8.81 billion.

Estimates of the quantity of fresh produce moving directly from farms to consumers vary considerably. The reported total tends to be underestimated because of the reluctance of direct marketers to reveal data on cash sales. Direct sales of 2.5 billion pounds represent about 3.7 percent of total production. The retail value of fresh fruits and vegetables marketed directly is set at $0.6 billion.

The marketing system for fresh fruits and vegetables, given the nature of the commodities and the great distances they travel to market, is truly impressive. This is especially so considering the fragmented nature of many firms involved in growing and marketing these crops. Until fairly recently, except for the large retailing organizations that also provided wholesaling services, most firms in the business were relatively small and operated at only one stage in the system. Except at the final stage of retailing and food service the majority of firms are still family owned and operated, although in recent years larger firms have entered and extended their operations to several stages. Some appreciation of the system can be gained from an overview of the structure and organization at each stage and the mechanism for coordination.

FIRMS INVOLVED IN MARKETING

Farm Production

Fresh fruits and vegetables are widely grown across the country on many farms. Even so, as we have seen, production does tend to be concentrated in major growing areas on specialized farms. There is considerable range in size of operation. A few large businesses now produce a major share of many commodities, while at the other extreme there are numerous small or parttime operations of which some are involved in selling directly to consumers. The value at the farm gate of fruits and vegetables grown for fresh market sale, domestic and export, was estimated at about $9.4 billion.

The Census of Agriculture is the acknowledged source of information on the number, size, and type of farms in this country. Unfortunately, the Census does not distinguish between the growing of fruits and vegetables for fresh market and for processing, which in some cases may involve quite different operations. The harvesting of many fruits and vegetables for processing is highly mechanized, while harvest of the same items for the fresh market must be performed largely by hand labor.

The Census has other limitations. The Census definition of a farm includes very small businesses. This deficiency can be overcome by considering only the larger farms. But the Census of Agriculture, unlike the Census of Manufacturers, counts operations in noncontiguous counties or in different states as separate businesses even though owned and controlled by the same firm. Many farm businesses now have widely scattered operations that the Census does not consolidate. This also tends to result in an underestimate of the average size of farm businesses.

Even with these limitations the Census does give a good indication of the large number of farms growing fruits, vegetables, and potatoes, the relatively small average size, and the few large operations that grow a major share of the crop. In 1987, for example, the Census counted 14,782 farms growing Irish potatoes, 60,819 farms growing vegetables for sale, and 120,434 farms with orchards or vineyards (Table 5.3). Farms with potatoes had an average of about 90 acres, farms with vegetables for sale an average of 60 acres, and farms with land in orchards and vineyards an average of about 40 acres. But averages mean little. Almost half the farms growing Irish potatoes had less than 5 acres and these had less than 0.1 percent of the total acreage. In the case of farms with vegetables harvested for sale and farms with land in orchards and vineyards, about 38 percent had less than 5 acres and these had about 2 percent of the acreage. At the other extreme the farms with 500 acres or more of these crops numbered less than 4 percent but had from 30 to 40 percent of the acreage.

The size of farm enterprises varies across the country. In 1987 there were

Table 5.3. Distribution of Farms with Potatoes, Vegetables for Sale, and Land in Orchards and Vineyards By Number and Acreage, 1987.

	IRISH POTATOES		VEGETABLES HARVESTED FOR SALE		LAND IN ORCHARDS AND VINEYARDS	
ACRES	FARMS	ACRES	FARMS	ACRES	FARMS	ACRES
			percent of total			
Under 5.0	48.3	0.1	38.1	1.3	37.5	2.1
5.0–24.9	10.6	1.4	29.2	5.8	37.1	11.3
25.0–49.9	6.9	2.8	11.6	7.0	10.9	10.0
50.0–99.9	9.6	7.5	9.6	11.4	7.1	12.7
100.0–249.9	13.8	24.2	7.1	18.8	5.0	19.6
250.0–499.9	7.2	27.6	2.4	14.5	1.4	12.3
500.0 and more	3.6	36.4	1.9	41.2	1.0	32.0
Total	100.0	100.0	100.0	100.0	100.0	100.0
			Numbers of Farms and Acres			
Farms	14,782		60,819		120,434	
Acres (1,000)		1,301.0		3,467.6		4,560.2

Source: USDC, Bureau of the Census, 1989.

3,787 farms in California harvesting vegetables for sale with an average of 233 acres (Table 5.4). The 2,822 farms in New York reporting that crop had an average of 53 acres each. These were vegetables destined for processing as well as sale on the fresh market. California had almost as many farms with less than 50 acres of vegetables as did New York, but more farms with larger acreages. California had 487 farms with 500 or more acres of vegetables, and these farms had over 70 percent of the acreage. New York had only 59 farms with 500 acres or more, and these farms had 38 percent of the vegetable acreage.

The Census also reported that 1,215 farms in California had 75 percent of the sales of vegetables, sweet corn, and melons with total sales of these crops of $1.7 billion, or an average of $1.4 million per farm. Another 1,993 California farms accounted for 75 percent of the sales of fruits, nuts, and berries with total sales of $2.1 billion. In New York 768 farms had 75 percent of the vegetable, sweet corn, and melon sales with total sales of $131 million, while 574 farms had 75 percent of the sales of fruit, nuts, and berries in New York with a total of $107 million.

California has a long history of large land holdings. Liebman (1983) has traced the pattern of land ownership from the Spanish and Mexican periods

Table 5.4. Distribution of Farms with Vegetables for Sale By Number and Acreage, California and New York, 1987.

	CALIFORNIA		NEW YORK	
ACRES	FARMS	ACRES	FARMS	ACRES
		percent of total		
Under 5.0 acres	22.2	0.2	33.9	1.3
5.0–24.9	22.0	1.1	34.5	7.2
25.0–49.9	9.2	1.4	10.8	6.9
50.0–99.9	10.4	3.1	9.6	12.2
100.0–249.9	14.1	9.8	6.6	19.5
250.0–499.9	9.1	13.7	2.3	14.8
500.0 and more	12.9	70.8	2.1	38.1
Total	100.0	100.0	100.0	100.0
		Numbers of Farms and Acres		
Farms	3,787		2,282	
Acres (1,000)		882.7		150.1

Source: USDC, Bureau of the Census, 1989.

to recent years, and suggested reasons why large holdings developed in California. Another study (Vallarejo, 1980) attempted to identify California farm operators with more than 1,000 acres of cropland in 1978, and presented a directory of the 211 largest operators by cropland size.

The methods growers have used to secure control of land resources has varied. Large landowners like Prudential Insurance Company, Southern Pacific Land Co., and Standard Oil of California have leased to others, while some like Tenneco West farmed about half the land they owned and leased the rest to other farmers. On balance, though, the 211 large farm operators in California in 1978 owned about half the cropland they farmed and leased the rest.

The 1980 study also looked at the entry of nonfarm businesses into farming. Tenneco and Superior Oil were identified as two large corporations that had recently entered farming. But only 19 of the 211 very large farms were owned by nonfarm businesses. These 19 operated about 300,000 acres or only 12 percent of the total acreage operated by the 211 farms. The others, many of them family held operations, had expanded through farming operations. Reasons for the many large farms in California are further discussed by Carter and Johnson (1980).

Shipping Point Operations

Many different sizes and types of firms are involved in fresh fruit and vegetable marketing operations at shipping point. Some perform physical functions such as harvesting, sorting, packing, cooling, storing, and the like, while some facilitate operations by negotiating sales, arranging transportation, or providing financing or other services. Some firms perform many of these functions while others specialize. Some are independent operations while others are diversified, or integrated horizontally or vertically. Shipping point operations are estimated to add about $8.0 billion in value to the $9.0 billion worth of fresh fruits and vegetables received from farmers.

The predominant type of firm at shipping point is generally the grower-shipper. This firm grows a proportion of the product it markets, and obtains the remainder from other growers either through outright purchase, for a fee per package, or some other type of arrangement. They may act as sales agent for a grower's entire crop or handle the product of a particular field or orchard block. The popularity of the grower-shipper organization apparently stems from the need of the shipper to have a guaranteed source of supply but also the ability to market a larger volume effectively than can be grown by one operation efficiently. Shippers without their own source of supply also enter into several different types of arrangements with growers. Most packer-shippers are proprietary firms, but some are farmer-owned cooperatives. Many locally owned packing and shipping cooperatives join together in sales and merchandising federations like Sunkist in California or Sealdsweet in Florida. Brokers arrange sales or purchases (buying brokers), usually without taking title to the product or physically handling it.

Large grower-shipper operations have developed in different ways. Most have expanded through internal growth like A. Duda and Sons in Florida and Merrill Farms in California. Some large family farm businesses have been bought by outside corporations, such as the Bud Antle operation in the Salinas Valley that was purchased by Castle and Cooke a few years ago. Large landowners like Tenneco and Superior Oil turned to intensive farming and marketing operations when irrigation water became available. The favorable prospects for fresh produce sales a few years ago attracted large firms with interests in other food products such as Castle and Cooke and Campbell Soup, and conglomerates with major interests outside the food business such as R. J. Reynolds and Procter and Gamble. Campbell Soup took a giant step into the shipment of fresh produce when the company purchased the Mendelson-Zeller Co., a large privately owned shipper.

The expansion in fresh fruit and vegetable sales has also encouraged some farm operations to shift production away from other crops as did Salyer American, a large cotton grower in the Central Valley, a few years ago. New marketing firms have also been developed by experienced people who left other companies to form their own, as did the management of Sun World and Fresh Western Marketing. Until the tax laws were changed a few years ago individuals seeking tax shelter had an incentive to invest in limited partnerships to grow and market fruits and vegetables.

Some firms have apparently sought to obtain a large share of the sales at shipping point not only for more effective distribution but also to have some control over markets and prices. The Purex Corporation tried this in the late 1960s without success. To what extent this objective motivates larger firms to increase their share of the market is hard to say.

Transportation

Long distance transportation is an important component in the marketing of fresh fruits and vegetables. Transportation within this country is estimated to add about $3.39 billion to the value of fresh produce moving to market. Of this amount $2.6 billion is added in hauling U.S.-grown product to U.S. markets, about $0.25 billion for hauling U.S. products for export, and $0.32 billion in hauling imports from port of entry to final destination.

The organization and operation of the four modes of transportation used for fresh fruits and vegetables differs greatly, as we shall see in chapter 17. Truck transportation is by far the most important method of shipping fresh produce. Truck transportation of perishables has been exempt from the economic regulation that has often restricted potential carriers from handling other types of products. Much of the trucking consequently is still performed by a large number of owner-operators of individual trucks. In most cases these truckers negotiate the charge for each individual load separately. The rates reflect current supply and demand conditions and can fluctuate widely throughout the year. Individual owner-operators often lease their vehicles to fleet operators by the trip or for longer periods. Fleet operations are gaining in importance, as is the use of annual contracts with shippers or receivers. There is a definite lack of information, however, on the organization and operation of firms involved in truck transportation of fresh produce stemming largely from the lack of economic regulation.

Railroads dominated the long-distance shipment of fresh fruits and vegetables prior to the advent of refrigerated tractor trailers. A few major railroads handle most of the 8 to 10 percent of the fresh produce that is shipped by rail, of which the bimodal trailer-on-flat-car (TOFC) operations is a significant part. Except during periods of truck shortages the shipment of fresh produce by rail car is largely limited to bulky and less perishable items like onions and potatoes.

Airfreight represents less than 1 percent of domestic fresh fruit and vegetable shipments, mostly handled by the major passenger airlines on regular flights. Air shipments from overseas are also relatively minor, being largely restricted to highly perishable items or to those needed to take advantage of a strong market.

Ships are seldom used now to carry fresh produce between places on the mainland. Costs and time requirements are prohibitive. Overseas shipments are handled by many different types of carriers, some domestic and some foreign. Shippers may contract for space on a cargo vessel, or charter a whole ship. Shipping firms are exempt from antitrust provisions with respect to setting charges, and many have joined forces to establish conferences to set rates and make other regulations.

Destination Markets

Wholesaling is estimated to add $2.08 billion to the value of fresh produce, of which $1.30 billion is involved in supplying produce for off premises use or retail sale, and $0.78 billion added to produce destined for food service use. Firms wholesaling fresh fruits and vegetables to retail stores in major markets fall generally into two categories, and will be described further in chapter 18. These are the vertically integrated wholesale–retail operations and the independent or nonintegrated wholesalers.

At one time most wholesaling of fresh fruits and vegetables at destination markets to retailers was a highly specialized and independent business. Many small firms just handled one type of commodity such as potatoes, tomatoes, or citrus fruit and specialized in certain services such as repacking or ripening. Merchandise was handled on consignment, and customers came to the place of business to purchase the produce and carry it away. There are still many independent wholesale produce houses, but most now handle many commodities and perform many different services. Integrated wholesale–retail and food service operations mainly buy direct from shipping point, but still depend on nearby independent wholesalers for specialty items and to fill in when unexpected shortages occur. The independent wholesalers still have a large share of the nonintegrated retail and food service business.

The integrated wholesale–retail operations consist of corporate and voluntary chains. Corporate chains own and operate 10 or more retail stores, mainly supermarkets, and service them out of their own distribution center or centers. Kroger Company and Safeway Stores are leading corporate chains. Voluntary retail chains are groups of retail stores that are serviced either by wholesalers under a franchise or contract agreement, or by the retailers themselves who cooperatively operate one or more wholesale distribution centers. Super Valu Stores and Fleming Companies are the two largest wholesale companies that service retail foodstores.

Foodservice operations today are supplied with fresh produce in many different ways. At one time, however, the predominant foodservice wholesalers were specialized firms called purveyors who devoted their entire operations to obtaining fresh produce and providing it to hotels, restaurants, and institutions. The growth and change in the foodservice business coupled with the expansion of wholesaling into other commodities and the integration with distributors has resulted in several new types of firms. Many corporate and voluntary retail chains now supply produce to foodservice operations. Many large fast food, restaurant, and hotel chains have established their own wholesale suppliers that in many cases also handle fresh fruits and vegetables. Large firms that began by supplying foodservice operations with frozen foods or dry groceries have extended their operations into perishables. Sysco Corporation, the largest foodservice distributor, with annual sales of $5.3 billion, now handles fresh produce in many markets (USDA ERS, 1989d). Wholesale clubs constitute a hybrid wholesaler-retailer providing merchandise to consumers and small foodservice operations. These are expanding in many areas. And the full line service wholesalers that supply fresh produce to retailers also sell to foodservice operations.

The Census of Wholesale Trade reports data on five types of food wholesalers. These are specialty merchants that handle perishables, including frozen foods and meats, general line merchants engaged in the wholesale distribution of a general line of groceries, limited line establishments that embrace a narrow range of dry groceries, manufacturers' sales offices, and agents and brokers. In 1982 food wholesalers operated 38,500 establishments with 673,800 employees and a sales volume of $288.6 billion (Table 5.5).

Epps (1986) described the structure and performance of specialty wholesalers. These are firms that distribute perishable foods to supermarkets, restaurants, food service facilities, and institutions. Three types were iden-

Table 5.5. Establishments, Employees, and Sales Volume by Type of Grocery Wholesaler, 1982.

	ESTABLISHMENTS	EMPLOYEES	SALES VOLUME
	thousands	thousands	billion $
Specialty Merchants	18.4	284.8	88.3
General Line Merchants	3.3	127.6	60.7
Limited Line Merchants	7.3	91.0	25.7
Manufacturers' Sales Offices	4.7	123.0	63.9
Agents and Brokers	4.8	47.4	50.0
Total	38.5	673.8	288.6

Source: Adapted from Epps, 1986.

tified in this study: merchants who take title to the goods they handled, agents and brokers who link producer-sellers and buyers, and manufacturers' sales branches.

In 1982 there were 4,800 establishments operated by fresh fruit and vegetable merchant wholesalers with a total of 86,500 employees and a sales volume of $18.5 billion. There were another 895 establishments operated by agents, brokers, and commission persons dealing in fresh fruits and vegetables with sales of $5.7 billion. The merchant wholesalers had an average of 20 employees per establishment, but almost one-quarter of the fruit and vegetable specialty wholesalers had only 1 or 2 employees. There were no dominant national merchant fruit and vegetable wholesalers, but in some markets such as Boston, Cincinnati, and Milwaukee the four largest had over half the sales in 1982. The four largest fruit and vegetable wholesalers nationally had an average of 23 establishments in 1982.

Fruit and vegetable wholesalers tended to concentrate on their specialty. About 40 percent of their sales were to retailers, almost 40 percent to other wholesalers, and the remainder to foodservice operators, government, and other buyers. The largest proportion, 72 percent, were incorporated followed by almost 20 percent that were individual proprietorships. About 2 percent were cooperatives, but these had the largest average sales.

The general line merchant wholesalers are also an important link in the marketing of fresh fruits and vegetables. These firms are fewer in number and generally larger than the specialty wholesalers, but carry many other products beside fruits and vegetables. Many have cooperative or voluntary arrangements with the owners of retail food stores, or specialize in serving institutions, restaurants, or fast-food organizations. There are several that have sales of over $1 billion (USDA ERS, 1989d).

Terminal produce markets such as the Hunt's Point Market in New York or the South Water Street Market in Chicago exist in major cities where many nonintegrated wholesalers conduct their operations. Current trade reports would indicate, however, that about half the nonintegrated wholesalers and almost all the integrated wholesale-retail firms conduct their businesses apart from the terminal markets.

Retailing

The value added to fresh produce in retailing and foodservice exceeds by far the value added in any other major marketing function or in farm production. In the mid-1980s the value added in retailing amounted to $8.81 billion and in foodservice to $9.6 billion for a total of $18.41 billion compared to $12.20 billion in total farm and import value of these products.

In 1988 there were 234,575 foodstores in the country with annual sales of $331 billion (USDA ERS, 1989d). Of these 165,453 were grocery stores with total sales of $312 billion, and 69,122 were specialized food stores with total sales of $19 billion. Sales of specialized produce stores were estimated at $2 billion. Corporate chain stores accounted for 62.8 percent of grocery store sales in 1988, or from about 50 percent in 1965. Supermarkets, defined as having annual sales of $2.5 million or more (1985 dollars), constituted 16.4 percent of total grocery stores but accounted for 74.6 percent of total sales. Operations of these firms will be described further in chapter 19.

The two largest retail food chains, Kroger Company and Safeway Stores, each had sales of more than $15 billion in 1987. Sales of the four largest chains amounted to 17.6 percent of total grocery store sales that year, but this figure means little since none of the top retailers operated in all regions of the country. Local market area concentration is more significant and was higher. The average four-firm market concentration of food retailers across all Standard Metropolitan Statistical Areas (SMSAs) was 58.3 percent in 1982 (USDA ERS, 1989d).

The volume and value of fresh produce retailed directly by farmers to consumers is difficult to measure. The 1982 Census counted 143,353 farms that sold farm products directly to consumers having a total value of $498 million. This included 14,076 farms that sold $143 million worth of fruits and nuts, and 12,868 farms that sold $93 million worth of vegetables and melons directly to consumers. Few records are kept, and direct marketers are reluctant to reveal financial information. Making allowance for sales of potatoes and for underreporting it appears that fresh produce farm sales might amount to $0.4 billion. With a value added in marketing through pick-your-own, roadside markets, farmer markets, and other means of about $0.2 billion, this gives a retail value of $0.6 billion.

Foodservice

This country had 727,000 foodservice establishments in 1987 that sold $184 billion worth of meals and snacks, excluding alcoholic beverages (USDA ERS, 1989d). Foodservice sales can be divided into commercial sales of $136 billion, and noncommercial sales of $48 billion. Restaurants and fast food operations each accounted for over 40 percent of the commercial sales, the remainder divided between cafeterias, lodging places, retail hosts, recreation, and drinking places. Leading in value of noncommercial sales were educational institutions and hospital and care facilities, followed by plant and office buildings, military, vending, transportation, associations, correctional facilities, day care centers, and other providers (USDA ERS, 1989d). The operations of foodservice firms will be discussed further in chapter 20.

SYSTEM COMMUNICATION AND COORDINATION

The organization and operation of the marketing system for fresh fruits and vegetables is under continual change, and the intense competition in most sectors has in general resulted in a high level of performance. The impact of weather on production plans, the highly perishable nature of most commodities, the great distances much of the volume travels to market, the long storage periods for the less perishable items, the many separate firms involved in the marketing process from farm to consumer, and the interrelationship between consumer demand for individual products, all make it especially difficult to coordinate the system. Criteria for satisfactory performance would be to provide consumers with the quantity and variety of products they want at prices that reflect costs, while returning to those involved in production and marketing sufficient revenues to cover expenses and adequately compensate initiative and innovation. Such an analysis of performance is beyond the scope of this book.

Many firms still operate at only one stage in the production and marketing process: in farming, in shipping point operations, with transportation, as wholesalers on markets, or in retailing or foodservice. Furthermore many of these firms are small businesses with highly seasonal operations that might have an imperfect view even of what is happening in their own segment of the industry.

A large share of the product moving from shipping point to destination market is bought by verbal agreement sight unseen by the chain store or terminal buyer from the shipper for a negotiated price based on each party's understanding of current and prospective market conditions. Transportation is often arranged through a third party. To successfully complete such a transaction requires effective methods of communication, the confidence of each in the integrity of the other, and also knowledge that a procedure exists for arbitrating differences that might arise later. The fresh fruit and vegetable marketing system has developed several coordinating mechanisms that have significantly contributed to the improvement in the performance of the system.

Communication and coordination have improved in part through changes in the structure of the industry. Firms have grown in size internally or through merger or acquisition, and integrated horizontally and vertically. Large corporate chains have maintained buying offices in important shipping areas, and do a small share of their own packing and shipping. Shippers have tended to diversify, some even operating at several locations across the country or handling contraseasonal imports to maintain their label in front of their customers all year. More and more marketing organizations are operating sales offices in major markets to have direct contact with developments there as well as to serve their customers better.

A number of public and private services have been developed to aid in coordination. A system of federal and state grades has been developed that enable the product to be sorted into lots that can be described in relatively objective terms, and this will be described further in chapter 7. Such grades can be the basis for regulating the movement and merchandising of the commodity, but primarily aid in communication when buyers are not in a position to view the product. The official grades are supplemented by shipper's brands that become the basis for transactions between buyers and sellers who have dealt with each other long enough to establish confidence.

The USDA Federal–State Inspection Service is charged with the responsibility of enforcing regulations regarding the use of official grades, but also provides a service at cost to buyers and sellers as to whether or not a given lot meets a specified grade. This can fill the dual responsibility of providing official certification that the product in question is as represented by the seller to the buyer, and also whether the product at destination fails to meet agreed upon specifications.

The National Agricultural Statistics Service (NASS) (formerly the Crop Reporting Service) gathers and disseminates official information on the production and storage stocks and season average prices of major commodities. The public information on production and stocks is used as a basis for trade pricing and marketing decisions that is supplemented by many firms with information from their own private sources.

The Market News Service, jointly operated by the U.S. and state departments of agriculture, gathers information on prices and movement of product at selected shipping points and major terminal markets. This information is promptly disseminated by teletype to other shipping points and markets, and by postal service, telephone recordings, newspaper reports, and electronic mail to private users.

The U.S. and state departments of agriculture also administer legislation intended to protect consumers, wholesale buyers, and sellers from unfair or fraudulent practices. Such legislation includes provision for licensing and bonding produce buyers. At the federal level the Perishable Agricultural Commodities Act (PACA) establishes a basis for mediating disputes between wholesale buyers and sellers and establishing penalties against the guilty party. The PACA also now provides a mechanism for sellers to have first priority to recover moneys due from buyers who subsequently declare bankruptcy.

Private firms also offer services that improve communication and coordination within the system. The publishers of two of the larger trade papers offer business and financial rating services (called respectively the Blue

Book and the Red Book) that provide information industry wide for all firms on the type of firm, the commodities handled and approximate total volume, the net assets of the company, a rating of the business character, and for buyers an indication of promptness of payment. One of the major trade papers offers a computer based electronic information and communication service to subscribers. This makes available information from NASS and Market News as well as information from other sources, including reporters on the trade paper, and provides an electronic mail service so subscribers can communicate with each other and also post notices concerning products they wish to buy or sell.

REFERENCES

Carter, Harold O., Warren E. Johnston, 1980. *Farm-Size Relationships With An Emphasis on California.* Department of Agricultural Economics, University of California, Davis.

Epps, Walter B., 1986. *Specialty Grocery Wholesaling: Structure and Performance.* U.S. Department of Agriculture, Agricultural Economic Report No. 547.

Liebman, Ellen, 1983. *California Farmland: A History of Large Agricultural Landholdings.* Roman and Allanheld.

Marion, Bruce W., Ed., 1986. *The Organization and Performance of the U.S. Food System.* D. C. Heath.

Putnam, Judith Jones, 1989. *Food Consumption, Prices, and Expenditures, 1966-87.* U.S. Department of Agriculture Economic Research Service, Statistical Bulletin No. 773.

U.S. Department of Agriculture, Agricultural Marketing Service, 1989a. *Fresh Fruit and Vegetable Arrival Totals for 23 Cities.* FVAS-3. Calendar Year 1988 and earlier issues.

—— 1989b. *Fresh Fruit and Vegetable Shipments by Commodities, States, and Months.* FVAS-4 Calendar Year 1988 and earlier years.

—— Economic Research Service, 1989a. *Fruit and Tree Nuts Situation and Outlook Yearbook,* TFS-250, August.

—— 1989b, *Fruit and Tree Nuts Situation and Outlook Report,* TFS-252, November.

—— 1989c, *Vegetables and Specialties Situation and Outlook Yearbook,* TVS-249, November.

—— 1989d, *Food Marketing Review, 1988.* Agricultural Economic Report, No. 614.

—— 1989e, *Foreign Agricultural Trade of the United States Calendar Year 1988 Supplement,* July.

—— National Agricultural Statistics Service, 1989. *Agricultural Prices 1988 Summary.* Pr 1-3(89), Washington, DC.

Vallarejo, Don, 1980. *Getting Bigger: Large Scale Farming in California* California Institute for Rural Studies.

Chapter 6

Marketing Systems for Three Major Fruits and Vegetables: Oranges, Apples, and Tomatoes

Each different fruit or vegetable has its own particular characteristics which make it unique with respect to growing, storing, packing, grading, shipping, wholesaling, and retailing. There are some similarities, though, in the way different fresh fruits and vegetables are marketed, because of similar physical and economic properties. These similarities and differences will be illustrated by examining some features of the marketing systems for oranges, apples, and tomatoes.

ORANGES

Varieties and Types

Many different types of citrus fruit are grown in this country, but oranges are the most important in terms of production and value. Next in order of importance are grapefruit, lemons, tangerines, temples, tangelos, and limes. Florida grows about two-thirds of total U.S. citrus production, California one-quarter, and Texas and Arizona the rest. Florida leads in the production of oranges and grapefruit, while California grows most of the lemons apart from a small quantity produced in Arizona.

Citrus readily hybridizes and so there are a large number of species and varieties (Seelig, 1982). Oranges grown commercially in this country are the sweet orange type, of which there are four principal kinds. These are the common orange such as the Florida or California valencia, the navel orange such as the Washington navel of California, the blood or pigmented orange such as the Ruby, and acidless orange grown in some Mediterranean areas.

Production and Utilization

U.S. production of oranges has ranged from 175 to almost 275 million boxes in recent years. Frost damage in Florida has severely reduced the crop several times in the past decade. About three-quarters of the orange crop is processed each year, but the proportion varies depending on the size of the crop and the stocks of processed products on hand. Production and marketing of the fresh and processed products are closely related. Many growers ship part of their production to fresh market and part to processing plants. Frozen concentrated orange juice is believed to have taken part of the market for fresh oranges. The marketing of fresh oranges is thus an integral part of a total citrus industry that extends across international borders and is changing rapidly. Ward and Kilmer (1989) list six primary factors that are contributing to the dynamics of this industry.

In 1988–89 U.S. orange production amounted to 207 million boxes, of which Florida produced 147 million or two-thirds (Table 6.1). California produced almost all the rest, apart from small quantities reported from Texas and Arizona. In Florida early and midseason varieties constituted 58 percent of the crop, with Valencias the rest. In California the navels and miscellaneous varieties accounted for about 60 percent of production, and Valencias the remainder.

Table 6.1. Oranges: U.S. Production and Utilization, 1988–89.

STATE AND SEASON		PRODUCTION	UTILIZATION	
			FRESH	PROCESSED
			1,000 boxes	
Florida	Early, Midseason	85,300	5,436	79,864
	Valencia	61,300	3,020	58,280
	Total	146,600	8,456	138,144
California	Navel and Misc.	34,000	25,900	8,100
	Valencia	23,000	13,500	9,500
	Total	57,000	39,400	17,600
Arizona	Navel and Misc.	550	500	50
	Valencia	1,150	715	435
	Total	1,700	1,215	485
Texas	Early, Midseason	1,200	1,021	179
	Valencia	650	115	535
	Total	1,850	1,556	714
U.S. Total	Early, Mid., and Navel	121,050	32,857	88,193
	Valencia	86,100	17,350	68,750
	Total	207,150	50,207	156,943

Source: Adapted from USDA NASS, 1989a.

About 24 percent of the 1988–89 U.S. crop was sold fresh and 76 percent processed. Over 94 percent of the Florida crop was processed, and Florida processing oranges made up 88 percent of the total processed. About 82 percent of the oranges processed in Florida went into frozen concentrated juice. Only a little more than one quarter of the California crop was processed.

U.S. output of oranges represents about one-quarter of the recorded world total. But Brazil has been increasing production of oranges in recent years and now grows about twice as many as the United States, exporting a large volume of frozen concentrated juice to this country and other countries. Other countries with major orange crops are Italy, Spain, Egypt, Israel, and Mexico.

Grades, Sizes, and Packs

Orange grades vary from state to state and encompass both federal, state, and private grades in some instances. Although federal grades are in effect and are often used, state or private grades are the main governing factor. Oranges are shipped mainly loose in fiberboard boxes, or in mesh bags or film bags in master containers.

The size of cartons in use for oranges varies from one region to another. Cartons used in California and Arizona have a net weight of 38 pounds, compared to the 4/5 bushel carton used in Florida with a net weight of 43 pounds, and the 7/10 bushel carton used in Texas with a net weight of 42 pounds.

Packing and Sorting

Packing oranges for shipment to fresh market involves washing, waxing, treating, and sometimes precooling and degreening. Oranges do not store well, but Valencias in Florida and Texas may be stored up to 8–12 weeks under proper temperature and humidity. Ripe fruit may be left on the tree although at some risk from unfavorable weather. Under the Federal Marketing Order in effect in California and Arizona the shipment of oranges to fresh market is regulated and fruit is harvested as required.

Arrivals at Major Markets

Total arrivals at major markets tend to fluctuate from year to year and sources change. In 1988 California supplied almost 80 percent of the oranges arriving at 23 major U.S. markets, Florida 19 percent, with Arizona and Texas accounting for the rest apart from small quantities from Israel and Mexico (Figure 6.1). Total arrivals tend to peak in late winter and early spring, drop to about one-third of this volume in late summer, and then peak again in December. California ships year round, with most Florida shipments arriving from November through May. California Navels are on the market from October through May, and Valencias from April through October.

Shipping Point Prices

Daily shipping point prices during the season have been summarized into monthly average prices for major grades and sizes of California, Arizona, and Texas oranges by the Market News Service (Table 6.2). The grades reported are Shipper first grade and Choice, and the sizes are 88–113s and 138s. Sunkist established the Shipper first grade many years ago and as this cooperative is still a major factor in the orange market it has been able to maintain this grade as the highest standard in the west.

The spread in prices between different varieties and types and different grades and packs varies from month to month and year to year. The causes for this variation will be discussed in the chapter on prices. In 1988 the highest prices for top grade navels occurred in May at the end of the 1987/

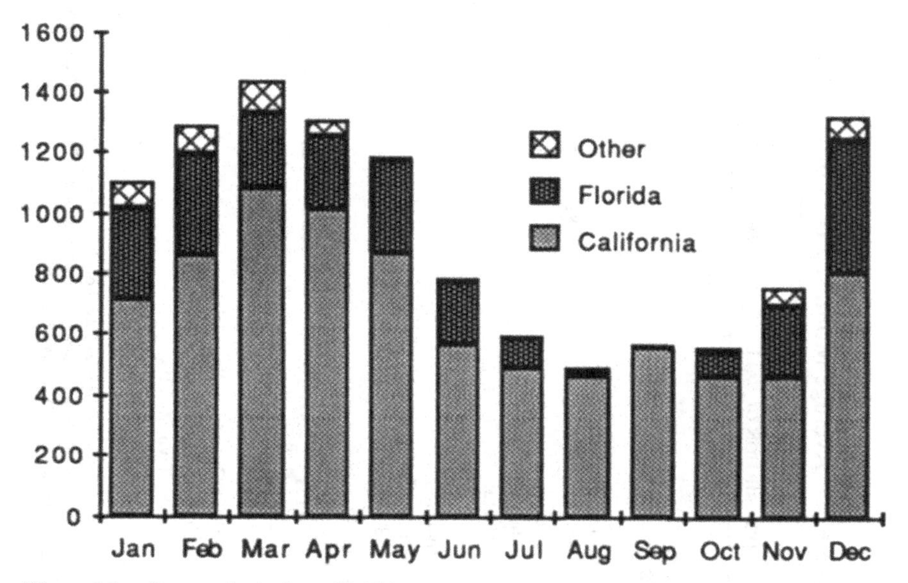

Figure 6.1. Orange Arrivals at 23 Cities by Source and Month, 1,000 cwt, 1988. *Source:* Adapted from USDA AMS, 1989a.

Table 6.2. Average Monthly Prices of Oranges at Selected Leading Shipping Points, 1988.

SHIPPING POINT AND DESCRIPTION OF SALE	JAN	FEB	MAR	APR	MAY	JUN	JUL	AUG	SEP	OCT	NOV	DEC
						dollars per unit						
Carton												
California including Arizona												
Navels												
Shipper first grade												
88-113s	9.81	8.54	7.92	9.81	11.88	—	—	—	—	—	—	7.59
138s	9.51	8.63	7.93	9.74	10.68	—	—	—	—	—	—	5.11
Choice												
88-113s	7.45	6.72	6.50	7.00	8.37	—	—	—	—	—	—	6.36
138s	7.50	7.01	6.61	7.09	8.01	—	—	—	—	—	—	4.67
Valencias												
Shipper first grade												
88-113s	—	—	9.13	9.57	10.65	8.52	8.69	9.18	12.28	9.85	8.81	—
138s	—	—	7.53	7.53	8.83	6.54	5.96	6.97	9.39	7.95	6.50	—
Choice												
88-113s	—	—	6.31	6.81	7.09	5.69	6.69	7.10	9.53	7.43	5.81	—
113s	—	—	6.12	7.00	6.73	5.50	5.24	6.35	8.46	6.60	5.27	—
Texas												
Lower Rio Grande Valley												
Combination U.S. One												
and U.S. Two 80-125s												
Early & Midseaon Varieties	7.82	7.75	7.39	—	—	—	—	—	—	10.00	8.58	7.75

Prices in this table are simple averages of the midpoints of the daily range and are only for months during which the market news offices were in operation in the specified districts. All sales are free on board at shipping point. Unless otherwise stated protective services are extra.
Source: Adapted from USDA AMS, 1989b.

88 season. That year the Shipper first grade oranges, as might be expected, sold at a premium over Choice but the premium varied. It tended to be larger for navels than Valencias, ranging for navels from $3.51 for size 88–113s per carton in May to $1.23 the following December. For Shipper first grade navel oranges the smaller sizes generally sold for less per carton than the larger sizes except in February and March. This year there was a substantial difference in the carton prices for Valencias between sizes with 88–113s selling for substantially more per carton than 138s. Some years the reverse is true. Of course $8.00 per carton for size 100 is 8 cents per orange while $10.00 for a carton of size 138 is only $7\frac{1}{4}$ per orange.

Retail Prices and Marketing Margins

The average retail price of a carton of California navel oranges sold in the Northeast during the 1980–87 seasons amounted to $16.76, increasing from $13.10 in 1980 to $20.01 in 1987 (Figure 6.2). The grower return fluctuated widely over this period, ranging from $1.87 to $5.79 and averaging $3.92. The margin for picking and hauling was relatively stable, rising from $0.70 to $0.84 and averaging $0.80. The packing house margin averaged $2.02

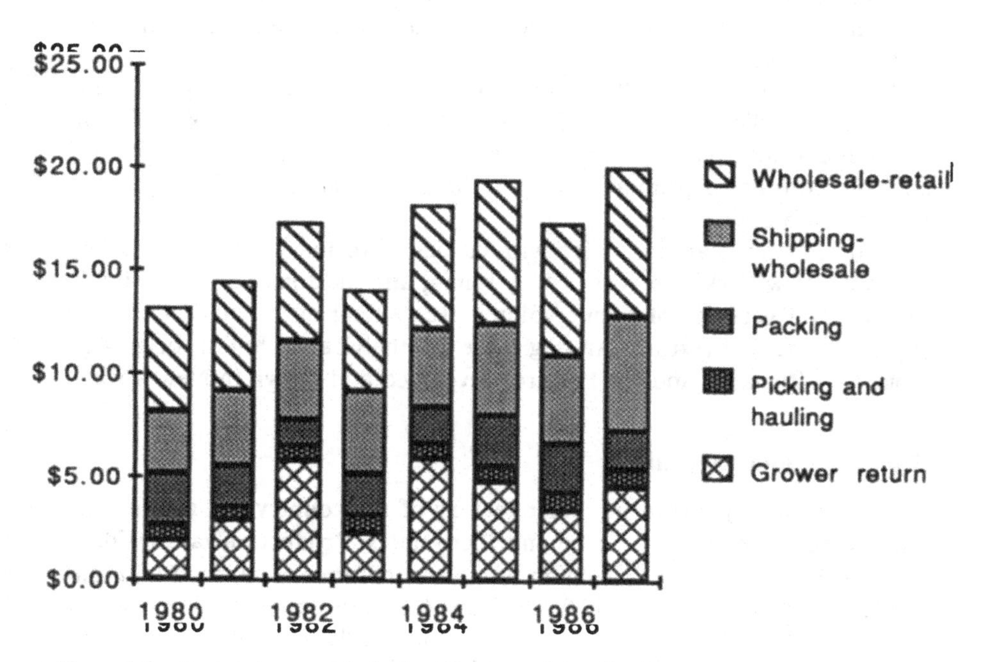

Figure 6.2. Season Average Marketing Margins and Retail Prices per Carton for California Navel Oranges Sold in the Northeast, 1980–87. *Source:* Adapted from Pearrow, 1988.

over this period without showing any strong upward trend. In contrast the margin between shipping point and wholesale prices rose from $2.99 to $5.51 per carton. The wholesale-retail margin, which accounted for the largest share of the retail price every year except one, increased from $4.99 to $7.24 over this period and averaged close to $6.00.

The grower return varied the most over this period, considering trend, and the return for picking and hauling the least. The variation in shipping-wholesale and in wholesale-retail margins was largely associated with the upward trend.

APPLES

Varieties

Many different varieties of apples are grown for commercial sale, and different strains or sports of each major variety (Ricks and Pierson, 1983). Red Delicious is the major apple variety. Production is rapidly increasing and now accounts for about 30 percent of the apple crop. Growers in Washington State have increased acreage and yields of Red Delicious in recent years, and plantings have increased in eastern states also. Red Delicious has been favored in the market because of its shelf life and external appearance. Golden Delicious and Romes are the only other varieties increasing in share of total output.

Some varieties such as R.I. Greening are used primarily for processing into sauce or slices. Golden Delicious and Northern Spy are examples of varieties considered to be dual purpose. Red Delicious and some other varieties are considered primarily fresh varieties but can be used for juice if blended with other varieties.

Certain varieties are identified with particular regions such as the McIntosh with New York, the York with Virginia, and the Jonathan with Michigan. Several varieties of minor volume are gaining in importance due to superior taste and texture. Among these are the Granny Smith, originally from New Zealand, and the Empire, developed and grown in New York.

Production and Utilization

Apples are widely grown in cooler regions of the country but production tends to be concentrated in certain northern states. Production varies widely from year to year due to changing weather conditions. The trend in apple production has been upward, from about 6 billion pounds utilized in the early 1970s to around 8 billion pounds in the mid 1980s. The previous record crop of 8.8 billion pounds, recorded in 1980, was exceeded in 1987 with a total production of 9.9 billion pounds.

Even larger crops are expected in the future. Anderson (1989) projected, based on tree surveys in major apple producing states, that apple production in Washington State would more than double the 1986 output by the year 2000. Production in New York and several other states is expected to increase by about 50 percent over this period. Plantings in South Carolina indicate that that state, from a relatively small base, is expected to produce about five times as many apples in 2000 as in 1986.

Total utilization of U.S. apple production in 1988 was about 9.1 billion pounds, with Washington state leading with about 3.9 billion pounds or over 40 percent of the total. New York produced 910 million pounds, followed by Michigan with 830 million pounds. Together these three states accounted for more than 60 percent of total production. Other important growing areas are California, Pennsylvania, Virginia, North Carolina, and West Virginia (Table 6.3).

Just under 60 percent of the crop is sold fresh, and a little more than 40 percent is processed. Growing for processing and fresh market tends to be carried out in the same regions and in many cases in the same orchards, although processing facilities tend to be concentrated in a few geographical areas. Juice and cider is the most important processed product, using about half the quantity of apples processed. Canned products like apple sauce and slices are next, taking about one-third. The remainder are used for dried and frozen products. Juice and cider is the most important product in Washington, while canned products lead in Pennsylvania and Virginia. Juice and cider production has been increasing, and now leads canning in utilization in New York and Michigan.

There is an active foreign trade in fresh apples (Sparks, 1989). The U.S. exports apples to many different countries. Exports vary from year to year

Table 6.3. Apples: U.S. Production and Utilization, 1988.

STATE	UTILIZED PRODUCTION	UTILIZATION	
		FRESH	PROCESSED
		million pounds	
Washington	3,900	2,850	1,050
New York	910	405	505
Michigan	830	230	600
Pennsylvania	520	138	382
California	630	305	325
Virginia	417	200	217
Other States	1,901	1,135	766
Total	9,108	5,263	3,845

Source: USDA NASS, 1989b.

but amount to about 10 to 15 percent of fresh utilization. Canada and Taiwan share the lead in export markets. Other major customers have been Saudi Arabia, Hong Kong, and the United Arab Emirates. Imports of fresh apples have been on the increase, coming mainly from Canada, Chile, and New Zealand.

Grades, Sizes, and Packs

Apples are subject to five U.S. grades. These are: U.S. Extra Fancy, U.S. Fancy, U.S. No. 1, U.S. Utility, and Combination grades. Some of the more common shipping containers for apples are cartons and boxes loose packed weighing approximately 38 to 42 pounds, tray packed weighing 40 to 45 pounds, and cell packed 37 to 43 pounds. Carton counts with minimum diameters include 2 1/4 inches, 180 count; 2 1/2 inches, 160; 2 5/8 inches, 140; 2 3/4 inches, 120; 3 inches, 96; and 3 1/2 inches, 80 count. Smaller apples are often packed in 3- or 5-pound film bags and shipped in a master carton (Selig and Hirsh, 1978).

Packing and Storing

The major apple varieties are quite different in terms of optimum storage and handling conditions. Red Delicious and some other varieties can be handled easily without bruising, while McIntosh bruises very easily. Fall harvested apples can be held for a few months in common storage, for longer in refrigerated storage, and some varieties can be held until the following summer under controlled atmosphere storage. The atmosphere of higher carbon dioxide and lower oxygen than normal retards the maturing process.

Specialized computer controlled packing lines enable shippers to efficiently separate fruit into several size and grade categories. Storage and handling will be discussed further in the chapter on shipping point operations.

Arrivals at Major Markets

In 1988 apple arrivals at 23 U.S. cities originated in 28 states and six foreign countries (Figure 6.3). About 15 percent of the domestic supply arrived by rail, the remainder by truck. Of the total reported arrivals of 120 million hundredweight, 75 percent came from Washington State. California came in a distant second with 5 percent, and New York followed with 4 percent. Michigan, Pennsylvania, Oregon, Idaho, and Virginia were also important sources of apples in these markets. Imports amounted to about 3 percent

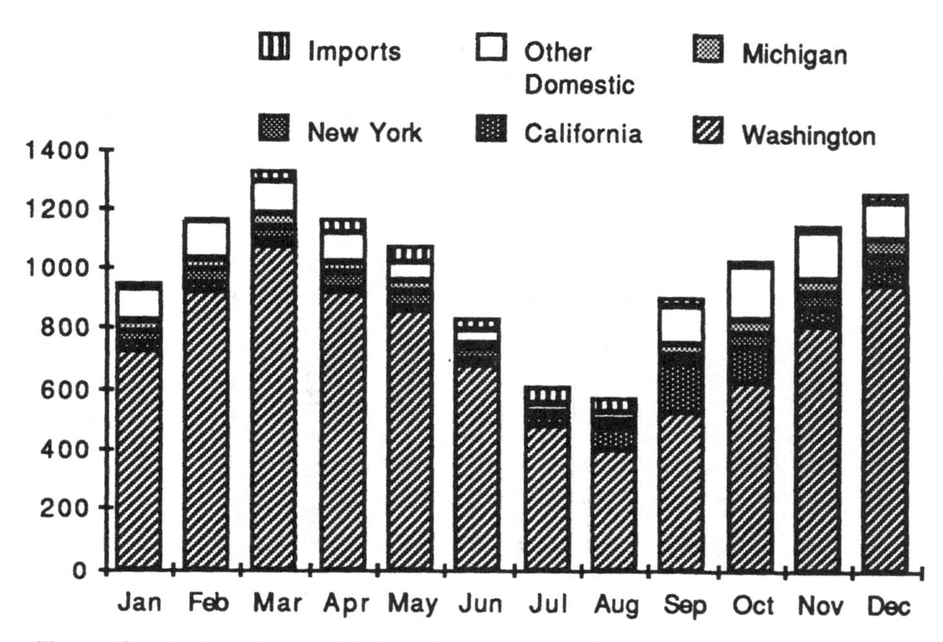

Figure 6.3. Apple Arrivals at 23 Cities By Source and Month, 1,000 cwt 1988. *Source:* Adapted from USDA AMS, 1989a.

of the total, led by Canada, New Zealand, and Chile with receipts also from France, Argentina, and Australia.

In 1988 apple arrivals peaked in March toward the end of the previous crop year, and then in December, early in the following crop year. Washington was in the market in volume every month. Imports from the southern hemisphere, mainly New Zealand and Chile, were greatest from March through August.

Shipping Point Prices

Daily shipping point prices of apples in 1988 were reported for Idaho, Michigan, North Carolina, and the Appalachian district of Maryland, Pennsylvania, Virginia, and West Virginia as well as for the New York regions of Lake Champlain–Mohawk Valley and Western and Central New York in addition to Washington's Yakima Valley–Wenatchee and New York's Hudson Valley. Prices were reported for Red Delicious, Golden Delicious, McIntosh, Cortland, Empires, Red Rome, Jonathan, Ida-Red and Winesap, Red Stayman, and Red York varieties. Prices were reported for tray pack, cell pack, and 12–3 lb film bags. Monthly average prices for

Table 6.4. Average Monthly Prices of Apples at Selected Leading Shipping Points, 1988.

SHIPPING POINT AND DESCRIPTION OF SALE	JAN	FEB	MAR	APR	MAY	JUN	JUL	AUG	SEP	OCT	NOV	DEC
	dollars per unit											
Washington												
Yakime Valley-Wenatchee District												
Wash. State or U.S. Extra Fancy												
Tray Pack, 88-113s												
Golden Delicious	8.95	—	—	—	—	—	—	—	14.38*	12.45*	11.99*	11.75*
controlled atmosphere	—	10.63	11.00	9.68	9.40	8.59	12.53	14.30	—	—	—	—
Red Delicious	7.75	—	—	—	—	—	—	—	20.45*	13.83*	12.15*	12.63*
controlled atmosphere		10.50	11.08	11.15	10.93	11.30	23.87	24.60	—	—	—	—
Granny Smith	10.63	12.58	15.78	16.00	15.45	—	—	—	—	—	16.00*	16.00*
New York												
U.S. Fancy												
Hudson Valley												
Cell Pack, 120s												
McIntosh	11.50	—	—	—	—	—	—	—	9.75	9.19	7.74	7.50
controlled atmosphere	12.33	12.31	12.50	12.50	—	—	—	—	—	—	—	—

Tray Pack 100s												
Red Delicious	—	—	—	—	—	—	—	—	—	14.00	11.50	11.00
controlled atmosphere	9.25	10.00	10.50	9.94	—	—	—	—	—	—	—	—
Film bags, 12 3-lb												
2 1/4″ min or up												
Cortland	—	—	—	—	—	—	—	—	—	8.92	8.17	7.56
McIntosh	7.75	—	—	—	—	—	—	—	8.88	8.88	7.50	7.38
controlled atmosphere	7.75	8.37	8.06	8.05	—	—	—	—	—	—	—	—
Red Delicious	7.50	—	—	—	—	—	—	—	—	9.00	8.63	8.50
controlled atmosphere	8.00	8.06	8.50	8.13	—	—	—	—	—	—	—	—
Golden Delicious	—	—	—	—	—	—	—	—	—	—	7.67	7.00
controlled atmosphere	—	6.58	7.50	—	—	—	—	—	—	—	—	—
Chile Dock Prices NYC/Philadelphia												
Tray Pack												
Granny Smith 80s	—	—	17.64	16.88	15.56	—	—	—	—	—	—	—
100s	—	—	14.79	13.94	13.44	—	—	—	—	—	—	—
113s	—	—	14.14	13.50	12.94	—	—	—	—	—	—	—

*Wash Extra Fancy

Prices in this table are simple averages of the midpoints of the daily range and are only for months during which the market news offices were in operation in the specified districts. All sales are free on board at shipping point. Unless otherwise stated protective services are extra.

Source: Adapted from USDA AMS, 1989b.

Washington State, New York's Hudson Valley, and for apples from Chile on the New York or Philadelphia docks indicate the degree of variability that exists (Table 6.4).

Apple prices, as do prices of other commodities, varied in 1988 from one region to another, and one variety, grade, and pack and month to another. Only careful study over many years would reveal how consistent such variation might be. Granny Smith, either from Washington or Chile, was a preferred variety at both the end of the 1987/88 season and the beginning of the 1988/89 season. In New York McIntosh cell pack apples sold at a premium over Red Delicious tray pack in the early months of 1988, while the reverse was true the following fall. Apples in 12–3 lb film bags generally sold for prices below those in cell or tray packs. Apples in film bags are often smaller than those in cartons, and sometimes of lower grade.

Retail Prices and Marketing Margins

The average retail value of a 42-pound carton of Extra Fancy Washington Red Delicious apples, size 138 or larger, sold in the Northeast during the seasons 1980/81–1988/89 was reported to be $26.55, rising erratically from $20.74 in 1980/81 to $32.96 in 1988/89 (Figure 6.4). The grower return averaged $5.88, varying from $4.07 in 1980/81 to $8.57 in 1985/86. The margin between the grower return and the shipping point price averaged $6.21, remaining fairly stable except for the unusually large amount of $8.52 in 1984/85. This margin reflects the costs of storage, packing, and other shipping point operations. The shipping–wholesale margin that covers transportation and wholesaling operations averaged $5.30. The wholesale-retail margin, largely reflecting retailing charges, was the largest item most years and averaged $9.16 for this period. This margin showed a strong upward trend, increasing from $7.82 to $14.37.

TOMATOES

Types and Varieties

Tomatoes are usually identified by type rather than variety. They come in many different colors, shapes, and sizes. In addition to the usual round red type there are yellow varieties, small cherry tomatoes, and the Italian or plum shaped types (Magoon, 1969). Growers selling direct to consumers will sometimes promote particular varieties, and at least one major shipper has been merchandising the special attributes of a patented variety. Consumers are generally more critical of the quality of tomatoes available in supermarkets than they are of any other fresh fruit or vegetable, faulting them mainly for lack of flavor.

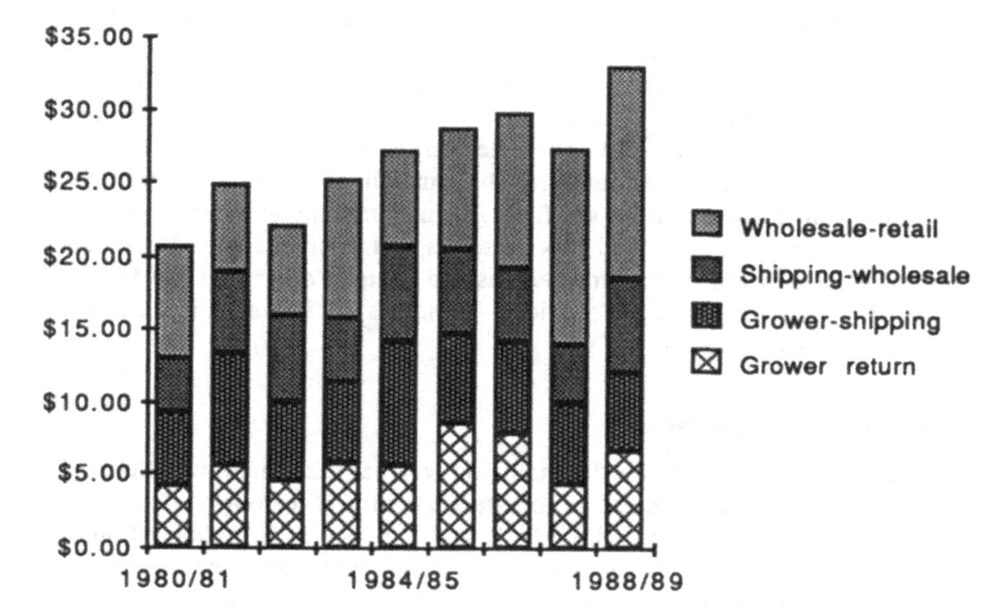

Figure 6.4. Season Average Marketing Margins and Retail Prices in Dollars per Carton for Washington Red Delicious Apples sold in the Northeast, 1981/82–1988/89. *Source:* Adapted from Pearrow, 1989.

Tomatoes are often described in the marketing system as to whether they have been harvested mature green or vine ripe, whether they were greenhouse (or hothouse) grown, or whether they were grown hydroponically (Mongelli, 1984). Tomatoes will ripen if harvested at any stage of development from mature green to fully vine ripe, especially if ethylene gas is used to regulate the ripening or degreening process.

Production and Utilization

Tomatoes are an important vegetable crop both for processing and fresh market. Growing tomatoes for fresh market, unlike oranges and apples, is quite distinct from growing for processing, and the fresh and processed products compete with each other only to a very limited extent if at all. Tomatoes for processing tend to be grown in a few major areas of concentrated production where large canning plants can be operated efficiently. Since fresh tomatoes cannot be stored for any length of time and are in demand year-round, the growing areas for fresh production are scattered across the country and change with changing seasons of the year.

Production of tomatoes for fresh market in the United States has grown from about 18 million hundredweight in 1970 to almost 35 million hundredweight in 1988 (Table 6.5). Commercial production is reported for 19 states, but over half the total is grown in Florida and about one-fifth in California. Two other states, South Carolina and Virginia, each grew over 1 million hundredweight in 1988, and Tennessee is often included with this group. Fourteen other states accounted for the remaining 12 percent of reported production. Most tomatoes are field grown, but some are greenhouse or hothouse grown in Ohio and New Jersey as well as other states. Tomatoes are also widely grown in home gardens and in states other than the 19 for which official estimates are provided. Tomatoes are imported in quantity from Mexico.

Grades, Sizes, and Packs

Grades for tomatoes differ depending on whether the tomato was field or greenhouse grown. Field grown tomatoes are subject to four grades: U.S. No. 1, U.S. Combination, U.S. No. 2, and U.S. No. 3. Greenhouse tomatoes are graded U.S. No. 1 or U.S. No. 2. As with other crops there are differences of opinion as to how well grades reflect consumer preferences (Sun, 1987). Most greenhouse and cherry type tomatoes are shipped in baskets containing about 10 pounds net weight. Mature green tomatoes for gas ripening and repacking close to market are shipped in cartons with volume fill or bulk pack holding 30 or 25 pounds. Pink and vine ripe tomatoes are usually shipped in two-layer cartons holding 20 pounds.

Packing, Storing, and Ripening

Tomatoes are handled according to the stage of maturity at which they are harvested. Tomatoes harvested as mature green will ripen eventually and

Table 6.5. U.S. Production of Tomatoes for Fresh Market, 1988.

STATE	PRODUCTION	PERCENT OF TOTAL
	1,000 cwt	percent
Florida	19,363	55.7
California	7,522	21.6
South Carolina	1,480	4.3
Virginia	1,260	3.6
Tennessee	828	2.4
Other states(14)	4,523	12.4
U.S. Mainland Total	34,776	100.0

Source: Adapted from USDA NASS, 1989c.

can be handled easily. The ripening process can be hastened and made more uniform by holding them in an atmosphere high in ethylene gas. Tomatoes harvested at the pink stage, known as vine ripe, need to be handled more carefully and are usually not ripened or degreened with ethylene. Tomatoes are subject to chilling injury if held much below 50 degrees Fahrenheit. A substantial share of the tomato arrivals at major markets are still gassed, sorted, and packed in consumer packages but with improvements in storage, ripening, and handling methods more of this is being done at shipping point.

Fresh Tomato Marketing System in the U.S.

Fresh tomatoes follow much the same marketing channels as do other fresh fruits and vegetables (Figure 6.5). The operations involved in sorting and packing are different from other crops and particularly the ripening, or degreening as the trade prefers to call this action. The industry has tried to provide consumers with tomatoes that have as much flavor as locally grown or home garden tomatoes but without great success. Vine ripened tomatoes were once believed to meet this requirement but in spite of the added costs of growing and harvesting tomatoes at the pink stage the resultant retail product was not considered superior to tomatoes harvested at the mature green stage, and could not command the additional price necessary to cover costs. As a result the industry has moved away from vine ripened tomatoes to harvesting mature green, and degreening these at shipping point or at destination market. Much, if not most, of the tomato crop is now sold in bulk at retail and there has been a decline in repacking into consumer packages at destination markets, but the practice still continues on a more limited basis.

Arrivals at Major Markets

Arrivals of 12.7 million hundredweight of tomatoes were reported at 23 major U.S. markets in 1988 (Figure 6.6). They came from 26 states and Puerto Rico and seven foreign countries. Less than 5 percent of the total came by rail. Small quantities come by air from Israel, the Netherlands, and Spain. Florida was the most important source of fresh tomatoes, supplying 45 percent, followed by California with 25 percent. Three other states together accounted for a further 5 percent—South Carolina, Virginia, and Tennessee. Mexico was by far the most important source of imports, shipping 18 percent that year. Other states and foreign countries accounted for the remaining 5 percent. Since tomato production is so sensitive to

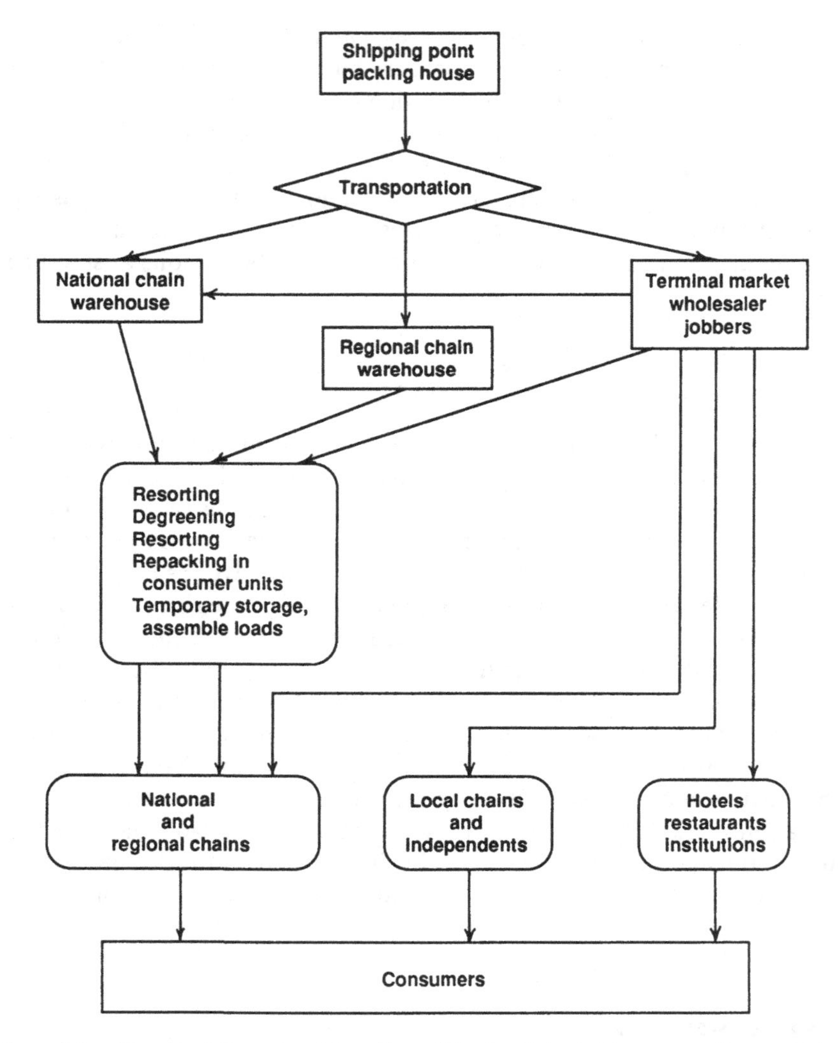

Figure 6.5. How Fresh Tomatoes Move From Shipping Point to Consumers. *Source:* Fahey, 1976.

weather conditions as well as economic forces regional production varies greatly from one year to the next.

Florida sources have a large share of the market from November through June, with peak arrivals in December and May. California is heaviest from July through November. Mexico is now in the market year round with largest shipments in February through April from the Culiacan area, and smaller amounts in the summer and fall from the northern Baja Peninsula.

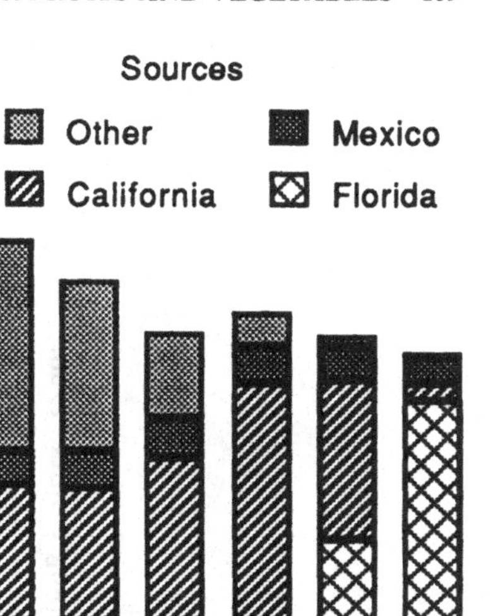

Figure 6.6 Tomato Arrivals at 23 Cities By Source and Month, 1,000 cwt 1988. *Source:* Adapted from USDA AMS 1989a.

Shipping Point Prices

Monthly average shipping point prices quoted for fresh tomatoes from Florida, for Mexican grown tomatoes f.o.b. Nogales, Arizona, for southern California tomatoes and Mexican tomatoes crossing through Otay Mesa, and for tomatoes from the northern Sam Joaquin Valley in 1988 give some idea of the diversity in tomato prices, grades, and packs (Table 6.6). Prices were also reported for the Gonzales–King City District of California, in New Jersey for the Swedesboro and Vineland Auction Sales, for western North Carolina, for the Charleston–Beaufort District of South Carolina, for Michigan, and for the Virginia–Maryland Eastern Shore. Packs for which prices are recorded are the 25-pound carton green, 2-layer flat pink, 3-layer lug pink, 20-pound carton pink, 3-layer lug place pack pink, and 2-layer flats place pack pink. The 25-pound carton green was the most commonly reported pack. Tomatoes were also reported as pink and turning-light red. Green tomatoes were most often quoted in sizes Extra Large, Large, and Medium, but occasionally in 5×6s and 6×6s. Pinks in place pack flats were sized in 4×5s, 5×6s, 6×6s, and 6×7s. The grade reported for Florida and South Carolina was 85 percent or more U.S. No. 1.

As one might expect there were considerable differences between prices for tomatoes by months, from one location to another, and from one size

Table 6.6. Average Monthly Prices of Tomatoes at Selected Leading Shipping Points, 1988.

SHIPPING POINT AND DESCRIPTION OF SALE	JAN	FEB	MAR	APR	MAY	JUN	JUL	AUG	SEP	OCT	NOV	DEC
						dollars per unit						
Florida 85% or more U.S. One Quality												
Florida												
25-lb cartons Green												
5×6s	10.90	6.63	8.59	11.21	7.08	5.96	—	—	—	8.98	12.16	6.90
6×6s	9.53	5.45	7.59	9.01	5.83	5.08	—	—	—	7.60	9.50	4.11
6×7s	8.15	4.78	6.64	6.27	4.83	5.00	—	—	—	—	5.95	3.69
Mexico												
Duty and crossing charges paid												
FOB Nogales, AZ (Point of Entry)												
3-layer lugs place pack Pink												
6×6s	8.27	4.16	7.14	4.58	—	—	—	—	—	—	—	—
6×7s	7.42	3.67	6.57	3.89	—	—	—	—	—	—	—	—
2-layer flats place pack Pink												
4×5–4×6s	7.52	4.49	6.28	6.12	—	—	—	—	—	—	—	—
25-lb cartons Green												

extra large	11.90	6.61	6.00	—	—	—	—	—	—	—	—	—
large	10.40	5.51	5.00	—	—	—	—	—	—	—	—	—
California												
Southern California												
& Baja Crossings through Otay Mesa												
Place Pack												
2 layer flats Pink												
4×5s	—	—	—	—	—	—	7.33	6.13	7.25	5.31	7.49	—
5×5s	—	—	—	—	—	—	6.50	5.78	6.50	4.63	6.44	—
5×6s	—	—	—	—	—	—	5.67	4.44	6.30	4.35	6.00	—
3 layer lugs Pink	—	—	—	—	—	—						
6×6s	—	—	—	—	—	—	6.22	5.91	7.50	6.58	6.25	—
6×7s	—	—	—	—	—	—	5.92	4.58	—	—	—	—
Northern San Joaquin Valley												
25-lb cartons Green												
large	—	—	—	—	—	5.45	6.33	8.11	8.34	5.35	8.44	—
medium	—	—	—	—	—	5.27	4.54	4.77	5.72	4.05	6.50	

Prices in this table are simple averages of the midpoints of the daily range and are only for months during which the market news offices were in operation in the specified districts. All sales are free on board at shipping point. Unless otherwise stated protective services are extra.
Source: Adapted from USDA AMS, 1989b.

and type of pack to another. Generally a carton of large tomatoes sold for more than a carton of smaller ones, but not always. During the winter months f.o.b. prices of Florida tomatoes were generally higher than Mexican tomatoes at Nogales, Arizona. In the late fall the f.o.b. prices of Florida tomatoes were higher than similar prices in California, probably due in part to the shorter distance to the Northeast market.

Retail Prices and Marketing Margins

During the months of January through May from 1980 to 1989 the retail price of a 25-lb carton of large mature green Florida tomatoes in Northeast markets averaged $19.06 (Figure 6.7). During the same period the Florida average f.o.b. price for a similar quantity, adjusted for a 15 percent loss in marketing, came to $8.97. The grower–packer thus received 47.1 percent of the retail value. The shipping point–wholesale margin averaged $3.22. The

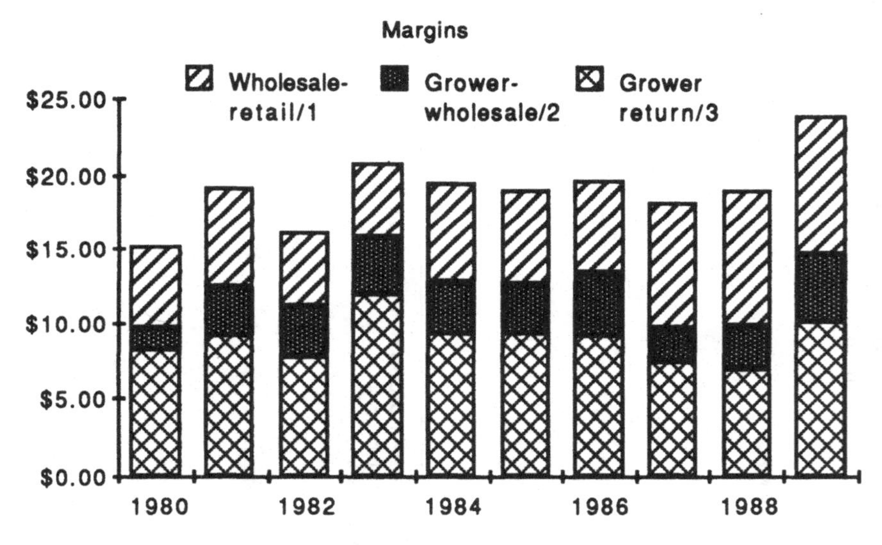

1. The dollar value per container of fresh market tomatoes received by retailers, intracity transporters, and secondary wholesalers.
2. The dollar value received per container of fresh market tomatoes by those who transport tomatoes from shipping point (Florida) to the wholesale market and by primary wholesalers in the wholesale market.
3. The dollar value for Florida tomatoes received at the f.o.b. shipping point. Packing may or may not be performed by growers.

Figure 6.7 Marketing Margins for Florida Fresh Market Tomatoes Marketed in the Northeast, January–May, dollars per 25-pound carton, 1980–89. *Source:* Adapted from Pearrow, 1990.

wholesale–retail margin came to $6.87 or 36 percent of the retail price.

During this period the retail value of a carton of Florida tomatoes varied from $15.17 in 1980, to $24.01 in 1989. Returns to all factors fluctuated widely during this period except for transportation charges.

SUMMARY

Some obvious similarities and differences, even in this broad overview, can be identified in the marketing of these three major fruits and vegetables. Each moves through the sequence of shipping point operations, long-distance transportation, and wholesaling at destination markets, and is mainly retailed through the produce departments of food stores or through food service operations. The physical characteristics differ and require different handling and storage methods. They come mainly from either California, Florida, or Washington State which are widely separated parts of the country. Retailing charges amounted to about one-third of the retail value, with minor variation from commodity to commodity. The return to the grower–packer varied much more than the retail margin.

Differences in physical characteristics that have led to different methods of handling, packing, and shipping point to the necessity for marketers to learn more about these special attributes of each commodity they handle.

REFERENCES

General

U.S. Department of Agriculture, Agricultural Marketing Service, 1989a. *Fresh Fruit and Vegetable Arrival Totals for 23 Cities.* FVAS-3, Calendar Year 1988.

—— 1989b. *Fresh Fruit and Vegetable Prices 1988, Wholesale Chicago and New York City, F.O.B. Leading Shipping Points.* April.

—— Economic Research Service, 1987. *Fresh Fruits and Vegetables Prices and Spreads in Selected Markets, 1975–84.* Statistical Bulletin No. 752.

—— National Agricultural Statistics Service, 1989a. *Citrus Fruits 1989 Summary* FrNt 3-1(89), September.

—— 1989b. *Noncitrus Fruits and Nuts* FrNt 1-3 (7-89), July.

—— 1989c. *Vegetables,* Vg 3-1 (1-89), January.

Produce Marketing Association, 1988. *Produce Marketing Almanac.* Produce Marketing Association, Newark, Delaware.

Oranges

Pearrow, Joan, 1988. U.S. Prices, Costs, and Spreads for California Fresh Oranges and Florida Frozen Concentrated Orange Juice, 1980–87. *Fruit and Tree Nuts*

Situation and Outlook Yearbook. U.S. Department of Agriculture, TFS-246, August.

Seelig, R. A., 1982. *Fresh Fruit and Vegetables Facts and Pointers: Oranges.* United Fresh Fruit and Vegetable Association, Washington D.C., March.

Ward, Ronald W., Richard L. Kilmer, 1989. *The Citrus Industry: A Domestic and International Perspective.* Ames, Iowa, Iowa State University Press.

Apples

Anderson, Bruce L., 1989. *Strategic Alternatives for the New York Apple Industry.* Cornell University, A.E. Res 89-15. September.

Pearrow, Joan, 1989. Washington Red Delicious Apples: Fresh Market Prices and Spreads. *Fruit and Tree Nuts Situation and Outlook Yearbook.* U.S. Department of Agriculture, TFS-250, August.

Ricks, Donald J., Thomas R. Pierson, 1983. *U.S. Apple Supplies: Trends and Future Projections.* Michigan State University, East Lansing, Michigan.

Seelig, R. A., Donald Hirsh, 1978. *Fresh Fruit and Vegetables Facts and Pointers: Apples.* United Fresh Fruit and Vegetable Association, Washington D.C.

Sparks, Amy Larsen, 1989. Situation and Prospects for Fresh Apple Markets. *Fruit and Tree Nuts Situation and Outlook Report.* U.S. Department of Agriculture, TFS-249, March.

Tomatoes

Fahey, James V., 1976. *How Fresh Tomatoes are Marketed.* U.S. Department of Agriculture, Agricultural Marketing Service, Marketing Bulletin No. 59.

Magoon, Robert L., 1969. *Fresh Fruit and Vegetable Facts and Pointers: Tomatoes.* United Fresh Fruit and Vegetable Association, Washington D.C., December.

Mongelli, Robert C., 1984. *Marketing Fresh Tomatoes: Systems and Costs.* U.S. Department of Agriculture, Marketing Research Report No. 1137.

Pearrow, Joan, 1990. Fresh Market and Canned Tomatoes: Prices and Spreads, 1980–89. *Vegetables and Specialties Situation and Outlook Report.* U.S. Department of Agriculture, Economic Research Service, TVS-250, March.

Sun, Theresa, 1987. *Quality Demand and Policy Implications for Florida Green Tomatoes.* U.S. Department of Agriculture, Technical Bulletin No. 1728.

Zepp, G. A., R. L. Simmons, 1980. *Producing Fresh Tomatoes in California and Baja California: Costs and Competition.* U.S. Department of Agriculture, Economics, Statistics, and Cooperatives Service, ESCS-78.

Part II

The Marketing Environment

Chapter 7

Market Information: Agricultural Statistics, Grading and Inspection, Market News, and Other Information Sources

Information as accurate and up-to-date as possible on supply, demand, and prices is essential for anyone directly or indirectly involved in the business of marketing fresh fruits and vegetables. Obtaining such information for these commodities is especially challenging because of the way they fluctuate in production, vary in quality, rapidly deteriorate, and also because many sales are made to buyers located long distances from production areas who are unable to physically examine the product. Most commodities are also produced in many different areas and sold in many markets, and supplies, demands, and prices are all interrelated.

Many firms and individuals rely heavily on their own private sources of information that they obtain either from branches of their own firm or from other buyers and sellers they have learned to trust. Buyers also sometimes visit sources of supply to obtain information, and sellers may periodically go to markets but this opportunity is limited. Some large retailers and food service operators have buying offices in major production areas but still need information from other sources. Some firms or individuals known as *bird dogs* will seek out sources of products for distant buyers to meet the latter's specifications. Onion and apple growers and other commodity associations have sometimes called on their members to develop crop production estimates. The most complete and accurate information on production, supplies, and prices still originates within the U.S. and State Departments of Agriculture, although increasingly such information is also being disseminated by private means.

The analysis of the basic market information to determine trends and relationships and future directions is also largely the function of the Departments of Agriculture and the Land Grant Colleges, but again the trade papers and major trade associations now play an important role in disseminating this information and informing the trade as to its implications.

CROP PRODUCTION, STORAGE STOCKS, DISPOSITION, AND VALUE

The National Agricultural Statistical Service (NASS) of the USDA in cooperation with each of the states is the primary source of estimates of acreage, yields, and production of major crops countrywide (USDA ERS, 1989b; USDA SRS, 1983). California, Florida, New York, and some other states supplement the national program with additional crop production estimates for their own individual state.

Survey methods used by NASS have changed significantly over the past quarter century. Prior to 1960 almost all published statistics were based on nonprobability mail surveys which depended on subjective responses and were prone to bias. Many growers failed to respond, and some of those that did reply underestimated production in an attempt to enhance prices. Since then NASS has increased the use of probability surveys which use objective and enumerative data collection methods.

The USDA vegetable estimating program has been continuously revised. In 1982, NASS reduced its estimating program for fresh market vegetables to nine crops (broccoli, carrots, cauliflower, celery, sweet corn, honeydew melons, lettuce, onions, and tomatoes) plus strawberries. Estimates for asparagus and cucumbers for pickles and reports of pickle stocks were reinstated in 1985. Estimates for nearly all fresh market vegetables include a combination of fresh market and processing production. However, where a large portion of the crop is processed, as for broccoli, cauliflower, asparagus, and carrots, separate estimates are shown for fresh market and processing. Estimates are only made for major producing states.

Fresh market vegetable statistics come from information collected from growers and shippers. Information on acreage to be harvested is released in January, April, July, and October. These forecasts are based on knowledgeable industry contacts and nonprobability mail surveys. Forecasts of celery acres are estimated monthly during the growing season for California, Florida, Michigan, and New York. Estimates of acreage for harvest and production are made during the growing season for strawberries and spring and summer onions. Acreage planted and harvested, yield, production, and value estimates for the entire year are made only at the end of the year, with preliminary estimates in December and revised data the following June.

Estimates of potato acreage planted and for harvest, coupled with forecasts of yield, are developed in a similar manner to those for other field crops. Acreage data are obtained by mail survey from growers and by stratified random sampling methods. To obtain objective yield forecasts sample fields are selected, two small plots are randomly located in each field, and the potatoes are dug just prior to harvest. Enumerators record the number of tubers and their weight from three sample hills, compute the number of hills per acre, and use this information to derive the gross yield per acre. Net yield is obtained by adjusting for harvest losses. This combined with acreage estimates and other information is the basis for production estimates. Production estimates are later checked with information on shipments to fresh markets and processing plants. Stocks estimates are based on storage reports from dealers and farmers, and on fresh shipments and processor utilization. Since these estimates are based on probability sampling their reliability can be determined.

Monthly reports issued on potatoes include information on intentions to plant, forecasts of yield and production, acres for harvest, stocks, and disposition and utilization of the crop.

The fruit and tree nut program provides estimates of production and utilization for 20 noncitrus fruits, seven citrus fruits, six tree nuts, and eight Hawaiian and miscellaneous fruits in states where there is significant production. At the end of the marketing season NASS estimates utilization of production, price for each type of utilization, and bearing acres and value for fruit and nut crops. Separate type or variety estimates are made for cherries, grapes, peaches, pears, plums and prunes, oranges, and pecans.

Except for citrus in California and Florida, most estimates for fruits and tree nuts are based on inquiries mailed to a sample of growers. NASS has incorporated objective measurement techniques to supplement mail surveys for several crops in California, Oregon, Washington, Michigan, Hawaii, and Florida. In Florida objective measurements and up-to-date tree inventories based on aerial photography have replaced mail inquiries in estimating citrus production. In other states, however, many production forecasts are based on judgments and rely on grower reported crop condition and expected yield or production.

For some crops NASS collects and publishes additional information such as fruit or nut size, percentage sound or marketable, grades and color, and production by varieties.

NASS initiated a program in 1965 to obtain regional estimates of the number of fruit trees, by type, variety, and age. Under this program, each state conducts a nearly complete tree census about every 5 years. Using this information, NASS can classify lists of growers by type of fruit and number of trees. This provides excellent frames for probability sampling.

Probability surveys for fruit are of two types. These include sample mail surveys allocated proportional to size, and objective yield measurements based on actual field counts and measurement. In the objective yield survey an up-to-date inventory of orchards and an accurate count of trees by variety, age, and acres, is required. The bearing surface of the tree is believed to be closely related to the cross-sectional area at its base. Sample limbs are selected with known probabilities to represent the bearing surface. Fruit on the sample limb are counted, weighed, and measured. The number of fruit per tree is calculated from counts on the selected sample limbs expanded by the bearing surface of the tree. Fruit size is projected to maturity by growth curves. In Florida, which has a very comprehensive program for oranges, an objective limb count survey is conducted in August and early September. Drop rates are projected for the coming months, and based on these data two forecasting models are used to estimate production.

Accuracy of production forecasts varies considerably depending on the crop, state, and season. The forecasts assume normal weather patterns and crop growth and development, and typical harvest losses which may or may not prevail. On those few occasions when the forecast exceeds the eventual production the growers often blame the USDA for depressing prices, but most studies have shown that these forecasts tend to underestimate rather than overestimate the crop size.

A list of the Crop Reporting Board estimates and the dates they are available is published annually. Estimates are released at 3:00 P.M. ET on the designated day in the South Building of the U.S. Department of Agriculture in Washington. Care is taken that economically sensitive information is not leaked in advance. State reports are issued from State Statistical Offices.

The Census of Agriculture conducted every 5 years provides a benchmark for production estimates but is also subject to error and not available promptly. The Census does collect much more information on farms and farming operations as well as farm production than does the NASS, but does not distinguish between production for processing and fresh market.

What estimates should be provided, when they should be made, how they should be obtained, how valuable they are, and who benefits from them are still debated (Armbruster, Henderson, Knutson, 1983). Many growers feel that the USDA tends to overestimate production and consequently depress prices. Careful examination generally indicates that although crops are sometimes overestimated the reverse is more common, and without official unbiased estimates growers would probably be at a disadvantage in dealing with brokers and wholesalers.

GRADES, GRADING, AND INSPECTION

Government grades serve several purposes, but chief among them is the impartial information they provide about the product especially to those unable to undertake a personal inspection. Private grades or brands also provide useful information, but may be variable and are usually less well understood. The first U.S. grade standards were established by the U.S. Department of Agriculture in 1917 for potatoes. Since that time the USDA has developed basic trading grades for about 100 fresh fruits, vegetables, and nuts.

The need for grade standards which would be applicable nationwide was brought on by the growth of long distance trading and the inauguration of the Federal-State Market News Service (Armbruster, Henderson, Knutson, 1983). Prices quoted by the Market News could only be meaningful in different markets if they were based on products of comparable quality. Other uses have also been found. Private procurement agencies as well as federal and state agencies use the grade standards in purchasing foods. Most U.S. grade standards are intended primarily for wholesale trading, but some consumers have come to recognize and use them.

Standards for fresh fruits and vegetables vary with the individual product. They usually define such factors as color, shape, size, maturity, and number and degree of defects. The range of grade names varies considerably from one product to another. For fresh apples the top three grades are Extra Fancy, Fancy, and No. 1, while for lettuce the grades are Fancy, No. 1, and No. 2, and for potatoes they are Extra No. 1, No. 1, and No. 2. The range in size or the count per carton may also be specified. For example, Size A potatoes are defined as having a minimum diameter of $1\frac{7}{8}$ inches with 40 percent of the potatoes $2\frac{1}{2}$ inches in diameter or larger or 6 ounces in weight or larger. When size is specified in terms of the number of potatoes in a 50-pound carton the potatoes in the 100 count size, which would average 8 ounces each, must be no less than 6 ounces or more than 10 ounces each. Special consumer grades have been developed for some products but they are seldom used. The wholesale grade standards for many fresh fruits and vegetables have been revised in recent years to provide for grading products in consumer packages.

The USDA has established a policy to gradually phase in four uniform terms—*U.S. Fancy, U.S. No. 1, U.S. No. 2,* and *U.S. No. 3*—to represent available levels of quality. These terms will be adopted, in whole or in part, during the normal process of revising existing standards or establishing new ones. The term *U.S. Fancy* will be adopted as the name of premium quality produce. U.S. No. 1 will represent the chief trading grade of a product—that good average quality found in the bulk of the crop. U.S. No. 3 will describe the lowest quality practical to pack under commercial conditions, and the U.S. No. 2 grade will be assigned to produce in the intermediate quality range, noticeably superior to No. 3.

The term *grading* may be used either to designate the act of sorting individual fruits or vegetables into relatively homogeneous lots that meet grade specifications, or to determine the grade applicable to a particular lot. The responsibility of sorting fruits and vegetables into lots meeting specific grades, either by selective harvest or in the packing shed, is that of the grower or shipper. Designating the U.S. grade on the wholesale carton may be optional under some circumstances and required under others. Regulations in many states require closed consumer packages to carry the U.S. grade as well as other information. Incorrect labeling or misbranding is, of course, prohibited by law.

The official certification of the U.S. grade of a lot of fresh fruits or vegetables may be made by federal inspectors or by state personnel supervised by federal inspectors. Such grading or inspection is voluntary, except in certain instances, and is provided on a fee-for-service basis. Grade certification is compulsory in those areas where producers have adopted federal or state marketing orders which establish minimum quality standards. When a request for official grading is based on U.S. grade standards the official certificate will show the USDA grade the product meets. If there are no U.S. grade standards for the product, or if a quality or condition inspection based on a State grade or a set of private specifications is required, the USDA graders can certify the product's quality or condition on that basis.

Grading services are divided into two main categories—shipping point grading and terminal market grading. The bulk of grade certification is done at shipping point as produce is being packed for shipment to market. This establishes the quality of the commodities at time of shipment, for sales purposes, or for verifying compliance with contract terms.

Receivers in terminal markets throughout the country can have shipments of produce graded for both quality and condition or condition only. Receivers use the grading service to determine whether or not a shipment meets contract terms at time of arrival. Public and private institutional buyers make extensive use of the grading service. The Federal-State Grading Service will help institutional buyers draw up specifications to describe the quality of produce they need, and then inspect deliveries to make sure they meet these specifications.

The USDA follows three basic principles in developing grade standards or making major changes. First, there must be a need for the standards or changes. Second, because their use is voluntary, there must be interest and support from the industry. And third, the new standards must be practical to use. If these requirements are met then proposed standards or major changes in existing standards are published in the Federal Register. Then if after a suitable time period there is found to be general support, the standards are published in final form in the Federal Register with a specified date on which they will become effective.

MARKET NEWS

The primary source of current information on markets for fresh fruits and vegetables is the Market News Service (MNS) of the USDA Agricultural Marketing Service (AMS) (USDA ERS, 1989c). Such information includes wholesale prices and market conditions at major shipping points and terminal markets across the country, and shipments from growing areas and arrivals at major markets. In cooperation with the states the MNS has reporters stationed at all major shipping points and terminal markets. Some states such as California and New York supplement the federal–state system with state supported reporters at additional markets or shipping points.

At shipping points during the marketing season reporters obtain information daily from local shippers on f.o.b. prices and shipments of commodities being marketed, and compile reports with information obtained by teletype from other MNS offices at other shipping points and major terminal markets carrying the same commodities. Reporters on terminal markets obtain information daily on prices from buyers and sellers and on arrivals in the market area. Prices reported in terminal markets are those received by wholesalers for sales of less than a carload or truckload. They issue reports that also include information from major shipping points and other markets.

The price ranges shown in market reports generally cover the high and low prices of fruits and vegetables in good quality and condition. Prices for stocks outside the usual range are appropriately designated. *Quality* refers to the physical properties affecting market value such as color, shape, texture, cleanliness, and freedom from defects. *Condition* refers to factors which may change with time such as maturity, decay, freezing injury, shriveling, and flabbiness. *Quality and condition* are described by the terms *fine, good, fair, ordinary* and *poor.*

Specific terms are used to describe market conditions. These include *the market,* defined as the price or price level at which a commodity is traded; *the trading activity,* the pace at which sales are being made, which can be active, moderate, slow, or inactive; *the price trend,* the direction in which prices are moving in relation to trading in the previous reporting period, which can be higher, firm, steady, or lower; *the supply/offering,* the quantity of a particular item available for current trading, which can be heavy, moderate, or light; and *the demand,* the desire to possess a commodity coupled with the willingness and ability to pay, which can be very good, good, moderate, light, or very light. *Mostly* refers to the majority of sales or volume, and *undertone* describes the situation or sense of direction in an unsettled (prices not established) market situation.

Shipments can only be reliably estimated in certain major shipping areas. With a substantial volume of direct delivery to major retail chain and institutional buyers it is not possible to count arrivals completely, but coverage is believed to record about 90 percent of the total. Both shipments and arrivals are now specified in 1,000-hundredweight units.

In addition to the daily reports from the Federal Market News Service a National Shipping Point Trends report is issued each Wednesday from the Philadelphia and Los Angeles terminal market news offices. A National Truck Rate and Cost Report is mailed each Wednesday from the New York City and San Francisco terminal market news offices. A marketing summary is prepared after the shipping season for each shipping point production area, and an annual summary of prices is assembled for each terminal market. Reports issued from Washington include weekly and annual transportation reports (rail and truck movement for fruit and vegetable commodities), and an annual summary of average monthly produce prices in principal areas compared with the average wholesale prices in Chicago and New York City.

Many state departments of agriculture also collect additional market information and issue reports.

DISSEMINATION OF PRIMARY INFORMATION

Information originating in the National Agricultural Statistics Service and the Market News Service is disseminated in many different ways. On some terminal markets the market reporter personally delivers a copy of the report to wholesalers on the market shortly after it is prepared. For a small fee subscribers can receive copies of crop and market news reports by mail. Telephone recorders at many shipping point and terminal market news offices provide information 24 hours a day. Many newspapers, radio stations, and television stations carry crop production and market news information. Weekly trade papers not only carry the crop and market reports but also considerable background information and are widely read in the industry.

A listing of where to write for fruit and vegetable terminal and shipping point market reports as well as a listing of mailed reports giving the frequency, the season, and the cost per month or per copy for the annual summary can be obtained from the Market News Service (USDA AMS, 1989). It is possible to obtain a connection on the USDA national teletype network by contacting the AMS Communications and Operations Branch. Fax subscriptions are available from selected offices. Telephone numbers are provided for those shipping point and terminal market offices that offer 24-hour market news information by means of telephone recorders.

In recent years crop and market information has become available through electronic data networks. As in the case of recorded telephone messages the transmission is instantaneous, but there are additional advantages in that the electronic service can provide information on specific crops or markets, and print out hard copy. Some networks carry information on all major agricultural commodities as well as other products, while at least one network specializes in market information on fresh fruits and vegetables. The network also provides late industry news, weather, credit ratings, and an electronic mail service in addition to commodity market information. To tie into the network requires a microcomputer, a telephone modem, and a printer, or instead of the computer and printer a fax machine. The growing popularity of fax machines has increased the use of this method of transmission. There is a monthly charge for the service that covers a specified amount of use and a small charge per minute or per hour for additional use.

MARKET ANALYSIS AND OUTLOOK

The complexity of the business and the tremendous flow of information that is available presents a real challenge to individual marketers to understand and interpret what is going on. There are several ways in which marketers can obtain help in recognizing what is happening in the market and what the implications may be for the future. Trade papers, magazines, trade associations, the U.S. and state departments of agriculture, and land grant colleges are all potential sources of help in understanding market developments. The USDA has used basic data to develop a comprehensive series on food utilization and prices (USDA ERS, 1989a).

The major weekly fresh fruit and vegetable marketing trade papers provide broad coverage of the industry and are widely read. They report developments from farm production through to supermarket retailing and food service for all different crops both in this country and in foreign sources of supply or export markets. They include political as well as economic developments of interest to their readers.

Several monthly magazines provide more specialized information. Some focus on the business of marketing fresh produce, while others serve fruit or vegetable growers nationally or regionally or those interested in certain phases of the industry such as retailing or transportation.

Trade associations provide educational services to members as well as representing their interests in state and federal governments. There are two major national trade associations that cover all commodities and all phases of the industry, and national or regional commodity associations and associations that work with marketers at different stages in the marketing process. The annual meetings and special seminars of the major fresh fruit and vegetable trade associations are strongly supported and provide excellent educational experiences.

The Economic Research Service (ERS) and Foreign Agriculture Service (FAS) of the U.S. Department of Agriculture conduct research on fresh fruit and vegetable markets and marketing and analyze market information. The results of this research and analysis is published in periodicals and special reports. The USDA holds an Outlook Conference every year to present information on the current situation for major commodity groups and the agricultural economy in general, and to forecast future directions.

Personnel at land grant colleges in the major producing states also conduct research, hold conferences, and issue reports on fruit and vegetable marketing from time to time.

Agricultural statistics, market news, and economic reports are available as follows:

National Agricultural Statistics Service Reports may be obtained by writing:

ERS/NASS
P.O. Box 1608
Rockville, Maryland 20850

Market News Reports can be obtained from shipping points or terminal markets or by writing:

USDA, AMS, F&V Division
Market News Branch
Room 2503-South Building
P.O. Box 96456
Washington DC 20090-6456

The following USDA Economic Research Service Periodicals

Fruit and Tree Nuts (4 per year)
National Food Review (4 per year)
Vegetables and Specialties (3 per year)

can be ordered by calling 1-800-999-6779 or writing to:

ERS-NASS
P.O. Box 1608
Rockville, MD 20849

The USDA Foreign Agricultural Service publication dealing with fruits and vegetables

Horticultural Products Review (monthly)

may be obtained by writing to:

Information Division
Foreign Agricultural Service
U.S. Department of Agriculture
Room 4644 South Building
Washington, DC 20250-1000

The following are the two major trade associations

The Produce Marketing Association
700 Barksdale Plaza
Newark, Delaware 19711

The United Fresh Fruit and Vegetable Association
North Washington at Madison
Alexandria, Virginia 22314

Among the major trade papers and magazines are the following

The Packer (weekly)
7950 College Boulevard
P.O. Box 2939
Shawnee Mission, Kansas 66210

Produce Business (monthly)
127 N.Y.C. Produce Co-op
Bronx, N.Y. 10474

The Produce News (weekly)
2185 Lemoine Avenue
Fort Lee, New Jersey 07024

REFERENCES

Armbruster, Walter J., Dennis R. Henderson, Ronald D. Knutson, Editors, 1983. *Federal Marketing Programs in Agriculture: Issues and Options*. Danville, Illinois, the Interstate Printers and Publishers.

U.S. Department of Agriculture, Agricultural Marketing Service, 1989. *The Market News Service on Fruits, Vegetables, Ornamentals, and Specialty Crops*, processed publication.

—— Economic Research Service, 1989a. *Major Statistical Series of the U.S. De-*

partment of Agriculture: Consumption and Utilization of Agricultural Products. Agriculture Handbook No. 671, Volume 5.

—— 1989b. *Major Statistical Series of the U.S. Department of Agriculture: Crop and Livestock Estimates.* Agriculture Handbook No. 671, Volume 7.

—— 1989c. *Major Statistical Series of the U.S. Department of Agriculture: Market News.* Agriculture Handbook No. 671, Volume 9.

——National Agricultural Statistics Service, 1989. *Agricultural Statistics Board Catalog.*

—— Statistical Reporting Service, 1983. *Scope and Methods of the Statistical Reporting Service.* Miscellaneous Publication No. 1308. Washington, D.C., September.

Chapter 8

Market Prices and Price Analysis

Keeping up on prices is one of the most challenging and difficult tasks facing marketers of fresh fruits and vegetables. Prices often fluctuate widely from day to day, and different prices usually exist for the same commodity at different locations or different stages in the marketing process, for different varieties, and for different grades and sizes. Reasons for these price fluctuations and these differences are often difficult to understand. This chapter will briefly discuss price theory, examine the process of price discovery or formation, describe the variability and differences in prices, and illustrate how knowledge of price response to changing supply and demand can be used in practice.

PRICE THEORY

Considerable study has gone into why prices are the level they are, why they vary over time, and how they differ from one commodity or service to another. A comprehensive theory has been developed which is too extensive to cover here, except for a few highlights. A more complete presentation may be found in books such as the one by Tomek and Robinson (1990).

In essence the theory postulates the existence of a *supply schedule* which is a schedule of quantities that would be offered for sale at various levels of prices, other things equal, and a *demand schedule* which is a schedule of quantities that would be purchased at various levels of prices, again other things equal.

The demand schedule for a product or service is determined by such things as the number of potential buyers, their taste and preferences, their purchasing power, and the quantities and prices of substitutes. Generally less of a commodity or service will be purchased at higher prices, other things equal. The demand schedule when plotted with price on the vertical axis, and quantity on the horizontal axis, slopes downward from left to right and has, in essence, a position or level and a slope. The position is defined by how much of the product or service will be bought at any specified price. A *shift in demand* occurs when more or less of the product will be purchased at a particular price. The slope is defined by the change in the quantity demanded for a given change in price. For comparison purposes the slope is measured in terms of *price elasticity of demand,* defined approximately as the percentage change in the quantity demanded for a one percent change in price.

The demand schedule is considered to be *elastic, inelastic,* or of *unit elasticity* with respect to price depending on whether the percentage change in the quantity demanded is greater than, less than, or equal to the percentage change in price. When the demand for a product at the farm is inelastic with respect to price, as it is for products like onions and potatoes, large crops bring very low prices and small crops very high prices. Under these conditions small crops are worth more to farmers as a group than large crops.

The supply schedule is determined by such things as costs of production, alternative opportunities, producers' preferences, and expectations with respect to yields. Since there is a delay between the time a decision is made to produce a crop and when it may be offered for sale the forthcoming supply largely depends on expected prices during the production period. Producers' intended supply response may also be influenced by weather or other factors beyond their control. Generally, however, more of a good will be offered for sale the higher the expected price, other things equal. A supply schedule also has a position and slope. When graphed the supply schedule slopes upward from left to right. The position is defined, as in the case of the demand schedule, by the quantity that will be offered for sale at various levels of prices. The slope is described by the *price elasticity of supply,* the change in quantity offered for sale with a 1 percent change in price, other things equal.

Prices tend to gravitate toward the level that equates the quantity demanded with the quantity supplied. This can be represented graphically as the intersection of the two schedules (Figure 8.1). At this point the quantity offered for sale and the quantity purchased will be in equilibrium. The theory is helpful in explaining the level and variability in prices even though it is not possible to observe actual supply or demand schedules, but only to estimate them by means of econometric analysis.

The price elasticity of demand reflects the response of sales to a change in price, but in many fruit and vegetable markets the more relevant concern is the response of price to a change in the quantity offered for sale. The measure used in such cases is termed the *price flexibility,* which is approxi-

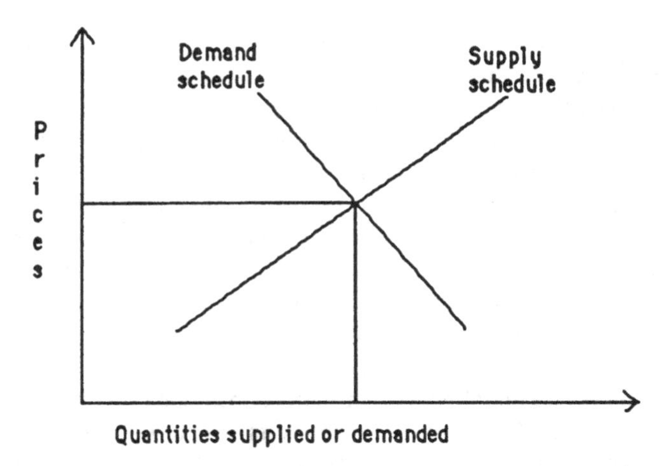

Figure 8.1. The Determination of Equilibrium Price.

mately the reciprocal of the price elasticity, or the percentage change in price for a 1 percent change in the quantity offered for sale.

Changes in supplies and prices of other products may affect the price of a particular commodity in either of two ways. If an increase in the price of lettuce increases the demand for spinach then these are competing crops. On the other hand, if a shortage of lettuce and an increase in the price is associated with a decreased demand for tomatoes then these are complementary crops. *Cross-price elasticity* is the change in the quantity associated with a 1 percent change in price of a competing or complementary crop, other things equal.

Changes in demand can come about through changes in consumer lifestyles, income levels, nutritional knowledge, and so forth. The purpose of product advertising is often to bring about a shift in demand. Increased income can result in an increase in demand for some products and a decrease for others. *Income elasticity* is defined in general as the percentage change in quantity demanded associated with a 1 percent change in income. Consumers demand less of certain goods and services with increased income, and these are called *inferior goods,* while those purchased in greater quantity when income increases are considered *superior goods.*

Market conditions or structure influence how the forces of supply and demand operate in practice. Markets range from conditions of *perfect competition* to those of *pure monopoly* on the selling side or *pure monopsony* on the buying side, though none of these extremes may actually exist. Perfect competition requires that there be many buyers and sellers in the market all with perfect knowledge of supply and demand conditions and trading a perfectly homogeneous product. Under such conditions no individual buyer or seller has any influence over prices, and prices are established at the intersection of the supply and demand schedules. In contrast a monopolist as the only seller would be able to dictate the selling price without reference to other prices, and hold that price indefinitely. If there were an adequate supply the quantity sold would be determined solely by the demand schedule. Pure monopsony represents the case of the single buyer who could dictate the buying price without reference to other prices. The nature of the supply schedule then determines the quantity that could be purchased.

In practice markets range between these extremes, with varying degrees of imperfection. Information is not perfect, commodities are not identical, some buyers or sellers may be large enough to influence the market but not completely control it. Shipping point markets for major products may at times approach competitive conditions, while at the other extreme some retailers may have a temporary local monopoly and be able to set prices weekly. In some markets growers may face a few large buyers and feel they are simply price takers, and look with envy at automobile manufacturers or other large companies who appear to have the ability to set prices at whatever levels they choose.

PRICE DISCOVERY AND ESTABLISHMENT

Who sets the prices of fresh fruits and vegetables, and how and where are they established? These questions are often asked, but there are no simple answers. Seldom if ever is a price set for a commodity or service without reference to another. Yesterday's price, the price at other locations, the prices of competing products, as well as changes in market supply and demand may all be taken into account.

Traditionally wholesale terminal markets have been the primary locus of price discovery for fresh fruits and vegetables. This was especially true when many shipments were made on consignment to terminal wholesalers, also known as commission houses. When merchandise was consigned the wholesalers then tried to get the best prices they could for their principals, less costs and commissions, in the face of current demand from retailers and institutions. Buyers walked the market, checking on supplies and asking prices, before striking a deal with a particular supplier. The prices reported on the wholesale markets were then relayed to other markets and discounts were deducted or premiums added to reflect location, stage in the marketing process, quality, or other factors. Price changes on wholesale markets were reflected forward to retailers and back to shippers.

Wholesale markets still play an important role in price discovery but the system has changed. Many chain retailers and other buyers purchase at shipping point and bypass the terminal market. Shippers are able to sell most of their products outright to terminal wholesalers rather than ship on consignment, except when markets are severely depressed. Shipping point price information, more difficult to gather, is becoming more important in relation to terminal market prices. Difficulty in moving supplies will still indicate when prices are higher than justified by current supply and demand, while active demand will signal the reverse. By this means prices tend toward equilibrium, although seldom achieving it.

Auction markets have had an important role in price discovery for fresh fruits and vegetables as well as performing other useful marketing functions. Auctions may be of the conventional or of the Dutch (reverse) auction type. A few auction markets still operate, but they are not as common today as they once were.

The cauliflower auction at Riverhead, Long Island, and the fruit and vegetable auction at Vineland, New Jersey, are examples of conventional shipping point auctions. Here buyers can acquire products from several growers each selling a relatively small volume, bidding against other buyers to fill their needs. Growers can be fairly sure of obtaining the going price and can learn about buyers' needs and preferences. As growers have become more specialized, with larger volumes for sale, there has been a tendency for them to bypass the auction markets.

The flower auction at Aalsmeer in Holland is a highly sophisticated operation where a tremendous volume of flowers is sold every day by the Dutch auction system. Unlike the conventional system where the prices start low and are increased until the last bid is made, prices start high in the Dutch system and decline until the first bid is made. Sellers generally believe that the Dutch system is less subject to collusion or manipulation than the conventional system. In any event sellers seem to prefer selling at auction rather than private treaty, while for most buyers the reverse is true.

Auctions were once used in many terminal markets primarily for citrus fruit from large cooperative and private shippers. Carload lots of fruit would be shipped to terminal markets to be sold on the auction block in smaller lots to many different buyers. Now with fewer buyers, each requiring larger quantities, the shipments go directly to retail chain buyers or large terminal wholesalers.

Means of establishing prices other than by private treaty or public auction have been suggested, and some have been attempted but generally without success. Setting prices on the basis of production costs or the prices of related crops seems logical but usually does not work for long since such prices often bear little relationship to conditions of supply and demand.

The use of differentials for quality, location, packaging, or other services over published market prices is often used either implicitly or explicitly and further emphasizes the need for accurate published prices.

Electronic marketing is believed by some to hold the promise of providing a means of establishing prices that would better reflect the true forces of supply and demand. The term has several definitions, one of which is the use of buy/sell notices on an electronic bulletin board. In this case buyers or sellers simply list the commodities needed or offered for sale, usually without any price specified. Deals may be consummated by electronic mail, or in the more usual case buyers and sellers get in touch with each other by telephone to work out the details.

For major commodities, such as cotton in Texas, electronic marketing consists of an auction conducted over a network of computers linked by telephone. Offerings are listed with reservation prices and buyers have an opportunity to bid on respective lots. Since buyers cannot see what they are bidding on, the product must be impartially graded and sized. Any possible benefits of electronic marketing cannot be realized fully until a large proportion of the buyers and sellers have the necessary equipment and participate in the service.

From time to time growers or major marketing firms attempt to secure sufficient control of the supply of a commodity to obtain a virtual monopoly and thus be able to set prices at higher levels. Experience with such projects has not been successful. Complete control of supply has generally been impossible, and the activity encourages outside production. Buyers search for substitutes and turn to other products.

Prices are linked together in such a manner that changes at one place or point in time are accompanied by changes at other places or times. Such changes are not immediate, nor are they necessarily equal. Ward (1982) found, for example, that changes in wholesale fresh vegetable prices led changes at retail and shipping point. He also found that decreases in wholesale prices were reflected more promptly at retail than were wholesale price increases, and that the same thing was true with respect to shipping point prices which apparently responded more rapidly to wholesale price decreases than to increases.

PRICE DIFFERENCES AT ONE POINT IN TIME

Origin, Variety, Quality, Condition, and Pack

Prices for a particular fruit or vegetable such as apples or potatoes may vary at any point in time for several reasons. One is that the product itself is highly variable, and may differ due to place of origin, variety or type, quality, condition, or pack. But products with identical specifications may still command different prices because of their geographic location or stage in the marketing process.

Price differences may also change from one time to another as noted for potato prices (Table 8.1). Prices for 50 pounds of potatoes on the New York wholesale market in March 1986, for example, varied all the way from $2.50 for potatoes from Long Island to $10.75 for the same quantity of Russet Burbank potatoes from Idaho.

The Idaho potatoes were a special baking type carefully graded, uniformly large in size, a specified number packed in a fiberboard carton, and shipped from a region with a reputation for good quality product. The potatoes from Long Island met the U.S. 1 grade standards but were the round white type mainly used for boiling or mashing with a wide range in size, and packed in 50-pound sacks. Relative prices were quite different two years later. Prices were substantially higher, except for the Idaho Russet Burbanks and the round whites from Minnesota–North Dakota. Prices for California Long Whites, a new crop of a special type just coming on the market, were almost double the previous level.

One way to understand these differences is to consider that there is a supply and a demand schedule for each type of pack identified by major characteristics which may differ, a little in some cases and a lot in others, from one pack to another. The changes in price differences from one season to the next between round red and round white potatoes, for example, may be due to the effect of changes in supplies. The widening spread in prices over time between Idaho baking type potatoes and Eastern round white

Table 8.1. Potato Prices on New York Wholesale Market, March 1986, 1988.

	MARCH 1986	MARCH 1988
U.S. No. 1 Size A		
50-lb sack washed		
Round Red		
Florida	$9.25	$14.28
Minnesota/North Dakota	5.56	6.13
Round White		
NY Long Island 2″ min	2.50	4.46
Maine Katahdin type unwashed	2.70	4.50
Canada PEI Canada No 1 $2\frac{1}{4}$″ min	3.13	4.69
50-lb carton		
California Long White	10.75	19.38
Idaho Russet Burbank 80-90s	10.13	10.03

Source: Adapted from USDA AMS, 1989.

potatoes are believed due to a shift in the demand schedules that increasingly favors the Northwestern potatoes.

These observations about price differences emphasize the need for price quotations to be specific with respect to the characteristics of the product particularly as to origin, variety, grade, size, and pack. Indeed, individual lots of fresh fruits and vegetables differ so much in market value that major buyers are beginning to purchase on the basis of product specifications that go beyond the normal attributes included in public price quotations and include additional characteristics, some of which may only be determined by laboratory analysis.

Geographic Location

Prices for commodities in surplus producing areas that ship excess supplies to other markets tend to be lower than prices in deficit areas that must depend on receipts from other regions to meet local demand. Under usual marketing practices the difference may be less but generally will not be greater than the cost of transferring the product from the surplus to the deficit region. A price difference greater than the cost of transferring the product between the regions will encourage shipments that will reduce supplies in the shipping region and increase them in the deficit region and thus bring prices closer.

Prices may differ between adjacent regions by less than the cost of transfer simply due to differences in regional supply and demand. In such cases local prices in different regions and the price differences may fluctuate independently of each other. There may also be highly perishable products for which transportation is not feasible and therefore no effective transfer cost to limit price differences.

An example of regional differences in prices is provided by comparing the price of California iceberg type lettuce at wholesale markets in the West, the North Central region, and in the Northeast (Table 8.2). Prices were lowest in the West, intermediate in the North Central region, and highest

Table 8.2. Wholesale Prices of California Iceberg Lettuce in Western, North Central, and Northeastern Markets, 1980 and 1986.

YEAR	WEST	NORTH CENTRAL	NORTHEAST	DIFFERENCE WEST TO NORTHEAST
		dollars per carton 24-head		
1980	6.60	8.30	9.73	3.13
1986	7.91	11.20	12.21	4.30

Source: Adapted from USDA ERS, 1989.

in the Northeast as might be expected. The difference between the annual average price in the West and in the Northeast was $3.13 in 1980 and $4.30 in 1986. This difference varies from year to year, reflecting mainly the costs of transportation from California to the Northeast, but also local market conditions.

Stages in the Marketing Process

Prices differ at different stages in the marketing process, increasing as the product moves toward the retail market. One can postulate a supply and demand for the marketing services performed at each stage. The demand for such services and their supply or availability does change, resulting in changing margins or spreads. The increase in price from one stage to the next represents the charge levied for the performance of such services. Over the long run the charges for services such as grading and packing, storing, selling, transporting, wholesaling, and retailing will tend to reflect the costs of providing such services just as farm prices will tend to reflect costs. But in the short run the amount actually levied will depend on marketing strategies and market conditions.

The difference between farm and retail prices for fresh fruits and vegetables—the marketing spread—has widened in recent years. Among the factors responsible may be the additional marketing services provided and the lower rate of increase in productivity in marketing than in farm production. Raleigh (1984) reported that over the period 1963 to 1983 the farm value of fresh fruits and vegetables increased 163 percent but the marketing spread increased 262 percent. As a result the farm share of the consumer's dollar spent for fresh fruits and vegetables declined from 33 percent in 1963 to 26 percent in 1983.

Prices paid Washington growers for Red Delicious apples, as was pointed out in chapter 6, varied considerably from the 1980/81 season to the 1988/89 season, but the total margin between the grower average price and the retail values in the Northeast rose fairly steadily from $16.67 per carton to $26.37 over this period.

PRICE VARIATION OVER TIME

Prices of many fresh fruits and vegetables fluctuate widely from day to day, month to month, and year to year, especially at shipping point or on wholesale markets. Price changes for fresh fruits and vegetables may be due to many factors such as changes in the supply of the product or of competing or substitute products, or changes in consumer demand. Some commodities such as lettuce experience violent fluctuations from week to week (Figure

8.2). Over the marketing year 1987–88 the weekly average f.o.b. price of a carton of California lettuce varied from $2.75 in May to $25.80 in November. Much of this variation in price was apparently associated with changes in weekly shipments which ranged from 1,470 thousand cartons to 2,837 thousand cartons.

The demand for western iceberg lettuce is considered to be highly inelastic since most of our lettuce comes from that region, there are no close substitutes, and consumers such as hotels, restaurants, and institutions have rather fixed requirements. Heaviest weekly lettuce shipments in 1987–88 were about double the lightest, but the highest prices were almost 10 times the lowest.

These price changes may be better understood by separating out the various factors responsible. These changes may be due to short-term or day-to-day fluctuations, seasonal variation, annual fluctuations, and trends or cycles. The importance of different kinds of price variation will differ for different commodities. Price changes for apples, for example, a storable crop with annual domestic supplies largely produced during one season of the year, will be different than for lettuce which is essentially nonstorable and produced throughout the year.

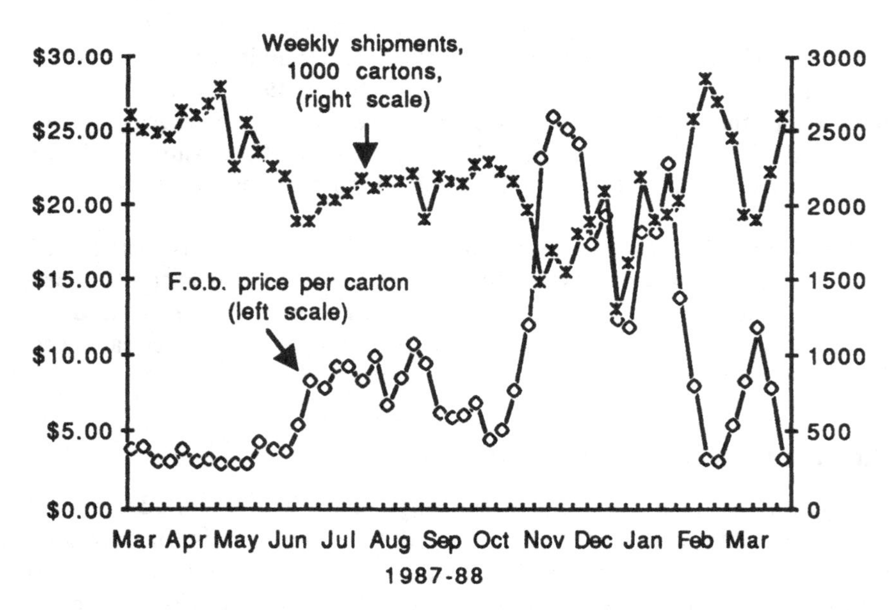

Figure 8.2. Fluctuations in California Weekly Lettuce Shipments and F.O.B. Lettuce Prices, March 1987–March 1988. *Source:* USDA AMS, 1988.

Day-to-Day Variation

Changes in supplies, actual or expected, are generally thought to be mainly responsible for short-run changes in prices. For perishable crops like lettuce the change in supply may be very real, while for storable crops like apples or onions the day-to-day changes in prices may hinge on rumors of exports or imports or changing marketing strategies. Days may pass without major price changes in prices of storable commodities, until new information becomes available.

Seasonal Variation

Prices of many fresh fruits and vegetables exhibit fairly regular seasonal variation, although at times the seasonal pattern may be obscured by other factors. In major production areas prices often tend to decline at the beginning of the local harvest period, reach a seasonal low during the peak of harvest, then rise as the marketing season draws to an end. Prices of storable crops such as cabbage and potatoes generally increase during the storage period, some years more than others. Over an extended period of years the average price increase during storage tends to equal the cost of storage. In any given year the increase seldom equals storage costs because of imperfect information about future supplies and market demand, storage quality, and for some crops the timing of new crop harvest. For highly perishable crops such as tomatoes, lettuce, or fresh sweet corn the end-of-season price increase may be of short duration or nonexistent.

Seasonal variation in prices of fruits and vegetables may be revealed by removing trend and cyclical price movements. Buxton (1988) did this for fruit crops and Buxton and Hamm (1988) for vegetable crops by dividing each monthly price over the period 1981–87 by its centered 12-month moving average. The seven ratios of a given month's price to its corresponding moving average (multiplied by 100) were averaged to determine the seasonal index for that month. The index indicates if the actual price was generally above, below, or the same as the moving average for that month. The variability of the ratios, as measured by their standard deviation, indicates the regularity of price changes and whether the seasonal index was weak or strong.

U.S. monthly average potato prices showed a high degree of seasonality as indicated by the relatively wide fluctuation in the seasonal index (Figure 8.3). The index was lowest in October—the major fall harvest season—at 76 then rose gradually at first and then more rapidly to a seasonal high of 137 in July. The band of irregularity was relatively narrow, being widest just prior to the major harvest season.

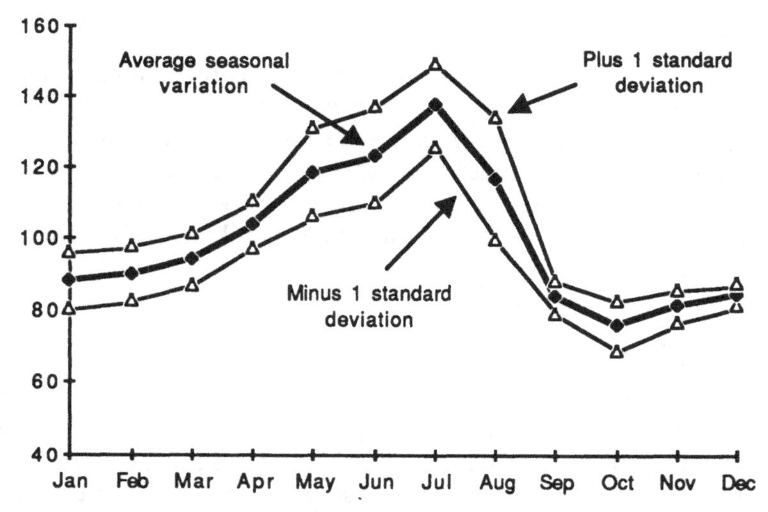

Figure 8.3. Seasonal Variation in Monthly Farm Potato Prices and Index of Variability, 1981–87. *Source:* Buxton, Hamm, 1988.

Strong seasonal price patterns were also recorded for apples, oranges, lemons, strawberries, pears, and sweet corn during this period. Several vegetables such as lettuce, tomatoes, onions, celery, carrots, broccoli, and cauliflower were found to have weak seasonal price patterns.

Year to Year Variation

The annual average prices of storable crops such as potatoes and onions often fluctuate widely from year to year. Changes in supply are usually the primary cause. Such changes may be due to growers' production decisions, or to factors beyond their control such as weather and pests. Indications are that growers make production decisions based on market conditions prior to planting rather than what might be expected at harvest time. While many growers may not make any changes in acreage from one year to the next there are always a few who do. Weather and pest conditions sometimes offset and at other times amplify growers' acreage changes.

The U.S. season average farm price of potatoes has fluctuated widely since 1970 from $1.90 per hundredweight in 1971 to $6.55 in 1980 (Figure 8.4). Production over the same period varied from 296 million hundredweight to 407 million hundredweight. The chart indicates a strong inverse relationship between production and price. In years when potato production was high, as in the late 1970s, prices were low. The short crop of 1980 brought high prices, as did the moderate sized crop of 1983.

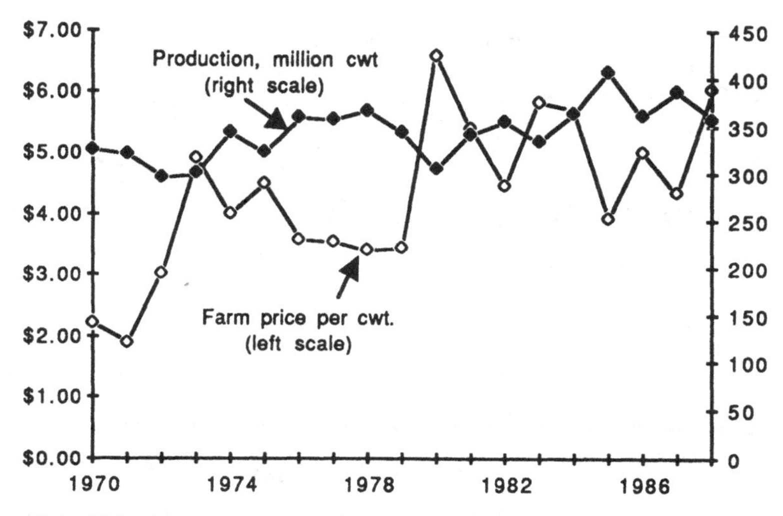

Figure 8.4. U.S. Potato Production and Season Average Farm Price, 1980–88. *Source:* Adapted from USDA ERS, 1989.

Annual average prices of orchard and vineyard crops also vary from year to year but are mainly due to changes in yields rather than changes in acreage. Production decisions take several years to come to fruition and are a major investment. Their impact is over a longer period of time.

Changes in demand from one year to the next may also be an important influence on annual average prices, particularly for crops that can be exported or that are subject to import competition. Such changes in domestic demand may arise from changes in foreign supplies or in barriers to international trade.

Price Trends

Major changes in factors affecting the supply or demand for a product will sometimes trigger an upward or downward trend in prices over a period of years until adjustment is completed. Such trends are often easier to identify than to explain.

Buxton (1988) reported that over the period from January 1981 through June 1987, the U.S. monthly average farm prices for apples varied from 10.7 cents to 30 cents per pound (Figure 8.5). A linear trend line fitted to these observations indicated a definite upward movement over this period. Prices of grapefruit, lemons, strawberries, and pears also showed an upward trend. In contrast the trends in prices of several vegetable crops at

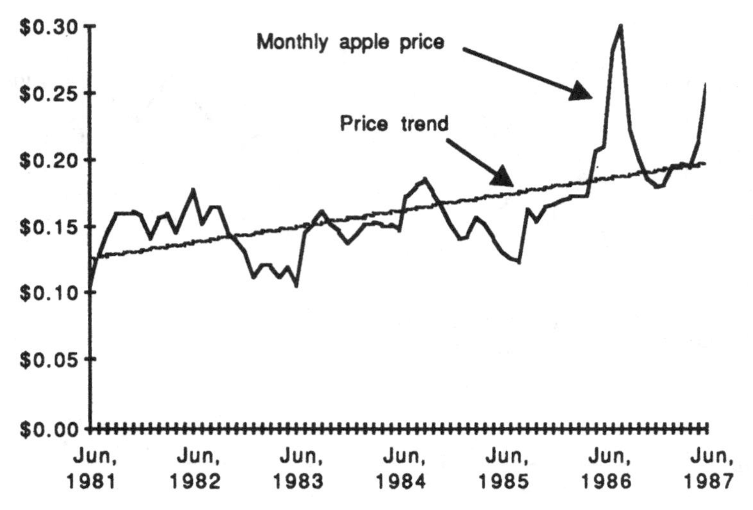

Figure 8.5. U.S. Monthly Farm Price of Apples and Linear Trend, June 1981-June 1987. *Source:* Adapted from Buxton, Hamm, 1988.

this time were relatively flat, except for prices of potatoes, broccoli, and cauliflower that trended down at this time (Buxton and Hamm, 1988).

These studies pointed out that the factors that determine trend—the longer term upward, downward, or constant direction of prices—may include general inflation, changes in production efficiency due to advances in technology or management practices, relative expansion or contraction of demand due to changes in population or per capita income, government price support programs, federal marketing orders, and competitive pressure from imports.

Upward or downward trends in prices obviously cannot continue indefinitely. Such trends can signal the need for adjustments in the industry, some of which may have to be severe. Upward trends often signal the opportunity to employ more resources in production, while downward trends reflect an excess supply situation.

Price Cycles

Cyclical price movements are defined as periodicities that repeat themselves regularly for periods of more than a year. To be predictable they need to be self-generated by conditions within the industry and not be the result of exterior shocks. They may occur because of intervals between the time a decision is made to produce a commodity and the time the crop is harvested, and because growers tend to make production decisions based on current prices or on prices in the recent past. Cyclical price patterns of several years in length often emerge for many fruit crops due to the length of time between planting decisions and actual production.

Some perennial crops, however, such as pistachio nuts and tart cherries, tend to bear heavily one year and lighter the next. Such alternate bearing is usually accompanied by corresponding fluctuations in farm prices. Vegetable crops planted annually might be expected to show shorter cyclical patterns than perennial crops such as apples or oranges where several years are required from planting to harvest.

Buxton (1988) and Buxton and Hamm (1988) constructed 12-month moving averages of monthly fruit and vegetable prices over the period June 1981 to June 1987 to look for cyclical patterns. Observation of the data revealed a 3-year cycle for potatoes (Figure 8.6). This is a relatively short period over which to observe cycles, and without some indication of the underlying causes it would be hazardous to predict that this pattern might continue in the future.

A California study of the lemon industry (Kinney et al., 1987) noted that that industry is subject to long-run production cycles. These cycles are apparently caused by the long lags that occur between changes in economic conditions and the associated changes in bearing acreage, and are influenced by factors both within and outside the industry. Changes in total lemon utilization influence the net return per acre received by lemon grow-

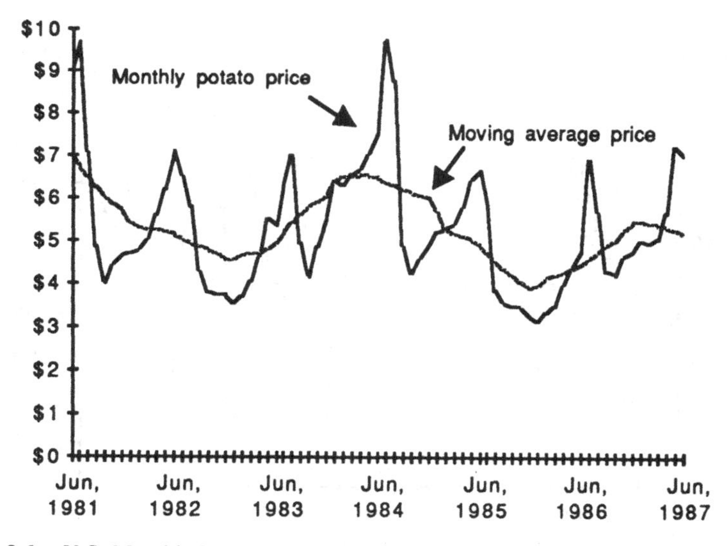

Figure 8.6. U.S. Monthly Farm Potato Prices and 12-Month Moving Average, June 1981–June 1987. *Source:* Adapted from Buxton, Hamm, 1988.

ers (Figure 8.7). Growers in turn respond to changing economic conditions by increasing or decreasing plantings (Figure 8.8). New plantings begin bearing in 5 or 6 years, and reach full production a few years later. Change in the rate of removals that typically occur after about 30 years was found to be associated with short-run changes in real per acre revenues and in price variability.

The two major expansions of the industry since 1950 have been associated with the growth of new market outlets. The development of new processed products in the 1950s and the export markets of the late 1960s and early 1970s brought increased plantings. This eventually led to oversupply causing declines in returns relative to costs.

Producers may respond differently to increases in prices than they do to decreases, resulting in asymmetrical cyclical patterns. Growers of annual crops appear to delay reducing production when faced with lower prices, yet promptly increase output following a period of favorable prices. Consequently prices may exhibit a pattern of several years at depressed levels followed by 1 year at higher level. On the other hand growers of tree fruit and vine crops require several years to increase output compared to the decrease that can be accomplished in a single season.

PRICES RELATIVE TO OTHER COMMODITIES AND SERVICES

The price of a commodity or service in relation to other commodities and services depends, as has been pointed out, on current supply and demand

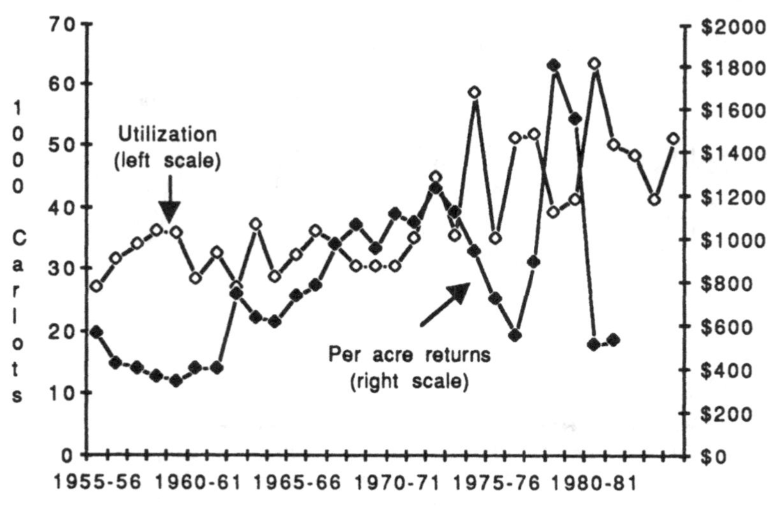

Figure 8.7. California-Arizona Lemon Utilization and Per Acre Net Returns Received By Lemon Growers, Seasons 1955-56 through 1981-82. *Source:* Adapted from Kinney et al., 1987.

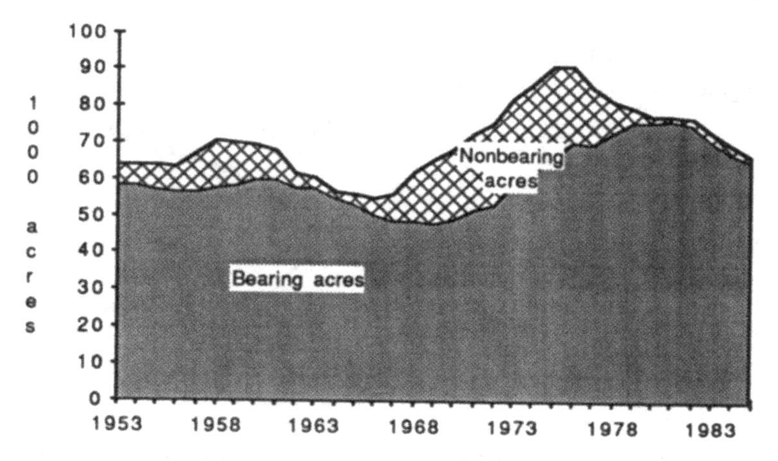

Figure 8.8. California–Arizona Bearing and Nonbearing Lemon Acreage, 1953–85. *Source:* Adapted from Kinney et al., 1987.

conditions. These, of course, can change. The price of round white potatoes has fallen relative to long russet potatoes as the demand for baking potatoes increased. The prices of fresh vegetables rose relative to canned vegetables as costs of harvesting and marketing fresh vegetables and demand for fresh vegetables increased relative to the canned product.

The prices of major fruits and vegetables tend to vary independently of each other over short periods of time, indicating that consumers do not consider them competing or complementary products. A simple correlation matrix of monthly prices of pears, apples, strawberries, grapefruit, lemons, and limes found that the strongest association was between lemons and limes, and that was relatively weak and only amounted to .21 (Buxton, 1988). In the case of fresh vegetables, the correlation matrix between prices of tomatoes, lettuce, celery, cauliflower, broccoli, sweet corn, carrots, onions, potatoes, and dry beans have similar results (Buxton and Hamm, 1988). There was a simple correlation coefficient of .49 between prices of broccoli and cauliflower, .41 between potatoes and dry beans, .38 between onions and potatoes, and .36 between onions and dry beans, but only .21 or less between the prices of the other 41 pairs of commodities.

Over a period of several years price changes for an individual commodity may be significant only as they differ from the general level of prices. Prices tend to change to a certain extent together. The index of retail prices of apples, for example, rose over the period of 1966–87 in much the same manner as food prices in general as represented by the Consumer Price Index (CPI) for food (Figure 8.9). To better observe the relative change in apple prices compared to all foods one can divide the index of apple prices by the CPI and multiply the result by 100 to obtain an index of deflated

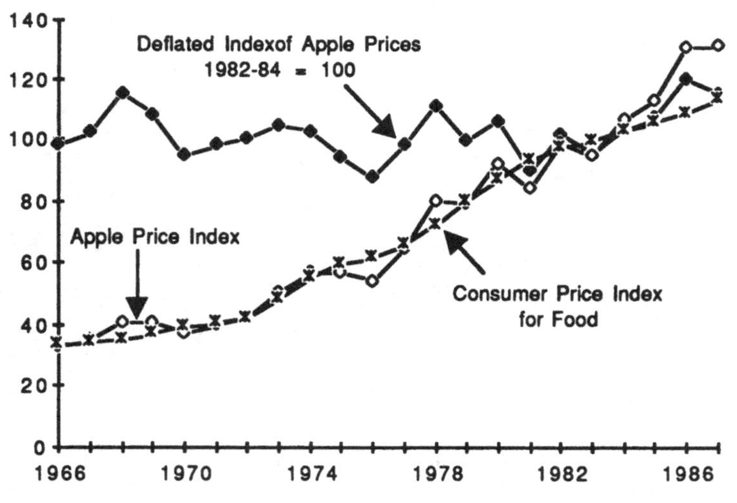

Figure 8.9. Index of Apple Prices, Consumer Price Index for Food, and Deflated Index of Apple Prices, 1982–84 = 100, 1966–1986

apple prices. This shows in one series of numbers what may be only vaguely apparent from a visual comparison of the original two series. The process is called *deflation*. Deflating the index of apple prices by the CPI clearly illustrates that retail apple prices have tended to follow the level of food prices in general, fluctuating a little above or below them from one year to the next.

The process of deflation can be used to study the relationship between changes of prices of a particular commodity or service and the general level of prices of competing products or services or of the general level of costs. The importance of deflation is the fact that it is often not the absolute level of a price that we are interested in so much as that price in relation to costs or the prices of competing goods or services.

ESTIMATING DEMAND AND SUPPLY RELATIONSHIPS

Demand and supply relationships are difficult to determine with confidence. Economists have spent considerable effort estimating such relationships. Knowledge of price flexibilities can be useful in improving price forecasts and in making or analyzing policy decisions such as marketing order provisions. Comprehensive reviews of past studies such as those published for California tree fruits, grapes, and nuts and for vegetables (Nuckton, 1978, 1980) illustrate the different results that can be obtained with different data, different estimation methods, and different hypotheses with respect to the way the market operates. Researchers and others, rather than relying on an existing store of estimates of economic relationships, tend now to develop their own coefficients as needed.

Huang (1985) used data on civilian food disappearance and retail prices for the years 1953–83 to estimate a complete system of price and expenditure elasticities for 40 food commodity categories and one nonfood category. The direct-price elasticities are estimates of the change in utilization that might accompany a 1 percent change in price, while the expenditure elasticities estimate the change in utilization that might accompany a 1 percent change in expenditures. Results of this study indicated that the demand was price elastic for grapes, was about unit elasticity for oranges, and was price inelastic for the other fruits and vegetables (Table 8.3). The expenditure elasticity was positive for all except four of these commodities, indicating that utilization would increase with increased consumer expenditures. According to these estimates the utilization of apples, bananas, carrots, and cabbage would decline with increased consumer expenditures. Relatively large standard errors of estimate for some of these expenditure elasticities suggest that these results may not be reliable.

PRICE ELASTICITY, SUPPLY MANAGEMENT, AND MARKET DIFFERENTIATION

Knowledge of the price elasticity of demand or the price flexibility for a particular product in a specific market has led to the development of mar-

Table 8.3. Estimated Direct-Price and Expenditure Elasticities for Selected Fresh Fruits and Vegetables, and Potatoes

COMMODITY	DIRECT-PRICE ELASTICITY	EXPENDITURE ELASTICITY
Apples	−.2015	−.3514
Oranges	−.9996	.4866
Bananas	−.4002	−.0429
Grapes	−1.3780	.4407
Grapefruit	−.2191	.4588
Lettuce	−.1371	.2344
Tomatoes	−.5584	.4619
Celery	−.2516	.1632
Onions	−.1964	.1603
Carrots	−.0388	−.1529
Cabbage	−.0385	−.3767
Potatoes	−.3688	.1586

Source: Adapted from Huang, 1985.

keting programs for commodities that might enhance returns to the industry. As we have seen, small crops and reduced supplies of commodities that have an inelastic demand with respect to price (a highly flexible price) result in greater total returns than do relatively large crops or large supplies. Under such conditions the natural tendency is to attempt to restrict production or marketings. Such strategies are often not successful for two general reasons. Most commodities with an inelastic demand at the farm are staples such as potatoes, onions, or lettuce that can be widely grown. Attempts to control supply are particularly difficult under these circumstances, because the interests of the individual firm are often at cross-purposes with the interests of the industry. Higher prices can only be achieved through restricting supply, but individual firms often have a strong incentive to maintain or increase production at these higher prices. And the higher the expected prices the greater the incentive. Imports also must be regulated to control total supply.

The other reason that supply management may not work is that consumer demand can change. If supply can be successfully controlled and prices raised above previous levels in the short run, the demand may decline as consumers gradually shift to alternative products. The necessary level of production and sales needed to achieve the desired prices would have to become increasingly restrictive and severely limit individual operations. This will be considered further in chapter 11 on "Marketing Orders."

Knowledge of differences in price elasticities of demand for a commodity in different markets has also been used to raise total returns through reallocating supplies between markets or establishing different prices in those markets. When price elasticities differ between markets the total net returns from a given supply can be increased by restricting supplies entering the market or markets with the more inelastic demand, and diverting the excess quantities to the market or markets with the more elastic demand. Higher prices in the market with the more inelastic demand will more than offset the effect on total revenue of the decline in prices in the market with more elastic demand. Means must be available to restrict the quantities going to certain markets in spite of higher net prices, which can present a real challenge.

In practice it has been found that for some commodities the demand for sales on the domestic fresh market is more inelastic than is the demand for sales for processing or for export. Programs have been developed to restrict supplies going for fresh sales, and divert the excess to processing or export. One example is the marketing program for California–Arizona lemons that has been operating for many years, which will be examined later. This practice results in higher net prices for fresh domestic sales both at the farm or shipping point and lower prices for sales for processing or export, which may be termed *price discrimination.*

Successful diversion under these conditions will, however, result in greater total net returns unless offset by higher costs of administering the program. But again pressures can develop to disrupt the operations. Incentives may grow for groups or individuals to ship to the market with higher net prices. The continually higher prices may bring about a shift in demand that results in smaller quantities being taken at current price levels, with all the adjustments and challenges that can bring.

REFERENCES

Buxton, Boyd M., 1988. Seasonal Farm Price Patterns for Selected U.S. Fruit Crops. *Fruit and Tree Nuts.* U.S. Department of Agriculture, Economic Research Service, TFS-246.

—— Shannon Reid Hamm, 1988. Seasonal Farm Price Patterns for Selected U.S. Vegetable Crops. *Vegetables and Specialties.* U.S. Department of Agriculture Economic Research Service TVS-246.

Huang, Kuo S., 1985. *U.S. Demand for Food: A Complete System of Price and Income Effects.* U.S. Department of Agriculture, Economic Research Service, Technical Bulletin No. 1714.

Kinney, William, Hoy Carman, Richard Green, John O'Connell, 1987. *An Analysis of Economic Adjustments in the California-Arizona Lemon Industry.* Giannini Foundation, University of California, Giannini Foundation Research Report No. 337.

Nuckton, Carole Frank, 1978. *Demand Relationships for California Tree Fruits, Grapes, and Nuts: A Review of Past Studies.* Giannini Foundation, University of California, Special Publication 3247.

—— 1980. *Demand Relationships for Vegetables: A Review of Past Studies.* Giannini Foundation, University of California, Special Publication 3247.

Pearrow, Joan, 1987. *Fresh Fruit and Vegetables: Prices and Spreads in Selected Markets, 1975–84.* U.S. Department of Agriculture, Economic Research Service, Statistical Bulletin No. 752.

Raleigh, Stephen M., 1984. Farm-to-Retail Spreads for Produce: 1963–83. U.S. Department of Agriculture. *National Food Review,* Winter 1984.

Tomek, William G., Kenneth L. Robinson, 1990. *Agricultural Product Prices.* Cornell University Press, Ithaca and London.

U.S. Department of Agriculture Agricultural Marketing Service, 1989. *Fresh Fruit and Vegetable Prices 1988 Wholesale Chicago and New York City, F.O.B. Leading Shipping Points,* unnumbered publication.

—— Economic Research Service, 1989a. *National Food Review,* Volume 12, Issue 2, April–June.

—— 1989b. *Vegetables and Specialties Situation and Outlook Yearbook,* TVS-248, August.

Ward, Ronald W., 1982. Asymmetry in Retail, Wholesale, and Shipping Point Prices for Fresh Vegetables. *American Journal of Agricultural Economics,* Vol 64, No 2, May, pp. 205–212.

Wescott, Paul C., 1983. Prices and Demand Responses in the Food Sector. U.S. Department of Agriculture, *National Food Review,* Spring 1983.

Chapter 9

Trade Practices, Credit Ratings, and Regulation of Trading (Perishable Agricultural Commodities Act)

The nature of fresh fruit and vegetable marketing is such that over the years trade practices and special types of businesses have evolved to support the system. These include private firms that supply fresh fruit and vegetable marketers with information on the products and services and on the financial worth and business character of companies with which they do business. Legislation and government regulations designed specifically for this industry have also been developed to prevent unfair trade practices, arbitrate disputes, and punish offenders. Procedures to protect sellers against bad debts in the event that buyers delay or default on payment are also provided. Marketers must still be aware of the special challenges they face in buying and selling produce and take precautions to deal with them.

TYPES OF FIRMS

In the early days the industry was highly fragmented, and most firms involved in fresh fruit and vegetable shipping and wholesaling were small, family owned businesses. These firms tended to specialize in performing only one or two marketing functions and sometimes handled only a small number of commodities. Some firms wholesaled carlot quantities of potatoes and onions, some ripened and repacked tomatoes, and others resold small lots to small independent retailers. There are still many such firms in operation, but more business is being done by firms that have expanded operations to include more services and more commodities, even though they remain family owned and operated. The operations of these firms consist of the performance of several functions or services, often for a fairly broad line of commodities. Knowledge of the individual functions, however, is still important.

Firms can be classified as to where they operate in the marketing system such as at shipping point or at the destination or terminal market. They can be classified also as to whether they physically handle the commodity or not, and whether they specialize or handle a broad line of commodities. The distinction can also be made whether the firm takes title to the commodity or handles the shipment on a commission or net return basis. Finally classification can be based on the services provided such as repacking, transportation, or retailing.

The definitions used to describe types of firms, the functions they perform, and the way they do business have been developed from experience with the Uniform Commercial Code and the regulations established under federal legislation. The following are some types of firms operating in this industry (USDA AMS, 1987, USDA ERS, 1989). More complete definitions may be found in the references.

Shipping Point

At shipping point many *growers* do not become involved in marketing but turn their products over to someone else to pack and sell. The grower may belong to a *cooperative* that handles the packing and shipping. Or product may be sent to a *grower–shipper* who also handles other growers' commodities on either a net return, flat fee, or outright purchase arrangement. There are also *shippers* who do not grow any commodities but may perform a wide variety of services such as financing, planting, harvesting, grading, packing, or also furnish labor, seed, containers, and other services and who rely entirely on obtaining products from growers on one of these three bases. *Sales agents* operate under contract with growers and do not physically handle the product but arrange for the sale. The contracts between growers and sales agents may limit the way in which sales may be made and how the proceeds of sales are allocated. *Brokers* operating at shipping point may be either *selling brokers* or *buying brokers* who arrange the purchase or sale of individual lots for sellers or buyers usually on a fee per carton basis, and do not take title to or physically handle the product.

Transportation

Firms that offer transportation generally specialize in providing just one mode such as trucking or railroad transportation and do not take title to the product. Truckers may either be *fleet operators* or *independent owner-operators*. For multimodal transportation such as trailer-on-flatcar, one firm often coordinates the service for the shipper or receiver. *Truck brokers* have traditionally arranged for truck transportation, while *freight forwarders* or *transportation brokers* arrange for transportation services involving other modes or multimodal shipments.

Destination Markets

At destination wholesale markets firms are often classified into whether they are primary or secondary handlers, importers or exporters, brokers or sales agencies, retailers, or food service operators. Primary handlers consist largely of *wholesale receivers* who purchase truck or rail carlots for their own account and resell mainly to other wholesalers or chain store accounts, *commission merchants* who mostly handle product on consignment or a net return basis, and *service wholesalers* who may provide several additional services for customers including delivery.

Secondary handlers include *jobbers* who purchase less than carload lots from primary receivers and sell to smaller retailers and food service operators. Jobbers may specialize in one commodity such as bananas or tomatoes, providing ripening and packaging services as well. *Purveyors* are jobbers who specialize in serving food service operations such as restaurants and institutions.

Brokers and *sales agencies* at destination markets usually do not physically handle produce although they may arrange for the physical handling. Brokers at wholesale markets do not take title to the product and may represent either seller or the buyer. *Distributors* buy carlots and sell either in carlots or less-than-carlot (l.c.l.) quantities. *Sales agencies* may represent cooperatives, importers, or shippers. In terminal markets that still have *fruit auctions* the auction company provides auction services, and *auction representatives* serve the interests of the fruit shipper on the auction.

Retailing

The major classifications of retail food companies include independent operators, voluntary chains, cooperative chains, and corporate chains. *Independent operators,* as their name implies, are not integrated with any supplier or other retailer, but are free to buy from any source. *Voluntary chains* are groups of independently owned retailers who are served under a franchise type of arrangement by a wholesaler who operates a distribution center. *Cooperative chains* are groups of independent retailers who have joined together to obtain wholesale distribution services. *Corporate chains* are integrated retailers generally with more than 10 stores who operate at least one wholesale distribution center. There are many different types of grocery stores but the main categories are supermarkets, convenience stores, and specialty stores. *Supermarkets* are defined as grocery stores, primarily self-service, that provide a full range of departments and have at least $2.5 million in annual sales (1985 dollars). A *convenience store* is a small grocery store selling a limited variety of food and nonfood products, typically open extended hours. A *specialty foodstore* is a store primarily engaged in the retail sale of a single food category such as fruits and vegetables, dairy products, or meat and seafood.

Food Service

Food service establishments are divided into two major categories, *commercial* and *noncommercial.* Commercial establishments are public establishments that prepare, serve, and sell meals and snacks for profit to the general public. They include full service restaurants, fast food outlets, hotels and motels that cater to the general public, and many other types. Noncommercial establishments are establishments where meals and snacks are prepared and served as an adjunct or supportive service to the primary purpose of the establishment. These include schools, hospitals, plants and offices, correctional facilities, military feeding, and transportation operations.

METHODS OF DOING BUSINESS

Transactions between buyers and sellers of fresh fruits and vegetables are handled differently than many other commodities because of the special nature of the business, and are important because ownership often changes hands several times between the grower and the consumer. Not only do financial arrangements vary, but settlement may be delayed and sellers may have difficulty recovering payment. The duties and responsibilities of persons involved in marketing fresh produce are too numerous to specify here, but the following is a very brief summary. Further detail may be obtained from the various trade publications and from regulations issued by the U.S. Department of Agriculture (USDA AMS, 1986; USDA AMS, 1987; *Blue Book; Red Book*).

Shippers usually purchase produce from growers in their own names, at various stages in the marketing process ranging from in the field ready for harvest to packed for wholesale distribution. Growers' agents perform a wide variety of services, usually distributing produce in their own names and collecting payment directly from consignees. They then render the net proceeds to their principals after deducting expenses and fees. Contracts between growers and sales agents specify what the sales agents are authorized to do.

Brokers may operate at shipping point or destination market. They carry on their business in several ways and are classified by their method of operation. The usual operation of brokers consists of negotiation for the purchase and sale of carlots whether of one commodity or several. In negotiating a contract a broker usually acts as agent of the buyer or seller but not as agent of both parties. Frequently the carlot brokers never see the produce they are quoting for sale or negotiating for purchase by the buyer. Generally under broker contracts the seller invoices the buyer, however the seller may invoice the broker who in turn collects from the buyer and remits to the seller. Buying brokers on the market may be authorized to view and appraise the merchandise being offered for sale before negotiating the purchase.

Wholesalers at destination markets may act as receivers and purchase commodities outright or serve as commission merchants and sell on consignment. The amount of the commission charged for commodities handled on consignment should be and generally is negotiated in advance. There is a great deal of latitude in the way commission sales might be handled, not all to the benefit of the seller, but government regulations restrict such activities as dumping the produce or purchase of the consignment by the merchant involved, and require that detailed records be kept.

Most produce shipped from major growing areas to destination markets is sold or consigned by the shipper and purchased by market wholesalers, retail chain stores, or the restaurant institutional trade. For outright sales the point of sale is usually the shipping point, and this means that the receiver pays for transportation, transportation services, and handling between shipping point and destination. But since the product is purchased sight unseen or may deteriorate unduly in transit, the means of handling such situations need to be specified. Trade terms have been developed that help to improve the understanding of each party's rights and responsibilities with respect to such transactions.

TRADE TERMS AND DEFINITIONS

Trade terms covering purchases and sales describe such factors as the point of sale, the opportunities for inspection and rejection, the time of shipment, and the promptness of payment. The following are brief descriptions of some of these terms. The references listed above provide a full explanation and should be consulted.

A frequent point of sale is *f.o.b.* which means that the produce quoted or sold is to be placed free on board the boat, railcar, or other carrier in suitable shipping condition and that the buyer assumes all risk of damage or delay in transit not caused by the seller. The buyer shall have the right of inspection at destination before the goods are paid for to determine if the goods complied with the terms of the contract at time of shipment. *F.o.b. steamer* means that the produce is to be placed free on board the steamer at shipping point in suitable condition and that the buyer assumes all responsibility of risk of damage thereafter. *F.a.s. steamer* means that the produce is to be delivered free alongside the steamer. *Delivered* or *delivered sale* means that the produce is to be delivered by the seller on board car or truck or on the dock if delivered by boat, at the market in which the buyer is located, or at such other market as is agreed upon, free of any and all charges for transportation and protective service. *C.i.f.* means cost, insurance, and freight, and c.i.f. sales are deemed to be the same as f.o.b. sales, except that the selling price includes insurance and the correct freight and refrigeration or heater charges to destination.

Suitable shipping condition, in relation to direct shipment, means that the commodity at the time of billing is in a condition which, if the shipment is handled under normal transportation service and conditions, will assure delivery without abnormal deterioration at the contract destination agreed upon by both parties.

F.o.b. acceptance or *shipping point acceptance* means that the buyer accepts the produce at shipping point and has no right of rejection. The buyer has recourse against the seller if the produce was not in suitable shipping condition by recovery of damages but not by rejection. *F.o.b. acceptance final* or *shipping point acceptance final* means that the buyer accepts the produce at shipping point and has no right of rejection. Suitable shipping condition does not apply under this trade term. The buyer does have recourse for a material breach of contract by recovery of damages, but not if the shipment has been rejected.

F.o.b. inspection and acceptance arrival means that the produce quoted or sold is to be placed by the seller free on board at shipping point, the cost of transportation to be borne by the buyer, but the seller to assume all risks of loss and damage in transit not caused by the buyer, who has the right to inspect the goods upon arrival and to reject them if, upon such inspection, they are found not to meet the specifications of the contract of sale at destination. The buyer may not reject without reasonable cause. Such a sale is f.o.b. only as to price and is on a delivered basis as to grade, quality, and condition.

Several general terms for the time of shipment have been specifically defined. These include *immediate shipment, quick shipment,* and *prompt shipment.* When *shipment soon as possible* is used the buyer has the right, at any time after 7 days from the date the order was given, to cancel the order provided the notice to cancel is received by the shipper before the shipment is made.

Joint account—split above means that the receiving joint partner will pay promptly the agreed cost of the shipment to the joint partner. After disposition of the produce the parties will divide equally the profits on the shipment after deduction of the cost of the shipment and proper expenses from the gross proceeds. The receiving joint partner will pay all expenses and cannot recover any loss resulting from the joint venture.

Buyers may refuse to accept goods which they believe do not meet the terms of the contract. If a shipment is rejected the seller can always demand proper and adequate proof that the goods did not conform to the contract terms. Government inspection is the way such proof is usually provided.

TRADE REFERENCES, CREDIT RATINGS, AND BUSINESS PRACTICE INFORMATION

Marketing firms frequently find the need to seek new sources of supply or new buyers with whom they have not dealt before. Entering into a verbal agreement on a commodity sight unseen with someone unfamiliar can be an especially risky business. One way to reduce the risk is to consult one of the trade reference manuals.

There are two major trade references widely used by wholesale handlers of fruits and vegetables not only in this country but also abroad. These are *The Blue Book* and *The Red Book* (see p. 162 for addresses). The material included in each book and the services provided by each publisher are generally similar, but many firms subscribe to both. They contain an explanation of the services provided as well as the text of federal regulations and federal legislation governing fresh fruit and vegetable marketing, a summary of state regulations, Canadian government legislation, as well as a description of customs and rules of the produce trade. The major portion of each book is the rating section that provides information on growers, shippers, receivers, commission merchants, brokers and jobbers, wholesale and retail grocers, and processors, who buy, sell, or use fresh fruits and vegetables in the United States and Canada and some in other countries. A section on truck transportation covers customs and rules in truck transportation, the names and rating of truck brokers, and a listing of truck stops. Also included is an index of firms that provide supplies and services used by the trade, an index of shipping dates and shipping points, and a name index of all firms listed in the book.

The business rating section in one book contains a listing by town or city of the name and address of each business, the names of the principals and their telephone numbers, and brief information about the business. The rating section consists of 6 parts. These are

1. a classification of the nature of the business such as shipper, receiver, commission merchant, broker, or jobber or any combination of these;
2. the annual wholesale volume bought or sold in truckload lots;
3. the commodities handled or specialties in order of importance;
4. a financial rating or credit worth ranging from less than $1,000 up to a maximum of $1 million or over;
5. a moral responsibility or integrity rating, ranging from X, poor, to XXXX, excellent; and
6. pay descriptions or payment promptness as measured between the time merchandise and invoices are received and checks are mailed, ranging, in the case of one book, from AA, within 5 days, to E, uncertain or after 45 days.

Ratings are based on reports obtained from members of the industry and are continually updated. The ratings for moral responsibility and pay promptness are based on a weighted average of industry reports. In addition to the publication of revised ratings books twice a year the services also publish weekly credit sheets and monthly supplements. Users can call in for the latest information on any other firms, and can leave a list of firms with the service and will be notified of any change in rating of any firm on the list. An arbitration service is provided to subscribers. Access to one of the ratings services can be achieved through an electronic information network.

A special category of subscriber is available to firms that meet superior standards with respect to moral responsibility and pay promptness. In addition to meeting these standards these firms must agree to settle disputes with other similar firms by arbitration. Recognition is provided by being listed in bold type in the rating book.

TRADE REGULATION AND DISPUTE RESOLUTION

Many years ago the prevalence of unfair or fraudulent trade practices in the marketing of fresh fruits and vegetables, particularly with respect to consignment sales, led to the passage of government legislation and the issuance of regulations intended to deal with this situation. In 1927 the Produce Agency Act was passed governing transactions between principals and agents involving the dumping or destruction of farm produce received on consignment, and required receivers to account correctly for consigned merchandise. The Act applied to all perishable farm products, but most action under this Act now involves only cut flowers since ornamentals are not covered by the Perishable Agricultural Commodities Act.

The Perishable Agricultural Commodities Act (PACA) was passed by

Congress in 1930 and has since been amended many times. The Act prohibits unfair and fraudulent practices in marketing fresh and frozen fruits and vegetables, sets penalties for violations, and provides for collecting damages from anyone who fails to live up to contract obligations. A few years ago the Act was amended to include trust provisions that would enable sellers to increase the likelihood of recovering the value of products sold to receivers who became financially insolvent.

Licensing is the key to PACA enforcement, although bonding may be required under special circumstances. The Act requires commission merchants, dealers, and brokers handling fresh or frozen fruits and vegetables in interstate or foreign commerce to be licensed (USDA AMS, 1983). Under PACA regulations dealers include shippers, wholesalers, certain retailers and truckers, and in some instances processors who buy or sell these commodities in wholesale or jobbing quantities (totalling 1 ton or more per day). A retailer is subject to licensing when the invoice cost of all purchases of fresh and frozen fruits and vegetables during a calendar year exceeds $230,000. All brokers negotiating sales of fresh fruits and vegetables must be licensed.

Growers who only sell the fruits and vegetables they grow do not need a license, but they do if they sell in interstate or foreign commerce any produce grown by other farmers. Truckers do not need a license if they only haul produce for freight charges, but do if they buy and sell produce in interstate or foreign trade.

License fees are used to pay the cost of administering the PACA. Regional offices are maintained throughout the country to administer the regulations and deal with infractions. Fines are levied for operating without a license. Violations of the PACA can result in suspension or revocation of a license. Information on licensing and PACA regulations can be obtained from the Regulatory Branch, Fruit and Vegetable Division, Agricultural Marketing Service, U.S. Department of Agriculture.

PACA regulations require traders to comply with the terms of their contracts (USDA AMS, 1983). Sellers must ship the quantity and quality of produce specified. Buyers must accept shipments that meet the contract and pay promptly after acceptance. Prompt payment means within 10 days unless there is a prior agreement to extend the time. Receivers who handle consigned shipments must issue accurate accountings and pay net proceeds promptly.

Regulations under the PACA include definitions of standard trade terms that are specifically designed to assist in the buying and selling of fresh fruits and vegetables. Proper use of these terms improves communication and enables shippers and dealers to draw up brief and accurate written contracts that will reduce misunderstandings.

Some of the trade practices defined by the PACA as unfair are:

Failure of a seller, without reasonable cause, to deliver produce sold or contracted to be sold
Rejection, without reasonable cause, of produce bought or contracted to be handled on consignment
Failure to pay promptly the agreed price of produce which complies with the contract terms
Discarding, dumping, or destroying any produce received by commission merchants on consignment without reasonable cause
Failure or refusal to account truly and correctly and to make full payment promptly for produce shipped on consignment or on joint account, and
Misbranding or misrepresentation of grade, quality, weight, or state of origin of fruits or vegetables in interstate or foreign commerce is prohibited.

PACA requires every licensee to prepare and maintain records which fully and correctly disclose all transactions involving the business. The regulations specify the records that must be kept, depending on the nature of the business. In addition to general records there are certain specific records that must be kept by grower's agents, shippers, brokers, and market receivers and commission merchants.

Informal or formal complaints may be filed with the USDA by anyone with a financial interest in a transaction, whether or not they hold a PACA license. To file an informal complaint simply requires sending a letter to the nearest PACA office explaining the nature of the complaint and the amount of damages claimed. Records will be reviewed, and if there appears to be a proper basis for the complaint an investigator will try to arrange an informal settlement. Complaints must be filed within 9 months of the alleged violation, but the sooner the better.

If the informal complaint cannot be settled or if the amount involved is more than $15,000, either party may request an oral hearing to present evidence. Cases involving smaller amounts simply require evidence to be submitted in writing. When the proceedings are completed the U.S. Department of Agriculture (USDA) decides the case. The person charged with the offense may be required to pay a reparations award. If that person does not pay their license is automatically suspended and he or she may not continue in the produce business or be employed by another PACA licensee. An appeal may be filed in a U.S. District Court within 30 days from the date of the order. If a reparation award is not paid the person who submitted the complaint may file suit in court to obtain a judgment on the award.

The most common violations calling for USDA disciplinary action are:

Failure to pay for numerous shipments of produce purchased
Slow payment
Failure of a commission merchant or grower's agent to render accurate accounting and pay promptly the net proceeds due to shippers and growers
Brokers authorized by shippers to invoice and collect for produce sold but who fail to remit the collected proceeds to the shippers
Flagrant misbranding or misrepresentation of produce shipped in interstate commerce, and
Refusal to produce records for examination in connection with complaints filed.

PACA Trust Provisions

One continuing problem in marketing has been the difficulty that shippers have had collecting unpaid bills from receivers who became insolvent. Other creditors with greater security generally had prior claim on the assets of the bankrupt firm. In 1984 the Congress approved legislation amending the PACA to require that under certain conditions receivers of fresh fruits and vegetables must establish and maintain a trust on perishable commodities received, or on products and proceeds derived from them, for the benefit of unpaid suppliers.

To preserve trust benefits, the supplier must file with the debtor and the Secretary of Agriculture written intent to preserve its rights within 30 days after default in payment by the debtor. In addition, trading contracts must call for payment within at least 30 days from receipt and acceptance of the commodity by the buyer. Suppliers who have preserved trust benefits become first secured creditors in the event of bankruptcy. Under this program many sellers have been able to recover from bankrupt buyers the proceeds of their sales. The program has also resulted in increased activity for the offices administering the PACA, however, which may in turn necessitate an increase in license fees or charges.

MINIMIZING DISPUTES AND LOSSES IN WHOLESALE TRANSACTIONS

PACA administrators suggest that the way to minimize disputes is for buyers and sellers to first check the financial status and reputation of anyone they plan to do business with, and second, when contract terms are agreed on verbally to get them down in writing as soon as possible. Fax machines can now be used to confirm transactions promptly. Knowledge of trading practices and means of recourse when encountering unfair or fraudulent practices is also essential.

Since trading in fresh fruits and vegetables is such a competitive and rapidly changing business one can never completely rely even on firms with which business has been conducted satisfactorily for many years. One common practice is to spread the risk by not concentrating a large share of the business with any single firm. By not putting all one's eggs in one basket the chance of suffering a crippling loss is reduced.

The Blue Book and *The Red Book* services can monitor firms and alert subscribers when unfavorable reports are submitted on companies with which they do business. Some firms find it beneficial to watch for changes in business practices on the part of their customers and to be cautious when they delay payment more than usual or even increase order size. Such changes may or may not signal deterioration in financial status.

The following are the addresses of the two main business rating services.

The Blue Book
Produce Reporter Company
315 West Wesley Street
Wheaton, Illinois

The Red Book
Vance Publishing Corporation
P.O. Box 2939
Shawnee Mission, Kansas, 66201

REFERENCES

U.S. Department of Agriculture, Agricultural Marketing Service, 1983. *The Perishable Agricultural Commodities Act: Fair Trading in the Fruit and Vegetable Industry*. Agricultural Marketing Service, Program Aid No. 804.

—— 1986. *Rules of Practice Under the Perishable Agricultural Commodities Act, 1930 (Other Then Formal Disciplinary Proceedings).*

—— 1987. *Regulations (Other than Rules of Practice) Under the Perishable Agricultural Commodities Act, 1930, and Perishable Agricultural Commodities Act.*

—— Economic Research Service, 1989. *Food Marketing Review, 1988.* Agricultural Economic Report No. 614.

Chapter 10

Cooperative Marketing

INTRODUCTION

Voluntary cooperation among growers plays an important role in the marketing of fresh fruits and vegetables. Cooperatives, because of the special nature of farming, have enabled farmers to achieve economic goals they may not have been able to individually. There are many different types of cooperatives, but this chapter will focus only on marketing cooperatives.

Cooperatives are business organizations that are voluntarily owned and controlled by members for their benefit as patrons or users (Abrahamsen, et al., 1977). Sexton and Iskow (1988) argue that an additional concept necessary to distinguish cooperatives from other forms of business is that they provide their members with a degree of vertical integration. This integration may be forward or upward as in the case of marketing cooperatives, or backward or downward as in the case of farm supply or service cooperatives. Members benefit from marketing cooperatives mainly from higher net returns for their products. Cooperatives are believed generally to be born of necessity, formed to correct deficiencies in the marketplace. Voluntary cooperation has been found to be beneficial for many types of fruit and vegetable marketing activity, including operating local packing houses and providing sales and merchandising services. While cooperatives can accomplish many desired objectives they may also be subject to inherent problems.

A BRIEF HISTORY

Fruit and vegetable growers have long considered themselves at a disadvantage in marketing their product. Individual farm production has often been too small in quantity to handle, pack, and ship efficiently. Changes in production from one year to the next have brought highly fluctuating prices.

The small number of buyers operating in the market have often been able to play off one grower against another and so apparently depress market prices. In shipping to distant markets growers have been at the mercy of commission wholesalers who had little incentive to secure higher prices. Through cooperation some growers have been able to develop new markets or new uses for their products.

Grower efforts to cooperatively market fruits, vegetables, and nuts in the United States began in the late 19th century (Hulse, et al., 1978). Early cooperative activity in the East involved adopting standard packages for fruit, shipping cranberries to England, and establishing auction markets. In the West fruit marketing cooperatives were organized to meet the problems of shipping long distances to Eastern markets. Local citrus cooperatives organized in the early 1890s were soon formed into district exchanges that each made its own sales. In 1895 the proposal to establish a central exchange with branch offices in important markets led to the formation of the Southern California Fruit Exchange, now known as Sunkist Growers, Inc. (Kirkman, 1978). The association launched its first advertising campaign in 1908 to popularize its new trademark. A dealer service unit was started in 1915 to make personal contacts with the retail trade. But early attempts by Sunkist to prorate citrus shipments were weakened by packers outside the organization. This eventually led to the development of marketing orders that could provide for compulsory support.

The system of locally owned grower associations federating into exchanges to supply nationwide and export marketing services also became the pattern for deciduous fruit cooperatives such as Blue Anchor in California and Diamond Fruit Growers in Oregon. Two California nut associations, Diamond Walnut Growers and California Almond Growers Exchange (now known as Blue Diamond, Inc.), also began as federated organizations, but like many fruit cooperatives they have changed to direct membership.

In Florida a number of citrus and vegetable cooperatives were organized about 1885, but it was not until 1909 that the Florida Citrus Exchange, now known as Seald-Sweet Growers, was formed. Originally patterned after the Southern California Fruit Exchange it remains a federated cooperative. Cooperative marketing of potatoes and vegetables was confined to small local operations in Florida and other production areas until the 1920s and 1930s.

The Sherman Antitrust Act of 1890 and state legislation provided farmers some protection against the domination of large scale businesses or trusts, but also hindered them from organizing cooperatively until limited exemption was secured through the Clayton Act of 1914 and the Capper–Volstead Act of 1922 (Manchester, 1982). Low farm prices during the 1920s fostered the growth of many farm cooperatives, some hoping to raise market prices through control of a substantial share of farm production. Their failure to attain success by this approach led others to attempt to increase returns through greater efficiency. The conflict between the bargaining and the operating approaches has still not been entirely resolved. Efforts to improve returns through cooperative bargaining, however, has not received much support from growers of fresh fruits and vegetables although from time to time cooperatives and other shippers will attempt to hold the line on prices or prevent price adjustments.

Cooperative Ownership and Control

Ownership and control of cooperatives differs from general business corporations in that voting by member/owners is not necessarily according to stock or other type of equity investment, but usually on a democratic or one member/one vote basis (Griffin, Volkin, Davidson, 1981). Prospective users of the organization become members by investing equity capital, on which they get only a limited return if any (Griffin, et al., 1980). Membership/ownership is generally restricted to users, and the aim is to keep member investment proportional to use. Use of a marketing cooperative means marketing farm products through the organization, and proportional investment would imply that a farm shipping twice as much product as another should have twice the investment in the organization.

Although most cooperatives operate on the basis of one member/one vote, some do take account of the variation in use by different members. In such cases provision is made for members who make more use of the organization to have more votes. Because use and ownership are so closely tied together, cooperatives have found it desirable to keep members informed of business conditions, and to educate them on impending issues.

METHODS OF OPERATION

Marketing cooperatives operate in most respects in the same ways as do other business corporations. A set of by-laws establishes the general scope and procedures, and a board of directors is elected to represent the members/owners in determining policies and hiring and supervising management to carry out such policies. Only in the way cooperatives seek to follow their goal of operating at cost, maintaining sufficient membership equity, and adjusting member investment to use do operations differ. These differences, however, make it hard to compare the success of a marketing cooperative in obtaining higher prices for members compared to returns that may be obtained from dealing with a proprietary corporation.

Marketing cooperatives cannot, of course, pay outright for members products on receipt and still follow the principle of operating at cost. Costs and net proceeds are not known with certainty until the product is finally sold. When there is a delay between the time the product is received and when it is sold members are sometimes provided with an initial payment representing a partial amount of the estimated final return.

Seldom is it possible for a marketing cooperative to return to members exactly what their product brought, less the costs of marketing it (Hulse et al., 1978). Costs are often pooled or averaged among products from different members that have similar characteristics. The same packing charge per box may be levied on fruit from different members, for example, even though packing charges may differ because of different quality. If quality and costs are very different, however, the difference in packing costs may be reflected by establishing several cost groups or pools according to fruit quality.

Pooling is the essence of cooperatives. Members pool capital and products. How members are paid depends on the extent to which costs, revenues, and risks are pooled or shared.

In order for sales revenue to reflect gross returns from each members' products the origin of the product must be identified all through the marketing process. This is easily done in some cases but difficult to do in others. And even though identity can be maintained it still may be desirable to pool or average returns if cooperative management makes the decision when and where to sell the product and net returns in different markets at different times are different, as they are likely to be. Good marketing may require the cooperative to provide customers as stable a supply as possible through periods of shortage as well as surplus. To preserve equity the general rule in pooling is to try to reflect to members those costs and returns that are due to factors under members' control or peculiar to their situation, and to pool those which are not.

Once net proceeds are determined marketing cooperatives usually make a patronage refund or dividend payment equal to all or part of the difference between the initial payment and the final amount due. Usual practice is to make only a partial payment at this time and to retain some net proceeds within the cooperative. Such retained proceeds are usually used to increase the members' investment in the cooperative or to replace previously retained proceeds that then can be paid to members to whom those retained proceeds are due. The practice of retaining proceeds for several years and then returning them out of current savings is known as operating a revolving fund. Having a revolving fund is a good way to keep member investment somewhat in proportion to use, although not the only one.

COOPERATIVE INCOME TAX EXEMPTION

Farmer cooperatives are often considered to have advantages over other business corporations not only because of favorable treatment under antitrust legislation but also under federal corporate income taxation. Income tax exemption for cooperatives actually is severely circumscribed, and limits operations to such an extent that many cooperatives voluntarily abandon exempt status.

All corporations operating on a cooperative basis may under certain circumstances (Subchapter T) exclude from taxable income the amount paid under patronage refunds or retained earnings (Griffin, Volkin, Davidson, 1981). If a cooperative or other corporation declares a patronage refund at least 20 percent must be paid in cash in order to secure tax exemption. In all cases, however, net margins or patronage earnings accruing to members of marketing cooperatives are subject to income tax payable by the member.

To qualify as a tax exempt or Section 521 association the organization must meet certain specific requirements. In general these include being a farmer association organized on a cooperative basis, with most stock owned by active members, with limited dividends on stock, less than half the business with nonmembers, and nonmembers treated like members with respect to such matters as pricing, pooling, and payment. Cooperatives that meet these and other requirements may deduct from taxable income, in addition to other normal deductions, the amount paid as dividends on capital stock during the taxable year, the amounts paid to patrons with respect to income derived from sources other than patronage such as rents or investment income, and the net margins paid as patronage dividends or that part retained and allocated to members (Griffin, Volkin, Davidson, 1981). A cooperative that does not qualify can exclude the patronage refunds and per-unit allocations from taxable income but not stock dividends or other income.

COOPERATIVE STRENGTHS AND WEAKNESSES

The concepts of service at cost, of democratic control, of open membership, and of benefits for users have had great appeal for farmers and have attracted many to organize or join farmer cooperatives. Being able to develop broad local support has often initially been a decided advantage for the cooperative organization compared to profit seeking proprietary competitors. Help in organizing is available from the U.S. Department of Agriculture (USDA) and the Cooperative Extension Service (Abrahamsen, Mobley, 1979), and debt capital at favorable interest rates can be obtained from the Farm Credit System.

In order to be successful the cooperative must have realistic or attainable goals, good management, and broad member support. Goals must be carefully evaluated to see whether on the basis of past experience or hard economic analysis they may be attained. Management must be carefully chosen, well paid, and given freedom to manage the business within the policy guidelines laid down by the board of directors. The organization needs the full support of members both in contributing equity capital and in using its services.

Once in operation, however, the cooperative is faced with challenges even greater than those facing other businesses. The main weaknesses of cooperatives, according to Sexton and Iskow (1988), are their difficulty in obtaining equity capital and their failure to adequately reward entrepreneurial activity. They may also be less flexible than other business organizations owing to their democratic nature.

Sexton and Iskow (1988) explored the key economic functions of cooperatives and described the market conditions amenable to a cooperative's presence. They also focused on the organizational, financial, and operational keys to the successful development of a cooperative. Finally they analyzed the experiences of 61 recently organized American agricultural cooperatives to discern the economic factors motivating each cooperative's inception and the organizational, financial, and operational keys to its subsequent success or failure.

Twenty-eight of the 61 cooperatives were considered by respondents to be major successes, while 11 were reported to be not successful. Statistical analysis was used to identify several factors critical to successful performance. These included initial involvement of a large number of members, growth in membership, use of full-time professional management, and acceptance of nonmember business.

Stern and Anderson (1986) studied a sample consisting of 14 well-established cooperatives that successfully marketed a variety of products including dairy products, fruits and vegetables, poultry and specialty products (nuts, raisins, etc.) They tested 12 hypotheses concerning the operation of these businesses and accepted 7. Significant features of these cooperatives were that their primary objective was to improve long-term returns of members, they carried out a thorough analysis of their own and competitors' strengths and weaknesses, they had an intimate knowledge of their customers and targeted their marketing efforts, they did a regular and systematic review of their entire marketing program, they fully exploited their advantages and managed their disadvantages, they had experienced marketing executives to manage their marketing programs, and their boards of directors were only involved in establishing policies, goals, and budgets.

CURRENT STATUS OF COOPERATIVES

In 1988, according to an Agricultural Cooperatives Service survey, there were 302 cooperatives specializing in fresh or processed fruits and vegetables (USDA ACS, 1989). These cooperatives had 51,650 members, net sales of $6.6 billion, and net income of $183.3 million. Compared to an earlier survey in 1975/76 this represented a decline of 25 percent in the number of associations and of 30 percent in membership but a more than doubling in the volume of business. Cooperatives are believed to handle about one-quarter of the total quantity of fruits and vegetables moving off U.S. farms. This share varies by region, commodity, and product form.

Most fruit and vegetable cooperatives are organized along commodity lines and specialized by commodity group (Hulse et al., 1978). A large number of small associations provide grading and packing services. There is a small number of large organizations and a few small ones that provide a full line of services including sales and merchandising.

Among the citrus cooperatives Sunkist Growers is the major factor in California. In late 1989 Sunkist reported (*The Packer,* 1989) that after 96 years of experience the cooperative handled 79 million cartons of fruit over a 12 month period from 5,950 growers through 65 packinghouses. Seald-Sweet has a similar operation in Florida. Diamond Fruit Growers in Oregon sells fresh apples and other deciduous fruit, and processes and markets canned and frozen products. A leading cooperative in strawberry marketing is Naturipe Berry Growers in California. Blue Anchor, also based in California, cooperatively markets grapes and other deciduous fruits for members in all major fruit districts in California and Arizona. Cooperatives also market other fruits grown by highly specialized growers in specialized areas. Members of Calavo Growers produce a large share of California's avocado crop, and the cooperative markets processed products as well as fresh fruit. Ocean Spray is a well-known cooperative marketer of cranberries in all forms.

Cooperatives that market fresh potatoes and other fresh vegetables are not as well known. In Florida, the Hastings Potato Growers and Florida Planters, market cabbage as well as potatoes for members. The Colorado Potato Growers Exchange markets onions, beans, and wheat as well as potatoes for local cooperative associations. Maine Potato Growers markets grain and dairy products as well as potatoes, and also buys equipment and farm supplies for members. Pioneer Growers and South Bay Growers in Florida each pack and ship a variety of vegetables for local growers.

Changes in farm production in recent years plus changes in marketing practices have brought major changes to farm marketing cooperatives. Some have been able to adapt well to these changes, while others have encountered difficulties. Cooperatives have many strong points, but also some attributes that can pose problems and must be handled effectively. The operations of some cooperatives have been closely tied to marketing orders so their future may depend to some extent on how orders survive.

REFERENCES

Abrahamsen, Martin A., J. Warren Mather, James R. Baarda, James Michael Kelly, 1977. *Cooperative Principles and Legal Foundations.* U.S. Department of Agriculture, Farmer Cooperative Service, Section 1, Bulletin 1.

Abrahamsen, Martin A., Howard Mobley, 1979. *Organizations Serving Cooperatives.* U.S. Department of Agriculture, Cooperative Information Report 1, Section 5.

Griffin, Nelda, David Volkin, Donald R. Davidson, 1981. *Cooperative Financing and Taxation.* U.S. Department of Agriculture, Agricultural Cooperative Service, Cooperative Information Report 1, Section 9.

—— Roger Wissman, William J. Monroe, Francis P. Yager, Elmer Purdue, 1980. *The Changing Financial Structure of Farmer Cooperatives.* U.S. Department of Agriculture, Farmer Cooperative Research Report No. 17.

Hulse, Fred E., Richard S. Berbich, Gilbert W. Biggs, Martin A. Blum, 1978. *Fruit, Vegetable, and Nut Cooperatives.* U.S. Department of Agriculture, Economics, Statistics, and Cooperatives Service. Cooperative Information Report 1, Section 13.

——. Gilbert W. Biggs, Roger A. Wissman, 1980. *Small Fruit and Vegetable Cooperative Operations.* U.S. Department of Agriculture, Economics, Statistics, and Cooperatives Service. Cooperative Information Report No. 27.

Kirkman, C. H. Jr., 1975. *The Sunkist Adventure.* U.S. Department of Agriculture, Farmer Cooperative Service, FCS Information 94.

Manchester, Alden C., 1982. *The Status of Marketing Cooperatives Under Antitrust Law.* U.S. Department of Agriculture, Economic Research Service, ERS-673.

Sexton, Richard, Julie Iskow, 1988. *Factors Critical to the Success or Failure of Emerging Agricultural Cooperatives.* Giannini Foundation, University of California, Information Series No. 88-3.

Stern, Ann G., Bruce L. Anderson, 1986. *An Analysis of U.S. Cooperatives With Successful Marketing Strategies.* Cornell University, Department of Agricultural Economics, A. E. Res. 86-19.

The Packer, 1989. *Sunkist Advertisement.* Vance Publishing Co. October 21 U.S. Department of Agriculture, Agriculture Cooperative Service, 1989. *Farmer Cooperatives,* November.

Chapter 11

Marketing Orders

Government enforced marketing orders have enabled growers to secure compulsory support for group action in marketing (Armbruster, Jesse, Nelson, Schafer, 1981). Under certain circumstances growers have needed to use the government authority provided by marketing orders to obtain greater participation in marketing programs. This chapter will discuss only federal marketing order legislation and programs even though legislation in California and several other states also authorizes marketing orders. The marketing order approach provides government authority to muster compulsory support for certain types of marketing activities provided a substantial proportion of the growers involved indicate their willingness to participate. Fruit and vegetable marketing order legislation does not permit the setting of prices directly, but the expectation is that returns will be stabilized or increased through such provisions as setting minimum standards for grades and sizes, regulating the flow to market, or raising money for advertising and promotion. The use of marketing orders for advertising and promotion will be discussed in a subsequent chapter.

HISTORICAL BACKGROUND

The number of fruit and vegetable cooperative marketing associations in this country grew rapidly during the late 1800s and early 1900s (Heifner et al., 1981). The depressed prices of the 1920s turned growers' attention to the need for programs to regulate the quantity and quality of fruit and vegetables marketed. Efforts were made to voluntarily organize producers into comprehensive sales agencies as a means of attaining more orderly marketing and higher prices. Some did achieve initial success in raising prices and improving market conditions, but all eventually failed. One limitation was the inability to induce or maintain a sufficient level of participation by growers and handlers. Since nonparticipants enjoyed the same benefits as participants without being subject to the regulations, there was little incentive to participate. This has been termed the *free-rider* problem, and it has threatened many voluntary self-help programs in agriculture.

The failure of voluntary programs led to government efforts in this country and others to become more involved in marketing. First efforts in the late 1920s and early 1930s by the U.S. government and the state of California consisted of entering into voluntary marketing agreements with first handlers and licensing them to handle commodities.

Experience in California and at the federal level led eventually to the passage of the Agricultural Marketing Agreement Act (AMAA) of 1937 which, in spite of significant amendments, still remains the statutory basis for federal marketing order programs. Fruit and vegetable orders employ different provisions than do milk orders, and operate in an entirely different competitive and institutional environment.

A marketing order is a legal mechanism under which regulations issued by authority of the Secretary of Agriculture are binding on all handlers of the product in a specified area. Marketing orders are initiated, and implemented only after approval, by the affected growers. Originally the stated goals were to provide a tool to establish and maintain orderly marketing and to achieve parity prices for farmers. The concept of parity has become outdated and orderly marketing is difficult to define, so the goals of marketing orders today are generally considered to be to obtain higher and less variable prices for farmers. The AMAA lists specifically those commodities that are eligible or ineligible for marketing order coverage and the types of regulations that can be enforced, and also requires that the geographical area covered be as small as practicable. Marketing order legislation for fruits and vegetables does not include the possibility of regulations to establish prices or control production directly.

TYPES OF REGULATIONS

There are three general types of provisions in federal marketing orders that can be used to solve production and price problems indirectly (Armbruster, Henderson, Knutson, 1983; Jesse, Johnson, 1981; Zepp, Powers, 1988). These are quality control, supply management, and market support. Quality control and market support provisions are the most widely used, with only a few marketing orders providing for supply management (Table 11.1). Enabling legislation in California and other states contains similar provisions.

Table 11.1. Provisions in Federal Fruit and Vegetable Marketing Orders, 1988.

	VEGETABLES	DRIED FRUITS, NUTS, SPECIALTY CROPS	FRUITS	TOTAL
	number			
Quality control				
Minimum grade	13	6	19	38
Minimum size	13	5	22	40
Supply management				
Producer allotments	1	1	1	3
Market allocations	0	5	1	6
Reserve pools	0	5	0	5
Prorates	2	0	5	7
Shipping holidays	4	0	5	9
Market support				
Research	9	6	19	34
Market development	9	6	19	34
Package standardization	10	2	15	27
Total marketing orders	13	7	23	43

Source: Adapted from Zepp and Powers, 1988.

Quality Control Regulations

Quality control provisions are the most generally used. These permit the setting of minimum grades, sizes, and maturity standards and are normally enforced through mandatory federal inspection. To the extent these regulations increase the quantity of lower quality material graded out they not only improve the average quality of the product moving to market and also reduce the supply, thus contributing to higher prices at least in the short run.

Legislation provides that if certain commodities are covered by a marketing order containing quality control provisions then imports of those commodities must meet the same or comparable standards. Quality regulations have seldom been set to significantly restrict sales, and so have had little impact on markets.

Supply Management Provisions

Supply management provisions represent the most powerful form of regulation permitted under marketing orders since direct regulation of supply has the greatest potential for affecting price. The two basic methods of quantity control are volume or sales management and market flow regulations. Volume management provisions attempt to influence price by reducing the quantity sold on the primary market. There are three types of volume management provisions. One is producer allotments that restrict the quantity of product a producer may sell. Producer allotments have been authorized for such crops as spearmint oil, Florida celery, and cranberries, but are seldom used now.

A second type of volume management is market allocation. Market allocation regulations restrict sales of a commodity in one of two or more different market outlets. Such a program enhances producer returns if the nature of demand (the elasticity of demand) differs among the markets, and buying in one market for resale in the other is difficult if not impossible. Allocation of product between the fresh and processed markets or between the domestic and export markets are examples of this approach. Restricting sales in the primary market (the market with the more inelastic demand), such as the fresh or domestic market, and diverting product to the secondary market, such as the processed or export market, has been shown to increase producer returns. Market allocations have been authorized in marketing orders for cranberries, almonds, walnuts, filberts, California dates, and raisins.

Reserve pools are a third method of volume management. These provide that at times of excess supply a portion of production may be placed in a set-aside or reserve pool rather than diverted immediately. Later if prices in the primary market improve then sales may be made out of the reserve pool, diverted to secondary markets, or disposed of in nonfood use. Reserve pools have been authorized in marketing orders for tart cherries, spearmint oil, almonds, walnuts, raisins, hops, and prunes.

The second form of quantity control authorized under marketing orders is market flow regulation. In principle all the production is eventually sold although in actual practice this may not be the case. Producer returns may be enhanced through market flow regulation by avoiding seasonal gluts or lost sales due to shortages. Market flow regulation is accomplished through prorates or shipping holidays.

Handler prorates specify the maximum quantity a handler may ship over a specified period of time, usually a week. Handler prorates have been authorized for citrus, Tokay grapes, Florida celery, and South Texas lettuce. Prorates are extensively used in the citrus industry where fruit may be stored on the tree for lengthy periods of time without significant quality loss. Use of the prorate throughout the whole season may force fruit otherwise suitable for fresh sale into processing outlets. Used in this way prorates become equivalent to market allocation provisions.

The provision for shipping holidays is another way of regulating within season shipment. A shipping holiday is a period when all commercial shipping is prohibited. The use of shipping holidays is a weak form of controlling market flow, and holidays are typically limited to periods surrounding calendar holidays. Shipping holidays have been authorized for Florida citrus, avocados and celery, California desert and Tokay grapes, Idaho/Oregon onions, and South Texas onions and lettuce.

Market Support Activities

The third category of provisions authorized under the AMAA have been called market support activities. These provisions do not directly affect the quantity sold but contribute to achieving legislative goals in general through reducing costs or expanding the market.

Several types of market support activities are used. Provisions may be made for standardization of containers and packs. Funds may be assessed to raise money for research or promotion. Advertising is permitted only for specific commodities listed in the Act. Orders may be used to require handlers to post minimum prices and to prohibit unfair trade practices, but these provisions are seldom used. All orders require handlers to provide shipping information and this is aggregated to provide data that is useful in decision making. Market support activities are widely used in all of the major commodity groups.

ORDER ADMINISTRATION

Fruit and vegetable marketing orders are an unusual form of regulation because they are initiated and largely controlled by growers and handlers while the terms are binding on the handlers. To initiate an order a proposal is developed by commodity producers in cooperation with the USDA which proceeds through legally defined public proceedings. If approved by the Secretary of Agriculture it is subjected to a public referendum of the growers concerned. A marketing order can become effective only if two-thirds of the producers voting, or producers with two-thirds of the production represented by the vote, approve the order. For California citrus, the necessary producer numerical majority is three-fourths. In producer referenda bloc voting by cooperatives is permitted by statute.

The procedure for amending an order is similar to initiating one. An order may be terminated by the Secretary of Agriculture if it is determined that the order obstructs or does not support the policy of the Act, or if more than 50 percent of the affected producers having at least 50 percent of the total volume of the commodity favor termination.

Orders are administered by committees composed of growers or growers and handlers and, sometimes, public members. The major responsibility of these committees is to recommend regulatory policy and specific regulations to the Secretary of Agriculture who is ultimately responsible for issuing regulations under the orders. Committees vary in size and method of election or appointment. Producers held a majority on all but seven committees in 1981; on those seven producers and handlers were equally represented. Seventeen orders specify some method of allocating committee membership among cooperative and independent producers and handlers. Sunkist and Sunkist-affiliated growers have not been permitted to have more than half the membership for the California–Arizona orange and lemon orders, even though the large cooperative has handled more than half the volume of the commodities affected.

ECONOMIC IMPACT AND PUBLIC POLICY

The number of federal marketing orders increased sporadically from 1937 to 1965 and has changed little since that time. In 1981 there were 47 federal marketing orders in effect covering production of 33 fruits and vegetables in 34 states (Armbruster, Henderson, Knutson, 1983). The estimated farm value of commodities marketed under these orders was $5.2 billion in 1980. In terms of crop value more than half the U.S. fruits and tree nuts and 15 percent of the vegetables were covered.

The economic impact of marketing orders on fruit and vegetable growers and on the economy in general has been the subject of repeated study, especially in recent years. Rising food prices in the 1970s caused marketing orders to be challenged for contributing to inflation. Marketing orders also came under scrutiny during the emphasis on deregulation in the 1980s. Attention has focused on whether growers as a group really have benefited, or whether some have benefited and others have not. Interest has also been directed as to the effect of marketing orders on the welfare of consumers, generally in terms of higher prices or misallocation of resources. Criteria have also been developed to help evaluate the performance of individual orders.

One study statistically analyzed commodity prices to determine the effect of orders on price levels and variability by comparing order and nonorder commodities (Jesse, Johnson, 1981). Prices for fruits and vegetables sold under federal orders were found to have not exceeded prices for similar nonorder commodities, nor were order commodity prices more stable than nonorder commodity prices. Historical price patterns for commodities covered by orders that limit quantity marketed were not different from price patterns for commodities covered by weaker orders. It was considered possible, however, that orders may tend to be instituted more frequently for products with extreme price variability and they may only be partially successful in reducing that instability.

Another study looked at the effect of orders on economic efficiency and came up with conclusions with respect to benefits and costs (Heifner et al., 1981). The benefits were considered to be that orders contribute to seasonal and interseasonal stability of producer prices and incomes, promote quality assurance, reduce trading costs, increase applied research, and increase the amount and availability of marketing information. Actual costs of fruit and vegetable marketing orders were considered to include the inefficient allocation of resources in the cases of the Arizona/California citrus and the filbert and walnut orders, restriction on firm growth, reduced price competition, and a smaller range of choice with respect to quality. Potential but uncertain costs were the diversion of wholesome food into nonfood use, restricted handling and packaging innovations, and inefficient trade flows. Orders were also considered to have had an effect on income distribution, entrepreneurial independence, and the number and size of farms by retarding the decline in farm numbers and helping small growers compete with larger growers.

The prorate provisions of the federal marketing orders for California/Arizona citrus have come under fire from some growers who claim these restrict initiative and penalize independent growers compared to cooperative members. Research has not been directed at these particular issues, but a study of the economic effect of terminating these orders found that while the orders have probably benefited consumers by lengthening the shipping season for fresh oranges they have likely had little long-run effect on prices and seem to have restrained the use of oranges for the fresh market and encouraged production for processing (Thor, Jesse, 1981). Terminating the orders, the authors state, would probably substantially reduce the processed orange shipments and lead to a sharp reduction in the Western orange processing industry. In the long run, however, producer prices and total seasonal fresh shipments would be comparable with present levels.

The handler prorate for California–Arizona navel oranges was temporarily suspended during the 1984–85 season due to a small crop and high prices. Powers, Zepp, Hoff (1986) found that only minor differences existed between the prorate suspension and the prorate periods in the stability of shipments and prices. Handler marketing practices changed very little.

Since, as we have seen earlier (Kinney, Carman, Green, O'Connell, 1987), the demand for fresh lemons in the domestic market is price inelastic while the export and processing markets are more price elastic the marketing order prorate for California–Arizona lemons is also used to restrict supplies moving to fresh domestic sales. The excess quantity is sold for export or processing. Kinney and others used a simulation model to project the impact of changes in the quantity of lemons allocated per capita to the fresh market. The immediate effect of increased allocation to fresh market was found to benefit consumers and reduce producer prices. It was concluded, however, that with the passage of time producer revenues would recover as the industry adjusted to the new level of prices.

These and other studies indicate how difficult it is to positively evaluate the costs and benefits of marketing orders to growers or the general public. It appears, however, that as firms involved in the production and marketing of fresh fruits and vegetables become fewer in number and larger in size, as methods of communication are improved, as more information on supplies and prices becomes readily available, that the role of marketing orders may change in the future.

REFERENCES

Armbruster, Walter J., Dennis R. Henderson, Ronald D. Knutson, Editors, 1983. *Federal Marketing Programs in Agriculture: Issues and Options.* The Interstate Printers and Publishers, Danville, Illinois.

Heifner, Richard, Walter Armbruster, Edward Jesse, Glenn Nelson, Carl Shafer, 1981. *A Review of Federal Marketing Orders for Fruits, Vegetables, and Specialty Crops: Economic Efficiency and Welfare Implications.* U.S. Department of Agriculture, Agricultural Marketing Service, Agricultural Economic Report No. 477.

Jesse, Edward V., Aaron C. Johnson, Jr., 1981. *Effectiveness of Federal Marketing Orders for Fruits and Vegetables.* U.S. Department of Agriculture, Economics and Statistics Service, Agriculture Economic Report No. 471.

Kinney, William, Hoy Carman, Richard Green, John O'Connell, 1987. *An Analysis of Economic Adjustments in the California-Arizona Lemon Industry.* Giannini Foundation, University of California, Giannini Foundation Research Report No. 337.

Polopolus, Leo C., Hoy F. Carman, Edward V. Jesse, James D. Shaffer, 1986. *Criteria for Evaluating Federal Marketing Orders: Fruits, Vegetables, Nuts, and Speciality Commodities.* U.S. Department of Agriculture, Economic Research Service.

Powers, Nicholas J., Glenn A. Zepp, Frederic L. Hoff, 1986. *Assessment of a Marketing Order Prorate Suspension: A Study of California-Arizona Navel Oranges.* U.S. Department of Agriculture, Agricultural Economic Report No. 557.

Thor, Peter K., Edward V. Jesse, 1981. *Economic Effects of Terminating Federal Marketing Orders for California-Arizona Oranges.* U.S. Department of Agriculture, Economic Research Service, Technical Bulletin No. 1664.

Zepp, Glenn, Nicholas Powers, 1988. Fruit and Vegetable Marketing Orders. In *National Food Review.* U.S. Department of Agriculture, Volume 11, Issue 3, July-September.

Chapter 12

Pesticide Use and Food Safety

INTRODUCTION

Surveys by the Food Marketing Institute, Vance Publishing Co., and Kansas State University, among others, have found that many consumers have become concerned about food safety. Their concern extends beyond the occurrence of natural components such as sodium or saturated fats to the presence of pesticide residues, additives, or contaminants. Fresh fruits and vegetables have come under particularly close scrutiny because of the chemicals used to control pests such as harmful insects, plant diseases, and noxious weeds. The problem is intensifying because of the emergence of new pests and of strains resistant to existing chemicals, while the development and testing of new control measures that meet safety standards is becoming more difficult and expensive. Concern is also growing about other effects of chemical pesticide use such as groundwater contamination and hazardous working conditions on farms.

The control of pests has long been a major problem in the production and distribution of fresh fruits and vegetables. Many types and kinds of pests prey on fruits and vegetables, adding to production costs, reducing yields and quality, or accelerating deterioration. Pest pressure differs from one crop to another, and for each crop may differ according to growing area and methods of production. There are three major types of pesticides used in farm production. These are herbicides that attack weeds, insecticides that control harmful insects, and fungicides that deal with fungus diseases of crops. The traditional reliance on chemical control is now being challenged, but alternative approaches such as bioengineering, irradiation, and organic farming have not proven completely satisfactory. Dissatisfaction with public food safety measures has led to the development of private pesticide residue testing services. Marketing firms are having to assess the costs and benefits of alternative strategies in regard to residue testing and the promotion of food safety, even though the danger to human health from pesticide residues is considered by toxicologists to be far less than that from other food-borne pathogens.

BACKGROUND ON CHEMICAL PESTICIDE USE AND PUBLIC CONCERN

Insects, plant diseases, and weeds have long presented a challenge to growers and handlers of fruits and vegetables. The Irish potato famine of the 1840s, for example, was caused by a disease known as late blight that is still common today. Early methods of pest control were largely primitive or nonexistent. Prior to World War II, Paris green, a mixture of copper sulphate and arsenic, was commonly used to control leaf eating insects. A mixture of lime and copper sulphate known as Bordeaux mixture was dusted on grape vines and other plants to protect against a fungus disease called downy mildew. Potato beetles were sometimes killed by knocking them off the plant into a can of kerosene. Pyrethrum, an extract of chrysanthemum flowers and an effective organic insecticide, was discovered over 150 years ago and is still in use. As early as 100 years ago the U.S. Department of Agriculture (USDA) brought in a few Australian vedalia beetles and achieved spectacular biological control of the cottonycushion scale that had threatened to destroy the California citrus industry (USDA ARS, 1989).

Following this initial success other methods of biological control were attempted but were not as effective, and few changes occurred in pest control measures until after World War II. At that time the recognition of the ability of DDT to control insects and the development of other chemicals led to the rapid adoption by U.S. farmers and food handlers of synthetic chemical pesticides. Chemical control was comparatively inexpensive and generally quite effective. The quantity of chemicals used in pest control increased rapidly during the immediate postwar period.

Soon, however, problems in the widespread use of chemical pesticides became evident (Carson, 1962; Bosso, 1987; Dunlap, 1981). Targeted insects in some areas developed resistance to DDT and other chemicals, and more toxic materials had to be used in order to obtain control. Pesticides sometimes harmed beneficial insects as well as other wildlife, and residues accumulated in the tissues of birds and animals as well as humans. The scientific community differed as to the potential impact of these developments on human health. Efforts by the USDA in the late 1950s to eradicate the gypsy moth in the Northeast and the fire ant in the South by the widespread aerial application of insecticides aroused the ire of many public officials and other citizens. The public was made aware of these concerns through the activities of various environmental organizations and by the publication of Rachel Carson's book *Silent Spring* in 1962.

The cranberry scare of 1959 also drew attention to pesticide use. Shortly before Thanksgiving that year the Department of Health, Education, and Welfare announced that some cranberries were believed to contain residues of an herbicide aminotriazole, thought to be a carcinogen. Sales of cranberries fell precipitously.

During this period the industry reacted in several ways. A small but dedicated group of fruit and vegetable growers adopted organic methods, and completely shunned the use of synthetic materials not only to kill pests but also to fertilize their crops. Cooperative Extension specialists began to promote the use of Integrated Pest Management (IPM) that involved using pest control methods only when necessary (the concept of the economic threshold) and relying on methods other than synthetic chemicals whenever possible.

State and federal legislation regarding food safety was first introduced in this country around the turn of the century, but the laws initially were relatively weak and dealt primarily with adulteration and contamination. The foundation of current food safety legislation was passed in 1938 in the form of the Federal Food, Drug, and Cosmetic Act (FFDCA). That Act defines conditions of adulteration and contamination, and provides the Food and Drug Administration (FDA) with considerable powers, including the authority to establish tolerance levels, to mandate safety testing prior to licensing, and to require a product's withdrawal from the market if evidence indicates lack of safety.

Regulations governing the manufacture, sale, and use of chemical pesticides have also been in place for many years. The passage in 1947 of the Federal Insecticide, Fungicide, and Rodenticide Act (FIFRA) was a major step. This Act required that all toxic chemicals for sale in interstate commerce be registered by the USDA against manufacturers' claims of effectiveness (Osteen, Szmedra, 1989). Pesticide registrations are essentially licenses that define permitted crops, livestock, methods of use, or locations of use for pesticide products. The FIFRA further required that the product label specify content and whether the substance was poisonous. The proliferation of responsibilities for regulating chemicals used in food production led to the establishment by executive order of the Environmental Protection Agency (EPA) in 1970, charged with administering the FIFRA.

In 1972 major amendments to the FIFRA were passed that required the government to obtain evidence from the manufacturers that new products could be safely used for the purpose intended and according to the methods recommended before they were introduced. The banning of pesticides was allowed for the first time, if evidence could be established that led directly to environmental damage (Osteen, Szmedra, 1989). Congress mandated that the EPA begin to assess the approximately 35,000 pesticide products previously registered by Federal and State authorities. The Animal and Plant Health Inspection Service (APHIS) was also formed in 1972 with responsibility for food safety through plant protection, animal health, quarantine inspection, and veterinary biology. In 1978 an amendment to the FIFRA imposed new requirements on what must be known about the chronic toxicity of pesticides before EPA can approve their use on specific crops.

Public concern with chemical pesticide use generally waned during the 1970s and early 1980s. The energy crisis of the early 1970s and the inflation of the later years in that decade turned consumers' thoughts in other directions, and the new legislation and regulations may have allayed their fears. Not until the late 1980s did chemical pesticide use and chemical contamination again arouse substantial public concern and severely affect the market for fresh fruits and vegetables.

CHEMICAL PESTICIDE USE ON FRUITS AND VEGETABLES

Large quantities of pesticides are used each year in the United States (Ferguson, 1985). The EPA estimated that 815 million pounds were used in agriculture in 1987 (quoted in Archibald, Winter, 1989). Herbicides constitute an important share and are widely used on field crops. Data are not readily available on the amount of pesticides used on fresh fruits and vegetables.

Growers have reported, as is generally recognized, that successful fruit production requires extensive pesticide use (Suguiyama, Carlson, 1985). Chemical programs to control pests on fruit have become exceedingly complex. Fruit crop pests have increased in severity and intensity due to more dense crop plantings, fertilization, and pest resistance to particular chemicals. The number of applications of pesticides on citrus and deciduous fruits per season differs according to the region, the pest species, and type of fruit crop. Deciduous fruit crops generally require more pesticides than do citrus fruits. The share of planted acres treated and the number of applications of pesticides to control diseases on oranges, grapefruit, and lemons are less in Arizona–California than in Florida. Pesticides to control insects and diseases on deciduous crops are usually applied fewer times in the West than in the Northeast, South, and North Central regions. For example, the control of diseases on apples in the West was reported to require fewer than 3 applications in 1978, compared to an average of almost 17 in the Northeast. One exception was that in the West insect control on pears involved an average of more than 18 applications of pesticides compared to about half that number in the Northeast.

The use of pesticides in vegetable production, as indicated by the percent of acres treated with active ingredients, is also different for different vegetables and different growing areas (Gianessi, Greene, 1988). Insecticides, for example, were reported to be applied to relatively few acres of watermelons and cucumbers, but to a large proportion of the acreage of lettuce, onions, and tomatoes.

Most growers of fall potatoes apply pesticides several times during the growing season, although the number of applications varies from one region to another (Table 12.1). Growers applied herbicides a little more frequently than one treatment per acre in each of the major states in 1988. The number of acre-treatments of insecticides varied from 1.23 in Colorado and 1.24 in Idaho to more than 3 acre-treatments in Washington, Pennsylvania, and Upstate New York. The greatest variation existed in fungicide applications which ranged from 1.14 acre-treatments in Idaho to 9.19 in Maine. The material used also differed from one region to another. Variation in fungicide applications were largely attributed to differences in weather conditions, mainly humidity.

PESTICIDE REGISTRATION AND ESTABLISHMENT OF TOLERANCES

Under the provisions of the Federal Insecticide, Fungicide, and Rodenticide Act (FIFRA) as amended in 1988 a pesticide must be registered with the EPA before it can be legally used on a food crop (Archibald, Winter, 1989). Potential registrants must submit extensive chemical residue data from actual field testing to the EPA to support their request for registration. Results

Table 12.1. Fall Potato Acre-Treatments of Herbicides, Insecticides, and Fungicides, Selected States, 1988.

		AVERAGE NUMBER OF ACRE-TREATMENTS		
	ACRES PLANTED	HERBICIDES	INSECTICIDES	FUNGICIDES
	1,000		number	
Idaho	350	1.28	1.24	1.14
Washington	115	1.17	3.11	1.53
Colorado	60	1.02	1.23	1.63
Minnesota	70	1.22	2.95	2.90
North Dakota	130	1.15	2.73	2.06
Michigan	32	1.31	2.34	2.32
Wisconsin	62	1.23	1.86	5.10
Maine	86	1.07	2.75	9.19
Pennsylvania	22	1.20	3.16	3.77
Upstate New York	24	1.17	3.44	2.34

Source: Adapted from USDA ERS, 1989b.

from toxicological studies must also be submitted as well as the pesticide's environmental fate. Data requirements are much more stringent now than they were 10 or 20 years ago. The EPA evaluates these data, weighing the safety of the residue level of the pesticide; the necessity of an adequate and economical food supply; and the other ways consumers may be affected by the same or related chemicals. This balancing of risks and benefits is the basis on which decisions to register the chemical are made. The registration if granted specifies the commodities to which the pesticide may be applied, application rates, and any other use restrictions. This information must be clearly carried on the label on the pesticide container, which has led to the statement that a product is or is not labeled for use on a particular crop.

Individual states have the authority to regulate pesticides more stringently than federal gencies, but are not allowed to authorize use of a pesticide not registered by the EPA for a given use unless under an emergency exemption (Archibald, Winter, 1989). The California Department of Food and Agriculture registers pesticides and regulates their application on fruit and vegetable crops. Chemical manufacturers must provide extensive information to have products registered in California, including data on application intervals, effectiveness, residue methodology, and pesticide chemistry. Commercial pesticide applicators must be licensed and are required to send usage reports on all applications to the agricultural commissioner of the county in which the chemical is used.

Establishing Tolerances

The EPA is required by the Federal Food, Drug, and Cosmetic Act (FFDCA) to establish a tolerance for each food item for which a pesticide is registered. This is the maximum level of residues allowed for the pesticide and also for any breakdown products that may legally appear in the food. Tolerances are set to exceed the maximum levels likely to result from the legal application of the pesticides, and are based on residue studies. Tolerances are primarily enforcement tools, and say little about the actual risks to human health posed by the pesticide residue (Archibald, Winter, 1989).

The determination of the tolerance level for a pesticide is a complicated process designed to take into account the cumulative residue of that pesticide that people are likely to encounter in all its food uses. As such it requires that the EPA determine the maximum expected human exposure based on residue studies and human food consumption levels. Residue levels and food consumption data become critical to these determinations. Maximum expected human exposure to the pesticide residue is then compared to certain standards. Standards differ for noncarcinogenic and suspected carcinogenic pesticides. For noncarcinogenic pesticides the residues are compared with acceptable daily intake (ADI) values developed by the U.N. Food and Agriculture Organization and the World Health Organization. An ADI is the daily intake of a chemical which, if ingested over a lifetime, appears to be without appreciable risk. For pesticides suspected of being carcinogenic the EPA has adopted a *negligible risk standard* of one additional cancer in 1 million persons, rather than a zero risk standard which would eliminate many useful pesticides (Roberts, van Ravenswaay, 1989). If risk estimates exceed the negligible risk standard a special risk-benefit analysis will be performed to determine if the pesticide should be registered for use on the specific commodity.

For agricultural use EPA tolerances specify residue limits for about 10,000 pesticide-food combinations, involving about 300 pesticide active ingredients. Roughly two-thirds of these active ingredients are in common use in the United States. Of these a large number have very specialized uses, some being limited to only one or a few crops. The EPA sets tolerances no higher than needed for the product's intended use, and these may reflect a very conservative margin of safety. This allows for the uncertainty inherent in calculating human risk on the basis of animal data, and the possibility that some people may be extra sensitive to a pesticide.

One limitation with the procedures used to set tolerances in the past is the fact that per capita consumption of some fruits and vegetables has risen since tolerances were set, causing potential understatement of residue intake (Greene, Zepp, 1989). In addition certain age groups eat larger amounts of some fruits and vegetables than the average on which residue ingestion is estimated, again causing understatement of potential residue intake. Toddlers, for example, are estimated to consume 11 times more fresh bananas and six times more apples per unit of body weight than the typical adult woman.

Reregistration and Cancellation

Since the current pesticide registration process is much more stringent than that previously followed there has been a procedure for reviewing the pesticides registered earlier against current standards and reregistering those that meet these standards. This reregistration involves a process now known as the Special Review, using current health and environmental protection criteria and the determination of the risks relative to the benefits. The review process, right of appeal, and other proceedings make this procedure one that requires many years to complete. From the inception of the review program in 1975 through November 1985 EPA completed 32 special reviews. As a result EPA canceled all uses of five active ingredients, canceled uses or imposed restrictions on certain uses of 26 other ingredients, and took no action on one ingredient (Osteen, Szmedra, 1989).

In early 1989 the EPA was in process of conducting Special Reviews on seven pesticides used in agriculture (Table 12.2). The public is informed of the initiation of a Special Review with the publication of risk analyses, Position Document (PD) 1. EPA presents its proposed regulatory decision in PD 2/3. After a period of public comment and scientific review, EPA's actual decision is published in PD 4.

The EPA can revoke (through either cancellation or suspension) a pesticide's registration if information indicates that the product presents an unreasonable risk of adverse effects on human health or the environment. If the EPA suspends the registration, use of the product is forbidden while the cancelation process is conducted. The EPA has invoked regulatory action against several major agricultural pesticide products in recent years involving restrictions on methods of application, uses, or complete cancelation (Table 12.3). The regulatory action may range all the way from complete cancelation for all uses to simply a label change requiring the use of protective clothing or extending the time field workers must wait before reentering a field treated with the chemical. The use of certain toxic chemicals may be restricted to certified applicators. Pesticide manufacturers sometimes voluntarily cancel the use of certain materials for various reasons.

Progress on the reregistration of chemicals registered before 1972 has

Table 12.2. Special Reviews Being Conducted by the Environmental Protection Agency of Pesticides Used in Agriculture, February 1989.

COMMON NAME	CATEGORY	MAJOR USE	POSSIBLE RISK	STATUS
Aldicarb	Insecticide	Peanuts, potatoes, nematicide	Acute toxicity cotton, citrus	PD 2/3
Amitrole	Herbicide	Noncrop areas	Carcinogen	PD 2/3
Captan	Fungicide	Apples, peaches, seed treatment	Tumors, birth defects	PD 4
Carbofuran	Insecticide	Corn, peanuts sorghum, sunflowers	Wildlife, bald eagles	PD 4
EBDCs	Fungicides	Apples, potatoes, tomatoes, citrus	Carcinogen, birth defects	PD 2/3
Parathion	Insecticide	Wheat, sorghum, fruits	Acute human toxicity	PD 1/2/3
Phosdrin	Insecticide	Vegetables, fruits	Acute human toxicity	PD 1

Source: USDA ERS, 1989a.

been very slow because of the high cost and time required (van Ravenswaay, 1989). The 1988 amendments to the FIFRA should help speed up the process. The new amendments set 1997 as the deadline for testing, reregistration, and possible cancelation of older chemicals. They allow the EPA to bill the manufacturer for part of the cost of studies needed to reregister the pesticide, relieve the EPA of the responsibility of reimbursing manufacturers for unused stocks of pesticides that have had their registrations suspended or canceled, and provide new rules for storage and disposal of pesticides taken off the market.

The rate at which restrictions are placed on the continued use of previously registered pesticides will probably increase. In February 1989 the EPA ruled to bar the use of captan on 42 fruits and vegetables (including broccoli, brussel sprouts, carrots, and cantaloupes), but to continue permitting its use on tomatoes, lettuce, and other commodities (Ferguson, 1989). In September 1989, four manufacturers of EBDCs announced that they were

Table 12.3. EPA Regulatory Action on Selected Pesticides, 1972–88.

ACTIVE INGREDIENT	ACTION OF CHEMICAL	REGULATORY ACTION AND YEAR[1]
Aldrin, Dieldrin	Control of soil insects	Canceled for all uses except termite control, 1974 Voluntarily canceled for termite control, 1987
Benomyl	Systemic fungicide for fruits, nut, vegetables, field crops, turf, and ornamentals.	Canceled for aerial spraying use, 1982
DDT	Broad spectrum insecticide	Canceled for all uses except control of vector diseases, health quarantine, and body lice, 1972
EBDCs (Mancozeb, Maneb, Metiram, Nabam, Zineb)	Fungicide for fruits, vegetables, and field crops	Label change including use of protective clothing and hazards to wildlife, 1982
Lindane	Seed insecticide treatment	Canceled for all registered uses except restricted use for commercial ornamentals avocados, pecans, Christmas trees, structures, and dog shampoos and dusts, 1984

[1] Restricted use pesticides must be applied by or under the direct supervision of a certified applicator.
Source: Adapted from Osteen, Szmedra, 1989.

voluntarily withdrawing a number of their registered uses on some 60 crops, including leafy vegetables, apples, carrots, celery, cucumbers, pears, squash, and citrus. This suspension leaves 13 crops on which EBDCs may be used: tomatoes, potatoes, sugar beets, wheat, sweet corn, grapes, cranberries, onions, bananas, figs, asparagus, peanuts, and almonds; EBDCs are fungicides that are particularly effective under warm humid conditions conducive to the growth of fungus.

The cost to the chemical manufacturer of reregistering pesticides under the 1988 FIFRA amendment is substantial. A fee of $150,000 will be charged for reregistering all except minor use chemicals. The cost to the manufacturer of filling data gaps such as long-term toxicology is believed likely to amount to $500,000 per study. As a result the manufacture of many products has been canceled and label uses reduced.

MONITORING USE AND INSURING COMPLIANCE

The FDA is charged with enforcing tolerances on food shipped in interstate commerce, except for meat and poultry for which the USDA is responsible (USDHHS, FDA, 1988b). The FDA collects samples from individual lots of both domestically grown and imported foods and analyzes them for pesticide residues. Violative residues occur when either small amounts of a pesticide are found on a commodity for which no tolerance has been set, or residues are found that exceed established tolerance levels.

When violative residues are found in domestic samples the FDA can initiate various sanctions such as seizure or injunction. For import samples shipments may be detained at the port of entry when illegal residues are found. When violative import samples are discovered for a particular shipper/country/commodity, the certification requirement may be invoked. When this requirement is in effect, the importer is responsible for having each shipment of the commodity in question analyzed and certified by a private laboratory to be free of the violative residues in question.

Domestic samples are collected as close as possible to the point of production. Sampling at an early stage of distribution provides the best opportunity to identify shipments of fresh produce that may contain illegal residues and effect timely regulatory action. Import samples are collected at the point of entry into U.S. commerce. Shipments from Mexico are given special attention because of the large volume of fresh foods that enter the United States from that country. Selective surveys are sometimes initiated to acquire information on specific pesticides, commodities, or pesticide/commodity/country combinations.

Samples collected, both domestic and import, are classified as either *surveillance* or *compliance*. Most samples are in the first category, that is they are collected without suspicion that the particular shipment contains illegal residues. Compliance samples are collected and analyzed as follow-up to the finding of an illegal residue or when other evidence indicates that a residue problem may exist in that shipment. The FDA cooperates with states such as California that also do extensive testing, and collects information on pesticide use in other countries to better monitor imported foods.

In 1987 the FDA tested for the presence of about 250 different pesticides in foods and found traces of about 100. Selective surveys were made for the presence of 9 different pesticides such as Aldicarb, Captan, and Daminozide on various fruits and vegetables.

Analyses of pesticide residues by the FDA in 1987 found that half or more of the samples of fresh fruits and vegetables both domestic and imported had no detectable residues (Table 12.4). Less than 1 percent of the samples had residues over tolerance. A small proportion of the fruits and vegetables had residues of pesticides for which there was no legal tolerance. Fruits or vegetables that were found to contain residues of pesticides for which no tolerance existed were primarily what are termed minor crops. Tolerances generally existed for the chemical on other crops but had not been approved for the crop in question.

The question whether imported fruits and vegetables are more likely than domestic samples to contain pesticide residues that violate current regulations is frequently raised. Archibald, Winter (1989) quote findings on nearly 20,000 samples of domestic and imported food analyzed by the FDA over a 5-year period which indicate a lower rate of violations under both the surveillance and compliance approach for imported as compared to domestic supplies. Results depend to a great extent on the sampling and analytical

Table 12.4. Results of FDA Analyses of Pesticide Residues in Fruits and Vegetables, 1987.

COMMODITY GROUP	TOTAL NUMBER OF SAMPLES	PERCENT SAMPLES WITH NO RESIDUES FOUND	PERCENT SAMPLES VIOLATIVE	
			OVER TOLERANCE	NO TOLERANCE
Domestic Samples				
Fruit	1,458	50	<1	<1
Vegetables	3,080	63	<1	2
Import Samples				
Fruit	2,720	51	<1	6
Vegetables	4,265	55	<1	4

Source: USDHHS FDA, 1988b.

procedures followed. Sampling procedures are not random, and screening methods differ in the chemicals that can be detected. But monitoring methods have improved. The FDA currently can use multiscreening methods capable of detecting 253 different chemicals at typical limits of 0.01 parts per million (Archibald, Winter, 1989).

The other major approach to monitoring used by the FDA is the Total Diet Study (USDHHS, FDA, 1988a). This effort is designed to estimate the dietary intakes of pesticide residues by 8 age/sex groups, from infants to senior citizens. Personnel from the FDA purchase foods from local supermarkets or grocery stores four times per year throughout the United States. Each of the four market basket samples is a composite of foods collected in a particular region. The cities of collection are changed each year. Each market basket contains 234 individual food items that have been chosen, based on national dietary surveys, to represent the diet of the U.S. population. The foods are prepared table-ready and then are analyzed for pesticide residues. The results of these analyses, coupled with data on the amounts of these foods consumed, allow for calculation of the dietary intakes of these residues. This provides an estimate of the actual amount of pesticide residues consumed in foods as they are actually eaten.

Of the over 200 chemicals that could be determined by the analytical procedures used in the Total Diet Survey, 53 pesticides were found in the 1987 market basket foods. These residues were compared with Acceptable Daily Intakes (ADIs). The comparisons show that the dietary intakes were consistently below established ADIs. For almost all pesticides the dietary intake was less than 1 percent of the ADI.

The ability of the FDA to enforce regulations with respect to pesticide residues is limited. The large number of pesticides and of fresh fruit and vegetable shipments, both domestic and imported, and the complex testing required makes thorough enforcement prohibitively expensive. A study by the U.S. General Accounting Office (GAO) (1986) found that the FDA sampled only a small percentage of imported food shipments and that the analytical methods used detected less than half the pesticides then available on world markets. The GAO also found that the FDA had been unable to prevent the marketing of about half the shipments of imported fresh fruits and vegetables that contained illegal pesticide residues, and had not usually collected liquidated damages for the distribution of such foods as it is authorized to do. Since that time, however, the FDA has tracked down and destroyed imports that were found under general surveillance to have excess pesticide residues, and under compliance has held succeeding shipments from the same growers until test results could be obtained.

Laws and enforcement programs to protect consumers from unsafe chemical residues on produce exist at the state as well as the federal level (Smallwood, 1989). Some states have recently enacted or proposed new pesticide regulations intended to limit exposure from all sources and require growers to keep records of pesticide use. Several states and over a dozen growers' associations across the country have implemented or proposed certification systems for organic produce to protect consumers from fraud.

California is a leader in establishing a strong pesticide monitoring program. California also conducted 13,500 residue tests in 1987 on samples from wholesale and retail markets, chainstore receiving docks, and ports of entry. More than 16,000 residue tests were expected to be made in 1989.

California voters have approved Proposition 65, The Safe Drinking Water and Toxic Enforcement Act of 1986, that is intended to provide warnings to residents about exposures—occupational, environmental, or from consumer products—to chemicals that cause cancer or reproductive toxic effects, and to prevent significant discharges of these chemicals into drinking water. Implementation of Proposition 65 has been slow because of limited scientific data on the carcinogenic effects of various chemicals, and because the law was not precise about acceptable levels of exposure.

PESTICIDE RISK ASSESSMENT AND RISK PERCEPTIONS

In 1987 the National Research Council (NRC, 1987) released a report on regulating pesticides in food. This contained the finding that 15 crops and animal products contributed nearly 80 percent of all estimated dietary oncogenic (cancer causing) risk, and 9 of these were fruits and vegetables. This study also made startlingly high estimates of the potential cancer risks from 28 pesticide residues in food, especially fresh produce. These estimates were based on the theoretical assumption, in the absence of complete residue data, that all foods contained the maximum residues legally allowed. In fact, as was determined some time later (Archibald, Winter, 1989) using actual residue data from the FDA and the California Department of Food and Agriculture, the risks were much less. For example, using actual data, the lifetime cancer risk from tomatoes dropped from the NRC estimate of 875 excess cancers per million to 0.33 excess cancers per million. The cancer risk posed for Captan, a fruit and vegetable fungicide, dropped from 474 excess cancers to 0.056 cancers per million for the highest population subgroup. But in the meantime many consumers and activist groups had become concerned.

In February 1989 the National Resources Defense Council (NRDC) issued a report alleging that children are exposed to dangerous levels of pesticide residues in fruits and vegetables (van Ravenswaay, 1989). In its 2-year study the NRDC looked at 23 pesticides used on 27 different fruits and vegetables. Twenty of the 23 pesticides were suspected of being carcinogens or neurotoxins. In this study the NRDC first determined from USDA consumption data how much of these foods children eat a day, and then combined these data with actual measurements of pesticide residues in these 27 foods. They then forecast the likely incidence of cancer, using a procedure that took into account the longer latency period that comes from early exposure, which EPA does not. Most of the risk could be traced to one pesticide, Alar, and its breakdown product UDMH. Alar is a growth regulator used on apples. According to the NRDC, preschool exposure to UDMH poses a cancer risk of 1:4,200, considerably higher than the 1:1 million lifetime risk that EPA strives for.

Alar had been selected for Special Review by the EPA in 1984. In 1985 the EPA announced its intention to cancel registration, but then reversed the decision. In 1986 some apple processors and supermarket operators informed suppliers they would not accept any Alar treated apples. Publicity surrounding the NRDC report caused near public hysteria. In early March public school officials in New York City, Los Angeles, and several other areas announced that apples and apple products were being removed from school lunch menus (Buxton, 1989). F.o.b. prices of Washington State apples, where most of the storage stocks were held, dropped sharply at a time when they might have been expected to rise. In May, 1989, the EPA increased its estimate of the proportion of apples treated with Alar from 5 percent of the crop to as much as 15 percent. The EPA also announced its intention to cancel registration, and recommended that Alar not be used until the Special Review was completed.

Conflicting reports about the extent of the use of Alar on apples arose in part because different analytical methods were used. Mass spectrometry, developed and commercialized in the 1970s, has enabled chemists to substitute electronic technology for time-consuming and expensive chemical methods. These machines are capable of identifying trace amounts of substances in very complex mixtures in concentrations as low as one part per trillion. Earlier versions of this equipment have been replaced with faster and more sensitive machines. But even though mass spectrometers can detect up to 200 compounds at one time, special analyses are often necessary to pinpoint the presence of certain chemicals.

In June 1989, the Uniroyal Chemical Company announced that it had reached an agreement with the EPA to voluntarily halt the sale of Alar for food uses in the United States. This halt of sales was treated as a "suspension" under the FIFRA. Suspension orders issued by the EPA allow for the use of existing stocks of materials according to label directions. Uniroyal agreed to voluntarily recall existing stocks of Alar. Growers were to be compensated for the value of returned stocks of Alar at the purchase price. Growers and distributors were strongly urged to comply with this recall.

Product tampering has also shaken consumer confidence and devastated the produce industry. In March 1989 the FDA discovered two grapes in a shipment from Chile that had been injected with small quantities of cyanide. The two berries were the only ones found to be contaminated in an examination of over 2,000 bunches of grapes, and the amount of poison was said to be less than would be required to make a small child sick, but FDA Commissioner Frank Young was reported to have advised consumers to discard all fruit in their homes unless they were certain that it was not from Chile (*New York Times,* March 15, 1989). U.S. stores refunded customers for recent grape purchases, and destroyed their inventories as did wholesalers. Boat loads en route from Chile were diverted. Early estimates of losses came to more than $240 million, including $50 million worth of fruit recalled in Chile and never shipped (*The Packer,* March 25, 1989). Losses were expected to mount as more fruit was recalled and destroyed.

For several years the Food Marketing Institute, among other groups, has been monitoring the attitude of consumers toward food safety. Consistently, consumers have reported greater concerns with residues such as chemical pesticides than with other hazards (Table 12.5).

A national survey conducted by the University of Florida in 1988 pro-

Table 12.5. Consumer Concerns About Selected Food Attributes, 1987.

RESPONSE TO QUESTION	SERIOUS HAZARD	SOMETHING OF A HAZARD	NOT A HAZARD AT ALL	NOT SURE
		percent		
Residues, Such as Pesticides and Herbicides	76	20	3	2
Antibiotics and Hormones in Poultry and Livestock Feed	61	32	4	3
Fats	55	40	3	1
Cholesterol	51	42	5	1
Salt in Food	43	49	6	1
Irradiated Foods	43	29	8	20
Nitrates in Foods	38	47	5	10
Additives and Preservatives	36	54	9	2
Sugar in Food	28	57	13	1
Artificial Coloring	24	53	20	3

Source: Adapted from AIC UC, 1988.

vides additional insight into consumer concerns about food safety in the marketplace (Smallwood, 1989). The chief food purchasers for 506 households were asked a series of questions about three general food safety concerns: additives and preservatives, pesticide and chemical residues, and bacterial contamination. They were asked to rank their concerns on a scale of 1 to 10, with 1 signifying no serious problem and 10 a serious problem. They rated residues as the most serious concern, followed by bacteria and additives (Table 12.6). The survey also found that the shoppers altered their food consumption to reduce perceived risks. Some 58 percent said they ate less of some foods because of safety concerns.

Any substance can be "toxic" under certain exposure conditions. Risk is the probability that the toxic properties will cause injury, disease, or death under actual or anticipated conditions of human exposure. Accurate risk assessment is difficult to achieve. Important assumptions must be made such as the amount of food people consume, the amount of pesticide remaining on food, and the extrapolation of data from animal experiments to humans.

When projecting cancer risks the potential increase is based on what may be called "worst-case" scenarios: that (1) all acreage of all crops are treated with all pesticides for which the crop has a tolerance; (2) residues are always present at the highest level; and (3) daily exposure to these residues occurs over a 70 year lifetime. Actual crop residues are usually considerably lower than the legal tolerances.

While attention has been focused on the risks inherent in the use of synthetic pesticides many point out that these risks are relatively minor compared to others that consumers face. Most toxicologists and food scientists believe that microbial pathogens are a much more serious hazard than chemical residues in the food supply (Roberts, van Ravenswaay, 1989).

Table 12.6. Consumers' Concern with Food Safety, 1988.

		LEVEL OF CONCERN[1]		
SAFETY ISSUE	NO OPINION	LOW (1 TO 4)	MEDIUM (5 TO 7)	HIGH (8 TO 10)
		Percent of respondents		
Pesticides and Chemical Residues	3	14	24	59
Bacterial Contamination	8	18	38	36
Additives and Preservatives	2	20	50	28

[1] Based on a scale from 1 to 10.
Source: Adapted from Smallwood, 1989.

ECONOMIC IMPACT OF PESTICIDE WITHDRAWAL

The publicity surrounding incidents of harmful chemicals allegedly found on fruits and vegetables and the banning of suspectedly dangerous pesticides has had an immediate and significant impact on the industry and consumers. Industry response, which will be discussed later, has mitigated the effects somewhat but only at considerable expense.

An economic evaluation of banning parathion, an insecticide widely used on vegetable, fruit, and field crops, illustrated the problems surrounding similar decisions regarding older pesticides currently being reviewed by the EPA (Ferguson, 1989). A preliminary review was concerned with whether generally lower control costs, higher crop yields for some crops, and other benefits outweighed the insecticide's potential hazard to pesticide applicators and its adverse environmental effects concerning some avian species. According to the FIFRA, the economic, social, and environmental costs and benefits must be considered before banning or reregistering uses.

The cost-benefit methodology used in deriving the short-term economic impacts of banning a chemical is based on a number of major assumptions and limitations. In general, however, the analysis concludes that pesticide bans will likely cause a net loss in economic efficiency and a redistribution of income from consumers to growers, with windfall gains to nonuser growers, and gains or losses to user growers depending on the crop's price elasticities of demand, and the supply and cost of alternative control.

Data came from a 1986–89 study by a panel of scientists on how the loss of parathion would affect crop yields and costs along with information from other sources. Sensitivity analysis, using three scenarios, took into account the limitations of data and lags that might be involved in the availability of pesticide inventories.

The economic study concluded that despite higher production costs on all crops and yield losses on some crops, some growers would reap greater returns due to the higher crop prices resulting from reduced production. Vegetable growers (both users and nonusers of parathion) would have an estimated increase in returns of $64–$90 million. Consumers, however, would pay $85–$109 million more annually in the short run. The combined impact of a parathion ban on growers and consumers would mean losses of $19–$22 million annually. These impacts have short-term implications of 1 year or less. In longer terms, competition would likely result in lower prices and loss of market share by those growers currently using parathion.

The reluctance of consumers to purchase apples and the subsequent drop in apple prices following the negative publicity surrounding the use of Alar resulted in a sharp drop in sales revenues for growers. Buxton (1989) estimated that this cost the producers about $140 million for the 1988/89 season. It was too early to tell whether the effects will continue into the following crop years.

ALTERNATIVE AGRICULTURE

One response to the increased concern with pesticide use has been greater efforts to reduce or eliminate pesticide use. Growers are searching for and trying various alternative methods of production. These are characterized by terms such as include organic farming, low input methods, integrated pest management (IPM), sustainable agriculture, and low input sustainable agriculture (LISA). Cook, Norris, and Pickel (1989) have defined these terms.

A report on Alternative Agriculture by a committee of the National Research Council (1989) concluded that federal programs have tolerated and often encouraged the use of unrealistically high-yield goals which have led to inefficient fertilizer and pesticide use, and unsustainable use of land and water. The report also argued that research at private and public institutions should give higher priority to the development and use of biological and genetic resources to reduce the use of chemicals, particularly those that threaten human health and the environment. The report reviewed a number of studies of alternative production practices, and presented case studies of 11 different farms or groups of farms using alternative methods. Economic data in the report on the profitability of these farm operations was scanty, and limited in the most part to general observations.

Organic Production

Growing crops organically is not a new phenomenon, but there has been an increase recently in this method of production. Growers claiming to use organic methods generally restrict themselves to naturally occurring rather than synthesized chemicals, although within this broad definition there is room for differences. In mid-1989 14 states had standards for defining what qualifies as organic produce, and Texas and Washington had implemented organic certification programs (van Ravenswaay, 1989). In addition, over a dozen private organizations across the country offered certification programs for growers. Because organic produce generally sells at a premium over conventionally grown items there is a temptation to substitute one for the other.

The California Certified Organic Farmers organization, one of the largest with over 400 members, uses laboratory analysis to verify that organic commodities meet certain standards. Most farms growing crops organically tend to be relatively small operations, but a few large farms in California were reported to be growing at least part of their production organically. Some growers have been highly innovative in developing new ways of pest control, as witness the vacuum cleaner type of machine used to remove harmful insects from lettuce.

Sales of organically grown fresh fruits and vegetables in the United States represent only a small proportion of the total, but according to most sources are increasing rapidly (van Ravenswaay, 1989). Unofficial surveys have found that prices for organically grown products are generally higher than for similar conventionally produced items.

The importance of the organic market is difficult to determine. In California the annual farm value of organic produce has been estimated at $50 million, a very small fraction of the total farm value of $15.5 billion. Industry analysts have estimated that sales were growing at 40 to 50 percent per year. Individual supermarket chains have reported substantial increases. It is recognized, however, that inconsistent quality and intermittent availability impede growth in this specialty market.

Integrated Pest Management

Integrated Pest Management (IPM) focuses on combining the use of biological and cultural controls, including plant resistance to pests and the augmentation of natural enemies of pest species, with chemical controls to manage pest populations rather than relying on a single method. Integrated Pest Management includes the use of pest monitoring and economic thresholds, instead of simply applying pesticides on a regular schedule as a precautionary measure (Osteen, Szmedra, 1989). Integrated Pest Management has been used mainly for insect control, although methods are also being developed to control of weeds and diseases.

Integrated Pest Management has generally been effective when followed on fruit and vegetable crops where the per acre use and cost of insecticides are high. Examples are available on the reduction in pesticide use from IPM for crops such as fresh market tomatoes, potatoes, and sweet corn (Gianessi, Greene, 1988). Since 1983 there has been a steady increase in the number of acres of vegetables grown under IPM.

Other Farming Alternatives

Low input farming is defined (Cook, Norris, Pickel, 1989) as involving methods that reduce the use of purchased inputs such as synthetic chemicals, and may include but is not limited to organic methods. Low input agriculture is most commonly found in Midwestern grain and livestock production rather than in horticultural crops.

Sustainable agriculture involves going beyond IPM by considering factors such as environmental impact, energy costs, effects on labor, profitability, and long-term effects of resource use. Organic methods may be appropriate for sustainable agriculture, but only if they are likely to prove viable over the long term.

Low Input Sustainable Agriculture (LISA) is simply a combination of the low input and sustainable agriculture definitions discussed already. The U.S. Department of Agriculture has instituted a program of grants for research and educational activities that might encourage the adoption of the LISA approach. These projects are designed to help farmers substitute management, scientific information, and on-farm resources for some of the purchased inputs they currently depend on for their farming enterprises.

Biological control methods are believed by many to have considerable potential (USDA, ARS, 1989). In recent years efforts have been intensified. One of the most effective introductions has been the bacteria *Bacillus thuringiensis* (Bt) used to control pests that attack cabbages and other vegetables. Growers initially prepared their own pesticide material from diseased insects, but now the product is commercially sold under two trade names. The use of genetic engineering techniques is also expected to result in an increased number of biological control methods being developed in the next few years.

Profitability and Potential for Alternative Agriculture

Whether alternative agriculture methods have a potential for providing acceptable fruits and vegetables with the use of less pesticide material will hinge largely on their ability to compete in the marketplace. In order to survive and grow the alternative farming methods must be profitable, which will depend on the relationship of revenues to costs. Organically grown products may be able to command greater returns per unit, but crops grown by other alternative methods may not. Costs will differ, as will product quality. The preponderance of evidence seems to point to a limited role for organic and low-input sustainable agriculture, but the jury is still out. Hard evidence is still insufficient.

One recent study compared costs and returns for Granny Smith apples produced under four different production methods in California (Table 12.7). Costs per acre for pest control materials and their application were lowest for low input organic methods, but yields were substantially lower and so also were net returns per acre. Yields for fruit grown under conventional, IPM, and high-input organic were the same. Compared to conventional methods the costs of pest control were lower for IPM and this re-

Table 12.7. Costs and Returns for Granny Smith Apples Grown Under Four Different Production Methods, California, 1989.

OPERATION	HIGH-INPUT ORGANIC	IPM	CONVENTIONAL	LOW-INPUT ORGANIC
Preharvest Costs:				
Pest Control Materials and Application	$455	$260	$427	$95
Other Cultural Costs	$1,554	$1,152	$1,152	$1,228
Additional Costs	$3,081	$3,081	$3,081	$2,922
Total Costs/acre	$5,090	$4,493	$4,660	$4,245
Yield (tons/acre)	20	20	20	8
Total Costs/Ton	$255	$225	$233	$530
Sales	$16,174	$12,051	$12,051	$6,470
Net Return/Acre	$11,084	$7,558	$7,391	$2,255*

*Could be reduced by 50 percent depending on codling moth damage in any given year.
Source: Cook, Norris, Pickel, 1989.

sulted in a higher net return. Pest control for high-input organic was a little higher than for conventional methods, but prices were about one-third more and so net returns were substantially greater.

Gianessi and Greene (1988) report that a study by the Office of Technology Assessment concluded that if pesticides were no longer available in the United States limited production of vegetables would still be possible. Some vegetables might become available only through imports, and of those that could be grown the yield and quality would be low and costs would be much higher.

INDUSTRY RESPONSE

Pesticide Records

Several states require that growers record their use of pesticides and be prepared to make these available to local authorities. Some buyers now ask their suppliers to certify that they applied only authorized pesticides in accordance with label directions. The legal implications with respect to product liability of such certificates is not clear and is under consideration.

Private Residue Testing

The development of better testing methods and the increased concern with pesticide residues has led to the use by members of the industry of private laboratories to test products for chemical residues. A testing company called NutriClean, based in Oakland, California, initiated the service, and by the end of 1988 was reported to be servicing nine chains with 420 stores scattered across the country (van Ravenswaay, 1989).

Retailers using the NutriClean services advertise two types of programs: dockside testing and certified produce. NutriClean's dockside testing program involves representative sampling of nine widely consumed fresh fruits and vegetables—apples, oranges, lettuce, potatoes, carrots, sweet corn, tomatoes, peaches, and grapes—at the loading docks of a retailer's central distribution centers. Samples are tested to see if they meet federal residue standards for 14 pesticides that the EPA has identified as potential carcinogens. Dockside testing is promoted as a supplement to the monitoring program conducted by the FDA.

NutriClean certifies produce from specific growers only after recording the grower's use of pesticides, collecting samples of produce from the field prior to harvest, sending the samples to a laboratory for testing, and determining that the samples have no detectable levels of pesticide residues. Testing includes the multiresidue tests used by EPA as well as custom testing for compounds used by growers that are not covered by the multiresidue test (van Ravenswaay, 1989). The cost per field varies depending on the type or variety of crop being grown and the type of pesticide applied.

Retailers who subscribe to NutriClean's services are granted the exclusive right to sell tested produce in their market territory. Growers who join the certification program must work with NutriClean associated retailers. In return they are assured of sales and larger market outlets.

Other firms are now offering services similar to NutriClean's. Grower-shippers on the West coast have also announced pesticide residue programs. Some plan to label their products, others say they will advertise, while some maintain they will test but not advertise. Members of the industry are still divided as to what extent they should draw customers' attention to the possibility of pesticide residues.

Information and Education Programs

Private as well as public agencies have stepped up their educational programs regarding the benefits and safety in the use of pesticides in the light of considerable misinformation and misunderstanding. The Center for Produce Quality was established by the United Fresh Fruit and Vegetable Association and the Produce Marketing Association using funds donated by industry members. The Center, as part of its public affairs campaign, is developing information kits on government and industry food safety efforts for use by retailers. Some retailers and other industry groups are also mounting education and information programs.

Along with private efforts, the FDA has taken steps to explain and expand its pesticide monitoring program. The Extension Service, the educational arm of the USDA and the Land Grant Colleges, is prepared to advise growers on the use of IPM practices.

PUBLIC POLICY

The National Research Council study (1987) referred to earlier, arrived at four basic conclusions. These were that pesticide regulation should be based on consistent standards, that a negligible rather than a zero risk standard should be applied, that the EPA should focus its energies on reducing risk from the most worrisome pesticides on the most-consumed crops, and finally that the EPA should develop improved tools and methods to evaluate regulatory actions.

The U.S. General Accounting Office (GAO) study (1986) referred to earlier concluded that the FDA program of surveillance and compliance was woefully inadequate and strongly recommended that the FDA increase its vigilance. No evidence was presented, however, that the FDA could expand its program under current budgetary constraints.

The leaders of five major trade associations representing all sectors of the fruit and vegetable industry wrote to President Bush in October 1989, urging him to take immediate and decisive action along the following lines (*The Packer,* October 14, 1989). The five organizations were the Produce Marketing Association, the United Fresh Fruit and Vegetable Association, the American Farm Bureau Federation, the National Food Processors Association, and the American Frozen Food Institute. Their requests were that the President:

Support the Environmental Protection Agency in seeking an accelerated pesticide cancellation process to assure public safety when sound scientific data demonstrate a significant diet-related health problem.

Simplify and expedite the reregistration of the older pesticides used on fruits and vegetables.

Direct the development and use of new, selective multiresidue methods for those pesticides that cannot be detected by the current multiresidue tests and which represent an enforcement problem.

Support and fund the necessary research to develop new pesticide registrations or to maintain the registration of older chemicals for use on fruits and vegetables (in instances where the commercial value of the compounds is insufficient to justify such expenditures by the chemical manufacturers).

Promote and fund expanded research and use of Integrated Pest Management and other alternative methods to minimize pesticide use.

Urge the Congress and the Office of Management and Budget to provide the federal agencies that safeguard food from pesticide hazards with the necessary funding to inspect, test, regulate, and, when necessary, to revise tolerances for pesticide residues.

Establish an interagency response team so that a mechanism exists for the Environmental Protection Agency, the Food and Drug Administration, and the U.S. Department of Agriculture to respond promptly to future crises as they may arise.

Establish a general crisis management plan in cooperation with the private sector so that in future crises public health is assured while market upheaval is minimized.

Communicate with and educate the public to the relative nature of risk in their daily diet.

REFERENCES

Agricultural Issues Center, University of California, 1988. *Regulating Chemicals: A Public Policy Quandary.* UC Agricultural Issues Center, Davis CA.

Archibald, Sandra O., Carl K. Winter, 1989. *Pesticides in Food: Assessing the Risks.* University of California Agricultural Issues Center. Preprint of chapter 1. In: *Chemicals in the Human Food Chain,* Carl K. Winter, James N. Seiber, and Carole F. Nuckton, editors.

Bosso, Christopher J., 1987. *Pesticides and Politics: The Life Cycle of a Public Issue.* Pittsburgh: University of Pittsburgh Press.

Buxton, Boyd M., 1989. Economic Impact of Consumer Concerns About Alar on Apples. USDA, *Fruit and Tree Nuts Situation and Outlook Yearbook,* TFS-250.

Carson, Rachel, 1962. *Silent Spring.* Boston: Houghton Mifflin.

Carter, Harold O., Carole Frank Nuckton, Editors, 1987. *Marketing California's Specialty Crops: Worldwide Competition and Constraints.* University of California, Agricultural Issues Center.

——, ——, Editors, 1988. *Chemicals in the Human Food Chain: Sources, Options, and Public Policy.* University of California, Agricultural Issues Center, Davis, CA.

Cook, Roberta, Kim Norris, Carolyn Pickel, 1989. *Economic Comparison of Organic and Conventional Production Methods for Fruits and Vegetables.* Unpublished paper, University of California, Davis.

Dunlap, Thomas R., 1981. *DDT, Scientists, Citizens, and Public Policy.* Princeton, Princeton University Press.

Ferguson, Walter L., 1985. *Pesticide Use on Selected Crops: Aggregated Data, 1977–80.* U.S. Department of Agriculture, Economic Research Service, Agriculture Information Bulletin No. 494.

—— 1989. The Economic Effects of Banning a Widely Used and Relatively Cheap

Pesticide: The Case of Parathion and Vegetables. In: *Vegetables and Specialties Situation and Outlook Report.* U.S. Department of Agriculture, Economic Research Service, TVS-248, August.

Gianessi, Leonard P., Catherine R. Greene, 1988. The Use of Pesticides in the Production of Vegetables; Benefits, Risks, Alternatives, and Regulatory Policies. In: *Vegetables and Specialties Situation and Outlook Report.* U.S. Department of Agriculture, Economic Research Service, TVS-245, September.

Greene, Catherine, Glenn Zepp, 1989. Changing Pesticide Regulations: A Promise for Safer Produce. *National Food Review.* U.S. Department of Agriculture, Volume 12, Issue 3, July–September.

National Research Council, 1987. *Regulating Pesticides in Food: The Delaney Paradox.* Washington, DC: National Academy Press.

National Research Council, 1989. *Alternative Agriculture.* Washington: National Academy Press.

Osteen, Craig D., Philip I. Szmedra, 1989. *Agricultural Pesticide Use Trends and Policy Issues.* U.S. Department of Agriculture, Economic Research Service, Agricultural Economic Report No. 622, September.

Roberts, Howard R., Editor, 1981. *Food Safety.* New York: John Wiley.

Roberts, Tanya, Eileen van Ravenswaay, 1989. The Economics of Food Safety. In: *National Food Review.* U.S. Department of Agriculture, Volume 12, Issue 3, July–September.

Smallwood, David, 1989. Consumer Demand for Safer Foods. *National Food Review.* U.S. Department of Agriculture, Volume 12, Issue 3, July–September.

Sporleder, Thomas L., Carol S. Kramer, Donald J. Epp, 1983. Chapter 9, Food Safety. In: Walter J. Armbruster, Dennis R. Henderson, Ronald D. Knutson. *Federal Marketing Programs in Agriculture: Issues and Options.* Dansville, Illinois; The Interstate Printers and Publishers.

Suguiyama, Luis F., Gerald A. Carlson, 1985. *Fruit Crop Pests: Growers Report the Severity and Intensity.* U.S. Department of Agriculture, Economic Research Service, Agriculture Information Bulletin No. 488.

U.S. Department of Health and Human Services, Food and Drug Administration, 1988a. Safety First: Protecting America's Food Supply. In: *FDA CONSUMER.* Volume 22, No. 8.

—— 1988b. *Residues in Foods—1987.*

U.S. General Accounting Office, 1986. *Pesticides: Better Sampling and Enforcement Needed on Imported Food.* Resources, Community, and Economic Development Division.

U.S. Department of Agriculture, Agricultural Research Service, 1989. *Agricultural Research,* March.

—— Economic Research Service, 1989a. *Agricultural Resources Inputs Situation and Outlook,* AR-13, February.

—— 1989b. *Agricultural Resources Inputs Situation and Outlook,* AR-15, August.

van Ravenswaay, Eileen, 1989. The Food Industry Responds to Consumers' Pesticide Fears. *National Food Review.* U.S. Department of Agriculture, Economic Research Service, Volume 12, Issue 3, July–September.

Vance Publishing Company, 1988. *The Packer,* March 25, Volume xv, No. 13.

Chapter 13

Nutritional Quality and Nutrition Marketing

INTRODUCTION

Recent years have brought a much wider recognition of the contribution that fruits and vegetables can make toward better health and physical fitness. The extent of this contribution is still being determined, but there is a general belief that consumer awareness of their nutritional attributes has been responsible in large part for the growth in consumption of this group of commodities, particularly in fresh form.

The many different fruits and vegetables vary greatly in their nutritive properties, and even different lots of the same commodity may contain different amounts of various constituents. The scientific community has made great strides in determining the properties of foods, in deciding the amounts of various nutrients we should eat, in providing guidelines for a healthy diet, and exploring the relationship between diet and health. The emphasis on nutrition has broadened from concern that we obtain sufficient nutrients to avoid dietary deficiencies to include care that we avoid an excess of certain other constituents.

Scientists, as evidenced by recent reports, agree that increased consumption of fruits and vegetables would contribute to a healthier population. Consumers are becoming aware of this, although surveys still reveal a considerable deficiency in nutritional knowledge. Marketers are challenged to take advantage of this situation through advertising and promotion as well as food labeling, but the complexity of scientific findings and the need of regulatory bodies to guard against fraudulent practices hampers their activity. Even so, progress is being made in bringing to consumers' attention the nutritive composition of fresh fruits and vegetables and the benefits that may be obtained from increased consumption.

THE NUTRITIVE COMPOSITION OF FRUITS AND VEGETABLES

Since the original publication of "The Chemical Composition of American Foods" in 1896 a long series of food composition tables have been issued. The most recent versions of fruits and vegetables are *Agriculture Handbook, 8-9* (USDA, FNS, 1982) which contains food composition tables on 263 fruit and fruit juice items, and *Handbook 8-1* (USDA, 1984) which contains information on 470 fresh and processed vegetable items. Each table lists the food composition of a 100-gram edible portion, and the composition of the edible portion of one pound as purchased.

Values are provided for refuse, energy, proximate composition (water, protein, fat, carbohydrate, and ash), nine mineral elements, nine vitamins, individual fatty acids, cholesterol, phytosterols, and 18 amino acids. The number of studies on which the mean values for components of the proximate composition and for the minerals and vitamins are based is listed, and the variability of these values is indicated by their standard deviation.

Fruits and vegetables, fresh and processed, make an important contribution to our diet (Table 13.1). This group constitutes about 30 percent by weight of the food we consume annually, but contributes less than 10 percent of the calories or food energy we get from fat, protein, and carbohydrates. Fruits and vegetables also contribute a major proportion of the essential vitamins and minerals needed for good health, especially ascorbic acid (vitamin C), vitamin A, vitamin B_6, and magnesium. Our food supply contains sufficient quantities of these nutrients so no one need go short, but the distribution is unequal and some segments of the population still need to increase consumption of the more nutritious commodities.

Recommended Dietary Allowances

Recommended Dietary Allowances (RDAs) are specific recommendations for the amounts of energy and certain nutrients that individuals in various demographic groups should consume. They are not considered minimum amounts necessary to maintain good health, but provide a considerable safety margin. The first RDAs developed by the NRC were for the intake of energy and eight nutrients and were adopted in 1941 (USDHHS, 1988). Recommended Dietary Allowances have been published periodically since that time; the tenth edition was issued in 1989 (NRC, 1989b).

Specific RDAs are established for protein, 11 vitamins, and seven minerals, and vary according to body size, gender, and energy consumption. For 10 additional nutrients research has been too limited to establish specific RDAs, and the NRC has proposed ranges of daily intake that are considered safe and adequate. The RDAs are now used for the development of

Table 13.1. Contribution of Fresh and Processed Fruits and Vegetables to Nutrient Levels Available for Consumption, 1984.

	CITRUS FRUIT	NONCITRUS FRUIT	POTATOES AND SWEET POTATOES	DARK-GREEN DEEP YELLOW VEGETABLES	OTHER VEGETABLES INCL. TOMATOES
			percent		
Food Energy	0.9	2.3	3.1	0.2	1.9
Protein	0.5	0.7	2.6	0.4	2.6
Fat	0.1	0.4	0.1	x	0.2
Carbohydrate	1.9	4.9	5.8	0.5	3.7
Calcium	0.9	1.4	1.1	1.4	4.1
Phosphorus	0.7	1.3	4.0	0.7	4.0
Magnesium	2.2	4.6	8.0	2.0	8.5
Iron	0.7	3.7	4.7	1.4	7.2
Zinc	0.5	1.0	3.4	0.5	3.5
Vitamin A Value	1.5	5.5	5.3	24.0	12.0
Thiamine	2.7	1.9	5.2	0.8	5.2
Riboflavin	0.5	1.8	1.6	1.0	3.9
Niacin	0.8	1.8	6.7	0.6	4.3
Vitamin B_6	1.3	7.5	10.9	2.1	9.0
Vitamin B_{12}	0.0	0.0	0.0	0.0	0.0
Ascorbic Acid	26.2	13.2	16.0	10.9	24.1
Cholesterol	0.0	0.0	0.0	0.0	0.0

[x] Less than 0.05 percent.
Source: Adapted from USDA, ERS, 1986.

standards for food assistance programs, for food labels, and for the evaluation of dietary adequacies, in addition to their original purpose of promoting nutritional health. The RDAs are designed to exceed the nutrient requirements of most individuals, but the allowances for energy are designed to reflect the average needs for people of different heights and weights, ages, and activity levels. Meal planners are advised that it is technically difficult and also unnecessary to try to design a single day's diet that contains all the RDAs for all nutrients, but it is suggested that menus meet all the RDAs over a period of 5 to 10 days.

The RDAs are intended to be used by experts in the field such as clinical nutritionists and physicians, and are not likely to be of help to the typical consumer.

DIETARY GUIDELINES AND RECOMMENDATIONS

Nutrition research in human populations on the role of diet in chronic diseases, such as coronary heart disease and cancer, is a relatively recent phenomenon dating back to World War II. Before that time nutrition research was concerned primarily with the role of essential nutrients in human deficiency diseases (NRC, 1989a). Although our knowledge is incomplete we are now deluged with information and given little guidance on how to separate fact from fallacy. To rectify this several U.S. government agencies and other expert groups have proposed dietary guidelines. An adequate diet was originally defined as one that included the basic food groups, and this would contain the essential nutrients sufficient to prevent deficiency diseases. Dietary adequacy now includes consideration of the most reasonable proportions of dietary factors for the prevention of chronic, as well as deficiency diseases (USDHHS, 1988).

The federal government has supported efforts to teach the public about nutrition for over 100 years (USDHHS, 1988). Since World War I, federally supported dietary guidance materials have been issued and revised regularly to meet the needs of specific target audiences and to reflect the emerging knowledge of nutritional science. From the U.S. Department of Agriculture's Five Food Groups of 1917 the recommendations have evolved to the joint statement issued by the Departments of Agriculture and Health and Human Services in 1985 that we should:

Eat a variety of foods
Maintain a desirable body weight
Avoid too much fat, saturated fat, and cholesterol
Eat foods with adequate starch and fiber
Avoid too much sugar
Avoid too much sodium
If you drink alcoholic beverages, do so in moderation

The USDA has also developed meal plans for consumers that not only would be nutritionally adequate but also take into account food costs. The diversity of food preferences, however, and the changing levels of prices make this program difficult to maintain effectively.

Dietary guidelines have also been issued by expert groups for specific audiences or for the prevention of specific diseases (USDHHS, 1988). The Department of Health and Human Services (DHHS) and the National Heart, Lung, and Blood Institute (NHLBI) in 1984 published Recommendations for the Control of High Blood Pressure. The DHHS and NHLBI combined to put out National Cholesterol Education Guidelines in 1987. The DHHS and the National Cancer Institute issued Dietary Guidelines for Cancer Prevention in 1988. Two major reviews of studies on the relation of diet or nutrition and health have been completed in recent years (NRC, 1989a; USDHHS, 1988). The staff who conducted these reviews were each advised by committees of noted nutritionists, and their findings are based on the consideration of a large number of research reports. Each of these reviews contains a wealth of information. The final documents are organized in somewhat different fashion, but the conclusions are in general quite similar. The importance of these publications warrants a summary of their contents.

The National Research Council Report titled *Diet and Health: Implications for Reducing Chronic Disease Risk* consists of 4 parts. The first part contains a summary and overviews of methodological considerations, dietary intake and nutritional status, genetics and nutrition, and the extent and distribution of chronic diseases. The second part looks at the evidence connecting dietary components such as fats, fibers, and vitamins to chronic diseases. The third part examines the impact of dietary patterns on eight individual chronic diseases or groups of diseases such as hypertension, cancer, and diabetes. The final part provides an overall assessment, conclusions, and recommendations.

The U.S. Department of Health and Human Services publication is titled, *The Surgeon General's Report on Nutrition and Health.* This Report summarizes research on the role of diet in health promotion and disease prevention. An introductory chapter provides a background and historical review of nutritional findings and public policy. Succeeding chapters present findings on the relationship of nutrition to each of 16 diseases or physical conditions such as cancer, high blood pressure, obesity, dental diseases, behavior, and aging. Two final chapters cover drug–nutrient interactions and dietary fads and frauds.

The tremendous amount of evidence, some of it conflicting, on the relationship of diet and health makes it difficult to develop concise yet unambiguous recommendations. The Surgeon General's Report makes the following recommendations for dietary changes for the general public:

Reduce consumption of fat (especially saturated fat) and cholesterol.
Achieve and maintain a desirable body weight.
Increase consumption of whole grain food and cereal products, vegetables (including dried beans and peas), and fruits.
Reduce intake of sodium by choosing foods relatively low in sodium and limiting the amount of salt added in food preparation and at the table.
To reduce the risk for chronic disease, take alcohol only in moderation (no more than two drinks a day), if at all.

Additional recommendations for some people are provided with respect to fluoride, sugars, calcium, and iron.

The National Research Council recommendations are somewhat more detailed and, as in the case of the Surgeon General's Report, are backed by considerable supporting evidence. They are as follows:

Reduce total fat intake to 30 percent or less of calories. Reduce saturated fatty acid intake to less than 10 percent of calories and the intake of cholesterol to less than 300 mg daily.
Every day eat five or more servings of a combination of vegetables and fruits, and especially green and yellow vegetables and citrus fruits. Also increase intake of starches and other complex carbohydrates by eating six or more daily servings of a combination of breads, cereals, and legumes.
Maintain total protein intake at levels lower than twice the Recommended Dietary Allowance (RDA) for all age groups.
Balance food intake and physical activity to maintain appropriate body weight.
The committee does not recommend alcohol consumption. For those who drink alcoholic beverages, the committee recommends limiting consumption to the equivalent of less than 1 ounce of pure alcohol a day.
Limit total daily intake of salt (sodium chloride) to 6 g or less.
Limit the use of salt in cooking and avoid adding it to food at the table.
Maintain adequate calcium intake.
Avoid taking dietary supplements in excess of the RDA in any one day.
Maintain an optimal intake of fluoride, particularly during the years of primary and secondary tooth formation and growth.

Both reports make specific recommendations with respect to the consumption of fruits, vegetables, and starches. In addition both reports strongly advise increased consumption of fruits and vegetables as a means of achieving other desired goals.

CONSUMER AWARENESS AND ATTITUDES TOWARD DIET AND HEALTH

Over the years consumers have become increasingly aware of the contribution of good nutrition toward fitness and health. Research studies have established the links between eating habits and disease, and public agencies and health organizations have provided dietary guidelines. Through the media this message has been widely disseminated with increasing frequency.

Food marketing companies have also contributed to greater awareness as part of their merchandising activities. The extent of consumers' nutritional knowledge and the impact this has had on their food purchases has been monitored by private and public agencies. One source of information is the health and diet surveys conducted by the Food and Drug Administration (FDA) every 2 years.

The FDA surveys have revealed not only the growing consumer concern with cholesterol but also the lack of knowledge (Lecos, 1988). In 1982 to study public perceptions of the major dietary causes of heart disease 29 percent of respondents singled out foods with fat, followed by 26 percent who identified cholesterol and eggs. Four years later the number citing fatty foods in the FDA survey jumped to 43 percent, and the number who now felt that cholesterol was a potential dietary hazard climbed to 40 percent. But the extent of nutritional knowledge was still limited. Many showed they were unfamiliar with the different types of fats, or what foods contained cholesterol. In 1986 only 29 percent knew that a product described as cholesterol-free could still be high in saturated fats. Almost half thought cholesterol was present in anything that contained fat or oil, and only one-third correctly indicated it is present only in foods of animal origin.

A major advertising campaign by the Kellogg Co. to promote its high-fiber cereal is credited with changing the public perception of the role diet can play in reducing the risk of cancer. In 1984, just before the company's campaign, only about one-third of those surveyed said they had ever heard of any dietary components that might help prevent cancer. Two years later this rose to 54 percent. And when asked what foods provided good sources of fiber over two-thirds of those responding answered breakfast cereals.

The response consumers give to surveys may not accurately reflect their purchasing practices but rather the answers they feel are expected or desired. But these answers may be useful if interpreted with caution. For example, those saying they believed consumption of fruits and vegetables prevented cancer jumped from 10 percent in 1984 to 22 percent in 1986. Many shoppers say they pay attention to ingredient lists, the percentage increasing from 78 in 1988 to 79 in 1986. There is no doubt that nutritional information on food labels has become more important to consumers than it was in 1976 and 1977 when it ranked far behind other food shopping aids, including USDA grades (Smith, Brown, Weimer, 1979).

Trade associations and other industry groups are also providing nutritional information to consumers and to retailers. The Produce Marketing Association (PMA) in its Nutrition Marketing Resource (1988) includes extensive information on consumer health concerns, and product nutrient sources in addition to information on labeling regulations and commodity nutritional promotional materials.

FOOD LABELING

Regulations to protect consumers against fraudulent misbranding of foods has been in place since shortly after the turn of the century (DHHS, 1988). In 1973 the Food and Drug Administration (FDA) authorized voluntary nutrition labeling and required nutrition labeling for fortified foods and for those for which nutritional claims were being made. In 1987 the FDA proposed a new policy for public health messages on food labels to permit health claims on package labels when the information is true and certain criteria are met.

Food fads and fraudulent claims have a long history and have not disappeared with the advent of scientific knowledge. Nutrition fraud flourishes in the United States today because of the diversity of cultures, the historical concern for health and the use of natural remedies, and the introduction of advanced communication technologies (USDHHS, 1988). Currently numerous government, medical, and consumer-oriented organizations are responsible for preventing and controlling fraud. At the federal level the FDA, the U.S. Postal Service, and the Federal Trade Commission (FTC) have authority to act against various kinds of illicit food and health-related practices. Misleading claims about food are, however, difficult to regulate because of the complexity of the scientific base. Only factual and nonmisleading information is allowed on food labels, so most false promotional claims appear in books, lectures, and the mass media where they are protected by constitutional rights.

In order to facilitate food labeling the FDA since the early 1970s has promulgated a simplified version of the National Research Council's Recommended Dietary Allowances (RDA). This consists of sets of what are called U.S. Recommended Daily Allowances (USRDA). Instead of labeling a food with the amount of the nutrient in a serving, the USRDA is used to indicate the proportion of one day's USRDA the serving will provide of each significant nutrient. The USRDA covers the amount of protein, 12 vitamins, and seven minerals recommended for four population groups.

The purpose of the USRDA is to provide a basis for nutritional labeling. There has been no legal requirement to provide nutrient information on food labels unless nutrients were added or nutritional claims were made. Fresh fruits and vegetables have been largely exempt from labeling regulations covering processed foods. This has been a somewhat mixed blessing as we shall see.

NUTRITION MARKETING

The nutritional benefits of fresh fruits and vegetables have been known for a long time, and are being confirmed by current research. To many consumers this appears to be an important consideration determining their purchases. Fresh fruit and vegetable marketers consequently have been eager to use claims for health and nutrition in their advertising and promotion programs. Their ability to do so has been restricted by confusing and complex regulations.

On the surface there would seem to be very little problem, but this is not necessarily the case (PMA, 1988). Fresh produce has been exempt from the FDA's complex nutritional labeling regulations, and the FTC has no specific regulations dealing with nutritional claims for fresh produce. But on the other hand, FTC legislation prohibits advertising that is misleading, and contains a broad prohibition against unfair or deceptive trade practices. According to the PMA (1988) failure of the FTC to act on some advertising claims has resulted in several states initiating investigations into deceptive practices.

Several efforts by the FDA to extend the nutritional labeling requirements to fresh fruits and vegetables have been rebuffed, but the FDA still has the responsibility to take action against any food if its labeling is false or misleading (PMA, 1988). Labeling is considered to include printed or graphic matter accompanying the product. This could include brochures, posters, and other point-of-sale material. Nutritional information on labels on fresh produce is not required by the FDA to be presented in any particular way, but the general recommendation has been to follow the specific regulations that apply to processed foods. The FDA has stated that a single nutrition labeling guide should be followed to reduce confusion and increase the potential benefits of nutritional labeling (PMA, 1988).

The FDA's nutritional labeling for processed foods requires a complete listing approach. All nutrient quantities are declared in terms of average serving size. The labeling should include the serving size per serving, number of servings per container, the number of calories per serving, the number of grams of protein, carbohydrates and fat per serving, and the percentage of the U.S. Recommended Daily Allowances (USRDA) (even if zero) for seven vitamins and minerals per serving. Several other nutrients must be listed if they are added, and may be listed if they occur naturally in the food. Requirements for cholesterol and sodium labeling are specified in detail.

Nutritional labeling of fresh fruits and vegetables is hampered by the fact that these products vary due to variations in climate, species, growing conditions, packing and handling methods, time to market, and other factors (USDA FNS, 1982; USDA FNS, 1984). Typical nutritive content has changed over the years with changes in varieties and methods of production. Statements by the FDA have led one industry group, the PMA, to conclude that data available from survey averages such as that published in the USDA's Handbook 8 are unsuitable as a basis for labeling claims. To overcome this deficiency the PMA has used data developed through the PMA's Nutrition Marketing Program. Under the PMA program, samples of the particular commodity are selected from supermarkets in two cities in each of six regions of the country. The sampling is repeated a sufficient number of times to reflect different growing and marketing patterns that occur during the year. The samples are flown overnight to a laboratory for immediate analysis. Results are statistically compiled to insure that there is a 95 percent chance that the produce items being displayed are at least equal to those set forth in any nutritional claim. Firms that wish to do so may use these data on labels and point-of-sale material without a disclaimer that these are average values and may not be found in the items displayed. For naturally occurring nutrients the PMA has followed the guidelines developed for food processors that nutritional values found in a product being displayed be no more than 20 percent below the nutritional values claimed on the product label.

The PMA, using this procedure endorsed by the FDA, has generated data on more than 20 commodities, and other commodities are continually being analyzed. There are some significant differences between the *USDA Handbook* data and the privately generated data, as is evident for potatoes (Table 13.2). The PMA data differ from the average analysis found in the USDA *Handbooks* because the PMA data are based on current information from a representative sample of product as purchased by customers in supermarkets across the country and, given the variability, are set at a level to insure that most product purchased will contain at least as much or more of the given nutrient. The California Almond Commission and the California Date Commission have developed data on their commodities using similar approaches. Data developed for a specific commodity are considered to be brand specific and allow consumers to believe that the nutrient content of the item is at least equal to that identified in the claim, rather than a generic claim based on average data for many different studies.

Specific nutritional information is still not available for many items. The PMA has stated that the FDA has indicated that it would accept claims based on generic data as long as this data is accurate and the claim based being made clearly states that it is based on generic data. Some shippers and retailers are still providing nutritional information based on USDA *Handbook 8* data and making claims based on such data. Others are using *Handbook 8* data without stating the source or making a disclaimer.

Under these confusing and complex regulations the Nutrition Marketing Task Force of the PMA has set forward guidelines for the nutrition marketing of fresh produce that it is hoped will not only avoid prosecution but may prevent the imposition of more restrictive legislation or regulations in

Table 13.2. Potato Nutritional Information, USDA Handbook 8-11 and Produce Marketing Association.

		USDA HANDBOOK 8-11	PRODUCE MARKETING ASSOCIATION
Serving Size	One average potato	7 oz	5.3 oz
Calories		218	110
Protein		5 g	3 g
Fat		0 g	0 g
Carbohydrate		50 g	23 g
Cholesterol		—	0**
Sodium		16 mg	10 mg
Potassium		826 mg	750 mg
Dietary Fiber		1 g	2.7 g
Percentage of U.S. Recommended Daily Allowance (USRDA)			
Protein		8	6
Vitamin A		*	*
Vitamin C		43	50
Thiamine		14	8
Riboflavin		4	2
Niacin		16	10
Calcium		*	*
Iron		10	8
Vitamin B_6		—	15
Folacin		—	8
Phosphorous		—	8
Magnesium		—	8
Zinc		—	2
Copper		—	8
Pantothenic Acid		—	4

*Contains less than 2 percent of the USRDA of these nutrients
**Information on cholesterol content is provided for individuals who, on the advice of their physician, are modifying their dietary intake of cholesterol.
Source: PMA, 1988; USDA, 1984.

the future. These guidelines suggest following the existing nutritional labeling format. Pictorial representation accompanying the nutrition information should display the raw produce, rather than some cooked dish, since cooking may change the nutritional quality. To be labeled a significant source a food must contain at least 10 percent or more of the USRDA in a single serving, and in a comparison must provide at least 10 percent more of the USRDA of the nutrient than the other product in order to be considered superior. An excellent source is considered to be one that provides 40 percent or more of the USRDA in a single serving, a good source one that provides 25 percent or more. For fiber an excellent source is one that provides 8 grams or more per serving, a good source one that provides 5 grams, and a source one that provides 2 grams or more per serving. Under FDA regulations low calorie refers to foods that contain not more than 40 calories per serving and contain not more than 0.4 calories per gram as consumed. The PMA (1988) also provides information on declaring and describing sodium content, and labeling cholesterol and fatty acid content. The FDA has not developed specific guidelines for serving size, so discretion is recommended.

Uncertainty still surrounds the use of medical and health claims. Under the Federal Food, Drug, and Cosmetic Act a food may be considered a drug if the claims on its label indicate that the food will be useful or effective in the treatment, prevention, or mitigation of disease, and so be subject to the FDA's extensive drug regulations. However the FDA has proposed amending nutritional labeling regulations to include health messages on food labels, and until the proposals are adopted the industry has recommended that health claims be limited to generic ones such as the claim that a certain food is a good source of fiber and fiber has been found to reduce the risk of some forms of cancer.

As part of its nutrition marketing program the PMA has prepared health messages for retailers and others to use in their advertising, and camera-ready copy on nutritional information, advertising statements, sample point-of-purchase signs, and recipes for a number of fruits and vegetables.

REFERENCES

Lecos, Chris, 1988. Getting the Message About Diet–Disease Links. In: *Safety First: Protecting America's Food Supply*. FDA Consumer Special Report. Food and Drug Administration, Washington, DC.

National Research Council (NRC), 1989a. *Diet and Health: Implications for Reducing Chronic Disease Risk*. Washington, DC, National Academy Press.

—— 1989b. *Recommended Dietary Allowances*, 10th ed. Washington, DC, National Academy Press.

Produce Marketing Association, 1988. *PMA Nutrition Marketing Resource*. Nutrition Marketing Task Force, Newark, Delaware.

U.S. Department of Agriculture, Economic Research Service, 1986. *National Food Review*, NFR-32, Winter.

—— Food and Nutrition Service, 1982. *Composition of Foods: Fruits and Fruit Juices, Raw, Processed, and Prepared*. Agriculture Handbook 8–9, rev. August.

—— 1984. *Composition of Foods: Vegetables and Vegetable Products, Raw, Processed, and Prepared*. Agriculture Handbook 8–11, rev. August.

U.S. Department of Health and Human Services (USDHHS), 1981. *Publication No. (FDA) 81–246*. U.S. Government Printing Office, Washington, DC.

—— 1988. *The Surgeon General's Report on Nutrition and Health,* DHHS Publication No. 88-50210. U.S. Government Printing Office, Washington, DC.

Smith, Richard B., Judy A. Brown, Jon P. Weimer, 1979. *Consumer Attitudes Toward Food Labeling and Other Shopping Aids.* U.S. Department of Agriculture, Agricultural Economic Report No. 439.

Chapter 14

Generic, Brand, and Private Label Advertising and Promotion

Advertising and promotion play an important role in marketing fresh fruits and vegetables. *Advertising* generally refers to the use of various mass media to deliver a message to potential buyers, usually but not always the final consumer. *Promotion* is the broader category of selling efforts that include issuing coupons, sponsoring contests, providing point-of-sale material, offering price discounts to retailers, building displays, and other activities that encourage product sales.

There are three major types of advertising and promotion programs directed specifically at buyers of fresh fruits and vegetables. One is *generic advertising and promotion* that is directed at expanding the market for a commodity or group of commodities grown in a specific area such as Washington state apples, California Iceberg lettuce, Jersey Fresh produce, or country wide, such as the national potato program. A second is *brand advertising and promotion* that seeks to increase sales and prices for a commodity or commodities sold by a specific company such as Sunkist oranges or Del Monte pineapples. The third is *private label advertising and promotion* that is used to expand sales of product under the label or brand of an individual company or group of companies such as a chain retailer or food service wholesaler.

Advertising provides consumers with information and an incentive to purchase the advertised product or service, and the successful advertiser receives increased revenue through greater sales or higher prices or both. There are many ways to reach a target audience, and the proper choice of media and message is essential. Advertising expenditures may or may not be cost effective, and planning and executing a successful advertising campaign is a challenging process. Accurate measurement of the impact is possible only with carefully designed research.

Food Advertising

The food system is the nation's largest advertiser, and advertising food and related products accounts for nearly 30 percent of all advertising expenditures, even though food purchases require only about 12 percent of consumers' total outlay (USDA ERS, 1989). In 1987 total expenditure on food and beverage related advertising reached almost $11 billion, of which $7.1 billion was spent on the seven major media (network television, spot television, cable television, magazines, newspaper supplements, network radio, and billboards), $1.8 billion on newspaper advertising, and $1.9 billion on coupons. Foodstores spent $1.6 billion on advertising of which 84 percent went for newspaper ads, although stores are relying more and more heavily on spot television.

The growth in food advertising seems to have slowed, as many leading advertisers have cut expenditures drastically. But promotion appears to be up sharply, according to industry estimates (USDA ERS, 1989).

GENERIC ADVERTISING

Funding

Generic advertising of farm products has a long history tracing back over 100 years (Nichols, Myers, Frank, 1988). Many early attempts to voluntarily enlist producers to advertise their fruits or vegetables were unsuccessful. There was an incentive not to participate, since those who stayed outside (often known as free-riders) benefited as much as those who contributed. Difficulties in getting all producers to support marketing programs and contribute to promotion led first to state and later to federal involvement. There are now many different types of generic promotional programs for fruits and vegetables (Table 14.1).

Some financial support for generic advertising has been available from government sources, but the bulk of the funding has had to come from the growers themselves. In 1986 $530 million was raised under government sponsored programs for commodity advertising and about $30 million by voluntary associations (Blisard, Blaylock, 1989). Of the total $530 million the sum of $435 million was spent on advertising and promotion and market research. The difference of $95 million went for administration, refunds to growers, and other expenses. Generic advertising and promotion expenditures aimed at the domestic market that year for fruit amounted to $58.2 million and for vegetables to $7.7 million. This total of $66 million is small relative to total food and beverage advertising of almost $11 billion. An additional $9.3 million was spent to advertise and promote fruits and vege-

Table 14.1. Characteristics of Generic Promotions for Fruits and Vegetables, 1986

CHARACTERISTIC	FRUIT	VEGETABLE
	number	
Total Promotions	82	50
Source of Legal Authority:		
Federal	15	5
State	67	45
Program Type:		
Producer-Funded	20	27
Marketing Order	43	23
Promotion Order	19	0
Refund:		
Yes	16	12
No	66	38
Initiating Referendum		
Yes	65	36
No	17	14
Periodic Formal Reapproval		
Yes	50	30
No	32	20

Source: Blisard, Blaylock, 1989.

tables in foreign markets, and $6.4 million for research involving these commodities.

Federal Programs

There are three types of federal involvement in generic advertising programs: legislated research and promotion acts, research and advertising activities under marketing orders, and joint promotion ventures with commodity groups or private firms to develop international markets (Armbruster, Frank, 1988).

Obtaining federal authority to assess producers of a specific commodity for generic advertising and research is a two-stage process (Morrison, 1984). Enabling legislation must first be obtained. This must include a provision, among others, establishing the percentage of producers that must support the program that incorporates the basic features of the act. The Potato Research and Promotion Act of 1971 applies to growers with 5 acres of potatoes or more, and requires that at least two-thirds of those voting who represent at least two-thirds of the production voting, support the program. In a referendum held in 1972 about 70 percent of those voting, who represented 20 percent of the eligible voters and 30 percent of the volume of production, voted in favor of the program (Connor, Ward, 1983). In 1979 an assessment rate of 1 cent per hundredweight raised $2.3 million, of which $1.7 million was spent on advertising and promotion. The Act requires all growers with 5 acres or more to pay the assessment, but does provide that it may be returned to those requesting a refund. About 10 percent of the growers involved here requested refunds. Similar federal programs have been in effect for wool, cotton, eggs, and wheat.

Some federal marketing orders also make provision for raising funds for advertising and promotion, but many are limited in the geographic coverage or the types of activities that can be undertaken. In 1982, $8.4 million was spent on promotion under 15 fruit and vegetable marketing orders covering commodities grown in either California, Texas, Florida, Washington, Hawaii, or Idaho-Oregon (Morrison, 1984). The money is usually collected by the first handler and deducted from payment to the grower. Some orders that do not permit paid advertising do include provisions for market research and development. Four federal marketing orders accept voluntary contributions for research and promotion, and the California almond, olive, and raisin marketing orders allow producers who engage in brand advertising to credit those expenditures toward their contributions to generic advertising campaigns. There are no provisions for refunds under federal fruit and vegetable marketing orders.

The federal government also supports advertising and promotion programs in foreign markets. Government foreign market development for specific commodities focuses on the Cooperator Program, the Export Incentive Program (EIP), and the Targeted Export Assistance (TEA) Program, all operated by the Foreign Agricultural Service (FAS) of the U.S. Department of Agriculture (Kinnucan, Williams, 1988). Both the FAS Cooperator Program and the Export Incentive Program were authorized in 1954 by federal legislation (PL 480). The Cooperator Program, under which most government expenditures for foreign market development are made, funds three types of activities: trade servicing, technical assistance, and consumer promotion. Trade servicing activities include bringing study teams from foreign countries to the United States, hosting trade conferences, and distributing promotional materials to foreign food buyers. Technical assistance programs seek to stimulate growth in the long-term demand for U.S. exports by pinpointing how U.S. commodities can be effectively used by foreign buyers. Consumer promotion under the Cooperator Program does not include branded advertising. The EIP provides market development in cases in which the FAS has determined that a Cooperator Program is infeasible or branded promotion would be more effective.

A relatively new federal program that encourages the development for export markets is Targeted Export Assistance (TEA). Enacted as part of the 1985 farm bill, TEA provides export promotion assistance to U.S. commodities injured by unfair foreign trade practices. Kinnucan, Williams (1988) reported that in 1986 total expenditures on U.S. government subsidized export promotion of agricultural commodities amounted to $102 million, of which the government contributed $42 million. Expenditures on fruits and vegetables amounted to $12.9 million. The USDA allocated $110 million to TEA to fund 36 projects in fiscal 1987 (USDA ERS, 1987). Of this amount about $55 million or half went to commodity organizations to promote the sale of fruits, vegetables, and nuts both fresh and processed in foreign markets.

State Programs

State legislated programs are also an important source of generic advertising. Several state promotion programs stem from marketing acts similar to those based on federal legislation. California and a few other states have also established boards or commissions that focus solely on demand expansion activities and are largely independent of the State Department of Agriculture. The California Iceberg Lettuce Commission, for example, carries out an extensive program of research and promotion to expand the market for California lettuce. Some states provide matching funds for commodity groups engaged in generic promotion in domestic and export markets. In 1979, $42.5 million was raised under state legislated programs in 48 states for fruit commodity promotion, and $4.3 million for vegetable promotion in 23 states (Morrison, 1984). Some of this was spent on foreign as well as domestic market development. About half the state programs allowed producers to request refunds.

Florida citrus growers, Idaho potato farmers, and Washington state apple orchardists have benefited from long-term advertising and promotion programs. In Washington, for example, the level of funding provides for a very comprehensive advertising and promotion effort, and the regulations require very strict grade standards. Not only is the top Washington apple grade, Extra Fancy, more restrictive than the comparable federal grade but the program mandates that apples meet high standards for firmness.

Several states have developed programs to provide generic advertising efforts for groups of commodities (Morr, 1989). The programs are different. Some states advertise all fruits and vegetables grown there. In others the growers of particular commodities must agree to contribute money and pack to a given specification in order to be eligible to participate in the program. New Jersey has a very active program featuring Jersey Fresh.

New York, which has compulsory state marketing order programs to raise funds to advertise apples and tart cherries, also initiated a Grown in New York program several years ago. Since many New York fruits and vegetables are sold in other states whose growers might object to the New York label the program was changed to emphasize a Seal of Quality. The state provides major funding to supplement the money contributed by individual producers who in order to use the Seal must voluntarily agree to pack a superior quality product. New York started the Seal of Quality program with eggs, onions, apples, and potatoes, and it now includes apple juice and cider, carrots, cheeses, maple syrup, sweet corn, and hydroponic lettuce.

According to the press report (Morr, 1989) 14 states had statewide generic promotion programs for fresh fruits and vegetables in 1989 and others were considering introducing such programs.

BRAND ADVERTISING AND PROMOTION

Advertising expenditures by marketing firms on fruit were estimated to amount to $51.9 million and on vegetables to $39.6 million in 1986 (USDA ERS, 1989). Much advertising by packers and shippers in the past has been directed to the wholesale and retail trade through trade papers. A small but increasing number of fruit and vegetable shippers are now able to undertake major media expenditures to promote their brands of fresh produce to consumers. Customer awareness of Sunkist oranges, Ocean Spray cranberries, Dole pineapple, and Chiquita bananas was only achieved after many years of effort. Now greater effort is being made by other shippers to reach consumers. While advertising to wholesale buyers is sometimes referred to as "push" advertising, that to consumers is considered "pull-through," and more effective in the long run.

Wholesale or Packer Brands

Advertising and promotion within the trade has long played an important role in communication and coordination betwen packers and shippers on the one hand, and retail affiliated or nonintegrated wholesale buyers on the other. Packers have sought to establish a reputation with wholesale buyers and retail chains for quality and service through advertising in trade papers, attending trade association meetings, hosting buyers in shipping areas, and visiting them in major markets. But an effective way of keeping their names and products in front of the buyer has been through the use of private brands on shipping containers.

Packer or shipper brands on wholesale crates or cartons carry their stamp of quality, often designating a superior pack than the customary U.S. grade. A packer may use different brands of the same quality when selling to different customers in the same market, or a second label may reflect a slightly inferior pack. Unlike U.S. grades the packers' brands may vary in quality from year to year with changes in the general quality of the crop, while maintaining the same relative position to other brands in terms of quality. Packers' brands are promoted in trade papers and, like federal or state grades, facilitate communication between buyer and seller.

Many packer brands have become very familiar to the trade. Reliance on well-known established packer brands has undoubtedly contributed to the ability of brokers to arrange sales and wholesalers to sell product in destination markets by telephone rather than have buyers physically examine the product.

Brand Consumer Advertising and Promotion to Consumers

Until recently much fresh produce lost the identity of the packer–shipper when it reached the retail store when it was sold in bulk or repackaged. Consumer bags of products like apples, potatoes, and onions packed at shipping point did carry the shippers' brand through to the consumer, but they were seldom supported with any significant advertising and promotion program. With the exception of a few products such as oranges, bananas, and pineapples consumers knew little about where the produce they purchased came from or who packed it.

A few firms, however, have been able to develop brand awareness for their products. Sunkist is one of these. Sunkist has been advertising to the consumer for more than 80 years (Kirkman, 1975), starting in 1908 with a full page ad in the *Des Moines Register and Leader.* In February 1914 the first national magazine ad for Sunkist appeared in the *Saturday Evening Post.* Sunkist and Kellogg joined in cooperative advertising in 1940, and in 1965 Sunkist first advertised in television commercials. Sunkist is able to carry the product identity to the consumer by stamping each orange. It is little wonder that consumers readily identify Sunkist with oranges.

Chiquita is another well-known brand. Chiquita Brands introduced Miss Chiquita to promote bananas in 1944, and the advertising jingle proved to be a great hit. Chiquita uses point of sale material or labels stuck on bananas to identify their products.

The potential profits from a well-established brand are widely known in the food business and represent a considerable incentive to fresh fruit and vegetable marketers. Three types of firms are now involved in developing new branded produce items. One type consists of firms with long-established brands like Sunkist, Chiquita, Dole, and Ocean Spray that originally were identified with oranges, bananas, pineapples, and cranberries and who are now extending their brands to other fresh items. Another group are firms with brands already well established in consumers' minds for other foods, but who have recently entered the fresh fruit and vegetable business; examples of these are the Pillsbury, Kraft, and Campbell companies. The third group are some of the larger fruit and vegetable shippers who are investing heavily to establish their brands in the minds of consumers. These include Sun World with patented varieties of tomatoes, yellow and red peppers, and seedless watermelon; Natural Pak Produce with tomatoes and avocados; the Nunes Co. with lettuce and cauliflower; and Frieda's Finest with some 300 new and different items.

Shippers who wish to develop a consumer franchise for their branded produce items have a challenging task ahead. They must develop quality control procedures to insure their pack will meet the highest standards all the way through to the retail counter. Major funding must be budgeted to advertise and promote the product. Funds must be carefully allocated among various advertising media and many different possible promotional activities to make most effective use of available resources.

PRIVATE LABEL ADVERTISING AND PROMOTION

Many retailers have long used their own brand or label to identify the quality of some produce items they offered for sale. This was relatively easy when many produce items were wrapped or bagged in the backroom of the store or at the distribution center. Now it is a little more difficult to arrange. Handling and transportation methods have improved to the point where less preparation is required at point of sale. More planning is required to obtain private label merchandise from shippers.

There are several advantages for the retailer in developing a private label. Use of a private label does not bind retailers to purchase branded merchandise from any particular seller, and so they can shop around for the best buy. The private label package does not have to support as heavy an advertising budget as branded merchandise, and so can be sold at a lower price. The specifications for the product can easily be set at the desired level. The package can be designed to promote the company logo and the store theme. On the other hand obtaining private label perishable merchandise packed to strict specifications by distant shippers can be a difficult task. Major firms may not be willing to pack for private label. Containers must be supplied in advance.

A recent development has been the entry of some food service wholesalers into private label packing. These wholesalers have a very large potential volume which provides considerable buying power. Consequently it may be possible for them to readily obtain merchandise packed to their specifications under their label. In effect, however, these wholesalers are in an intermediate position. To the packer they are obtaining private label merchandise. To the foodservice operator, on the other hand, they are providing branded merchandise even though the brand may not be heavily advertised and carry a premium price.

EFFECTIVE ADVERTISING

The object of advertising is to expand the market by increasing the volume of sales, obtaining higher prices, or both (Wills, Cox, 1988). Brand advertising can accomplish this by either increasing market share or the total market, while to be successful generic advertising must increase the total market for the commodity. Because of the limitations of the human stomach this generally means capturing sales of another food product.

An effective program may consist of television commercials, billboard advertisements, point-of-purchase posters, coupons, and tie-in sales with other firms. Some shippers are entering into cooperative advertising arrangements with retailers. With so many new products coming on the market the retail store shelf space is at a premium, but without it all advertising and promotion would be lost. In marketing dry groceries food manufacturers are often asked by retailers to pay for shelf space with what are called "slotting allowances." It is uncertain to what extent this will happen in marketing fresh produce.

Measuring the effectiveness of advertising is difficult at best. Many variables are involved, and the impact of advertising seems to be cumulative and distributed over time rather than taking effect immediately (Connor, Ward, 1983). Studies of the advertising of milk and citrus products have found a significant response, but there are few if any comparable studies for fresh fruits and vegetables (Forker, Liu, 1988). With such heavy expenditures on advertising and promotion by other commodity groups the ability to simply maintain sales may be a significant achievement.

Designing an Effective Program

An effective advertising campaign is believed to consist of several essential components (Armbruster, Myers, 1985). The product must lend itself to promotion in terms of consumer demand. Luxury or specialty products are more suited for promotion than staples. Sufficient money must be raised to fund a significant effort. The target audience must be selected with care, as must the most appropriate theme or message. The media mix used to deliver the message must be chosen carefully. Finally the product must be available in the quantity and of the quality necessary to meet consumer expectations.

The audiences targeted for generic advertising are often fairly broad groups such as young adults or mothers with small children, but can be even more narrowly focused with respect to ethnic background or educational level. The campaign may use such themes as taste and flavor, convenience, nutrition, food safety, low calorie, new uses, or a combination. In the case of well-funded programs the delivery often includes national television, while more modest efforts may be limited to promotional methods such as point-of-sale materials of store contests. The national potato program, for example, has stressed low calories, has been aimed at homemakers, and much of the budget has been spent on advertising in women's magazines. More recently the theme has been changed to convenience. Through research the California strawberry growers found that the heaviest users tended to eat strawberries without other fruits or foods, and so attempted to encourage others to adopt this practice.

In addition to the use of mass media many well funded generic programs also employ regional representatives to undertake promotional activities in major markets. These people distribute and set up point-of-sale material, conduct contests on merchandising and displaying their commodity, monitor quality, instruct on handling methods, and perform other activities.

Generic and brand advertising of fruits and vegetables has some inherent problems on the supply side. Production and quality often fluctuate widely from season to season or year to year. Quality control practices differ among shippers. There is greater difficulty in expanding the market for staples like potatoes and onions than for specialty items like strawberries or avocadoes. Any benefit from higher prices is likely to be lost in increased output. This may be the reason there has been greater support for generic advertising among specialty fruit growers located far from market than among those growing staple vegetables close to market.

REFERENCES

Armbruster, Walter J., and Garry Frank, 1988. Program Funding, Structure and Characteristics. In: *Generic Agricultural Commodity Advertising and Promotion,* Walter J. Armbruster, Robert L. Wills, Editors. Ithaca, NY, Cornell University, A.E. Ext 88-3.

—— Lester H. Myers, 1985. *Research on Effectiveness of Agricultural Commodity Promotion.* U.S. Department of Agriculture and Farm Foundation.

Blisard, William N., James R. Blaylock, 1989. *Generic Promotion of Agricultural Products: Balancing Producers' and Consumers' Needs.* U.S. Department of Agriculture, Economic Research Service, Agriculture Information Bulletin No. 565.

Connor, John M., Ronald W. Ward, Editors, 1983. *Advertising and the Food System.* College of Agricultural and Life Sciences, University of Wisconsin, North Central Regional Research Publication, 287.

Forker, Olan D., Donald J. Liu, 1988. Program Evaluation. In: *Generic Agricultural Commodity Advertising and Promotion,* Walter J. Armbruster, Robert L. Wills, Editors. Ithaca, NY, Cornell University, A.E. Ext 88-3.

Kinnucan, Henry W., Gary W. Williams, 1988. International Programs. In: *Generic Agricultural Commodity Advertising and Promotion,* Walter J. Armbruster, Robert L. Wills, Editors. Ithaca, NY, Cornell University, A.E. Ext 88-3.

Kirkman, C. H. Jr., 1975. *The Sunkist Adventure.* U.S. Department of Agriculture, Farmer Cooperative Service, FCS Information Bulletin 94.

Morr, Dianne, 1989. Tapping into State Loyalty. *Produce Business,* September.

Morrison, Rosanna Mentzer, 1984. *Generic Advertising of Farm Products.* U.S. Department of Agriculture, Economic Research Service, Agriculture Information Bulletin No. 481.

Nichols, John P., Lester H. Myers, Garry Frank, 1988. Its Role in Marketing. In: *Generic Agricultural Commodity Advertising and Promotion,* Walter J. Armbruster, Robert L. Wills, Editors. Ithaca, NY, Cornell University, A.E. Ext 88-3.

U.S. Department of Agriculture, Economic Research Service, 1987. *National Food Review,* Volume 10.

—— Economic Research Service, 1989. *Food Marketing Review, 1988.* Agriculture Economic Report No. 614 August.

Wills, Robert L., Thomas L. Cox, 1988. Economics and Impacts. In: *Generic Agricultural Commodity Advertising and Promotion,* Walter J. Armbruster and Robert L. Wills, Editors. Ithaca, NY, Cornell University, A.E. Ext 88-3.

Part III

Marketing Operations and Firms

Chapter 15

International Trade

INTRODUCTION

Buying and selling horticultural products across international boundaries is much more involved than carrying out the same transactions within the country. There can be higher transportation charges, tariff and nontariff barriers, and special documentation requirements. There are greater risks in terms of collection problems and exchange rate fluctuations. Just developing and maintaining the business is challenging, since language and business practices and customs may be very different, and supplies available for foreign trade may fluctuate widely from one season to the next.

In spite of all these problems the international trade in horticultural products has expanded tremendously in the past few years. The relocation of production to areas with favorable growing conditions that has occurred within this country has also taken place worldwide. This has been facilitated by improvements in communication and personal contact between buyers and sellers, in transportation with the development of jumbo jets and container ships, and in the growth of multinational firms that can overcome many institutional barriers. International treaties such as the General Agreement on Tariffs and Trade (GATT) have led to the reduction in tariff barriers. Cooperation between countries in programs such as the International Monetary Fund (IMF) and the World Bank have provided assistance to developing countries, enabling them to expand foreign trade.

Economic theory argues that free trade will result in a gain in aggregate welfare, and this has been the goal of U.S. trade policy in recent years. The prevalence, however, of what appear to be protectionist policies coupled with unfair or discriminatory trading practices on the part of some countries has recently led to considerable support for restrictions on trade that would protect affected industries. The direction that U.S. trade policy may take in the next few years is consequently uncertain at present.

RECENT CHANGES IN FOREIGN TRADE IN FRESH FRUITS AND VEGETABLES

This country has come to depend heavily on international trade in fresh fruits and vegetables. We desire products such as tropical fruits that cannot be grown economically here, and a continual supply of fresh fruits and vegetables during the winter when our production in limited. We also need markets for products that can be efficiently grown here during our main season.

Trade in fruits and vegetables fluctuates from year to year depending on supplies and prices in this country and in countries with which we trade. Often this variability is weather related, unplanned, and unpredictable. Trends do emerge, however, based on changes in consumer demand or on changes in production and marketing costs and methods. Exports of fresh fruits and vegetables from this country grew rapidly during the 1960s and 1970s, declined in the early 1980s, and then resumed their growth (Figure 15.1). Imports of bananas exceeded all other fruits and vegetables combined until just recently, but were relatively stable from 1985 to 1988. Imports of other fruits and vegetables have also increased, more than doubling between 1981 and 1988.

Banana imports come mainly from Central and South American countries. Only in recent years have official trade statistics revealed the quantities of bananas shipped from individual Central American countries. Prior

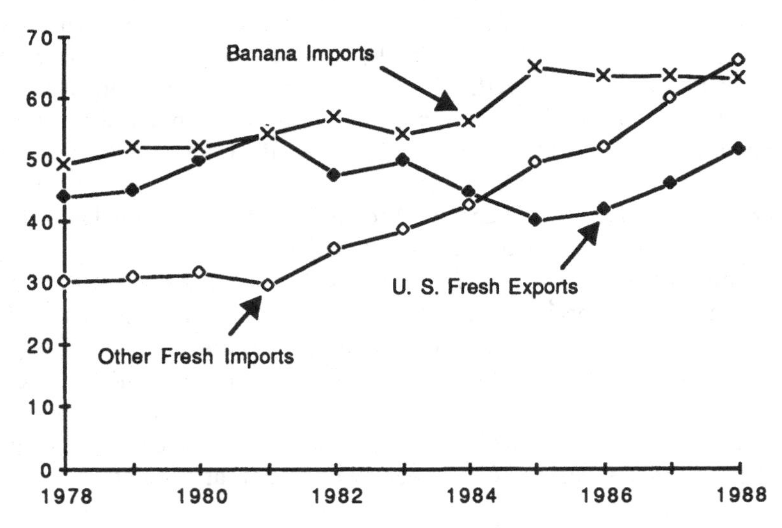

Figure 15.1. U.S. Imports of Bananas and Other Fresh Fruits and Vegetables, and Exports of Fresh Fruits and Vegetables (million cwt), 1978–88. *Source:* Adapted from USDA AMS, 1989; USDA FAS, 1989a.

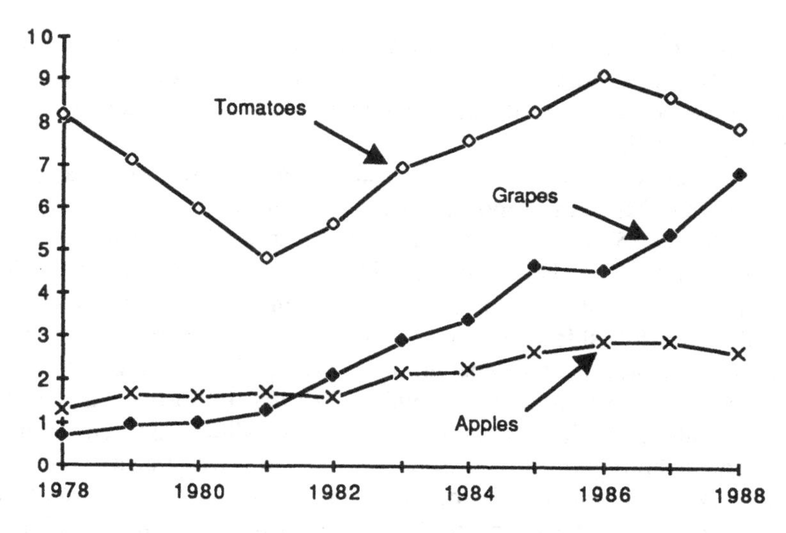

Figure 15.2. Imports of Apples, Tomatoes, and Grapes (million cwt), 1987–88. *Source:* Adapted from USDA AMS, 1989.

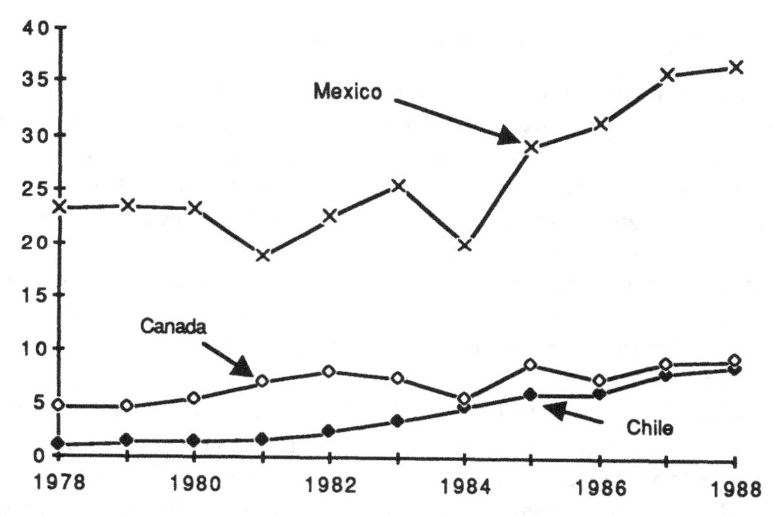

Figure 15.3. Imports of Fresh Fruits and Vegetables Other Than Bananas From Leading Countries of Origin, (million cwt) 1978–88. *Source:* Adapted from USDA AMS, 1989.

to that time the shipments were grouped because of the small number of firms involved in exporting bananas. Now as more firms are entering the banana trade the data for individual countries no longer reveals the operations of specific firms.

Tomatoes and other winter vegetables are the most important fresh products, other than bananas, imported into this country, and most come from Mexico (Figures 15.2, 15.3). This trade has fluctuated in recent years, increasing from 1981 to 1986 then dropping off in 1988. Imports of grapes have risen dramatically since 1981, coming mainly from Chile during our winter season. Chile ships other fruits and vegetables, but grapes have been by far the most important. Grape imports increased from less than 1 million hundredweight in 1980 to almost 7 million in 1988. Imports of apples rose from 1.3 million hundredweight in 1978 to 2.9 million hundredweight in 1987. Canada ships some apples to the U.S., as do several other countries. Canada is also a major supplier of root crops, especially potatoes and carrots. Imports from Canada more than doubled in the 10 years ending in 1988, to the distress of Northeastern U.S. growers with whom they compete directly.

TARIFFS, NONTARIFF BARRIERS, AND SUBSIDIZED EXPORTS

All countries, including our own, maintain some degree of protection for domestic producers against foreign competition. The industries protected and the degree of protection varies considerably from country to country and over time. Governments can impose many types of trade restrictions and these are generally classified into tariff and nontariff barriers. Tariffs are widely used to raise public revenues and protect domestic industry, but now are relatively minor deterrents to the international movement of goods. The nontariff barriers are much more subtle and insidious, and cause more distortion from what would otherwise be efficient patterns of production.

Tariffs are of two general types: the ad valorem tariff and the specific duty. The ad valorem tariff is levied as a specified percentage of the value that the customs places on the good for purposes of tariff assessment. A specific duty is a fixed charge per unit, volume, or weight. For example in 1979 the duty on carrots from Canada was changed from the ad valorem duty of 6 percent of value to the specific duty of one-half cent per pound (Bierlen, Blandford, 1987).

Unlike tariffs on industrial goods the tariffs on seasonally produced products like fresh fruits and vegetables are sometimes different at different times of the year. Tariffs on some products are low during our main growing season, higher when our off-season growers must compete against foreign imports, and lowest when there is no domestic production. There are many different types of nontariff barriers, and most of them are much more restrictive on the flow of goods than tariffs. These can be classed in different ways, but Hillman (1976) distinguished categories that included quantitative restrictions or quotas, licensing requirements, variable levies, state trading operations, health and sanitary regulations, marketing standards and labeling, and bilateral agreements as well as several others. Most of these are self-explanatory. The California study group (Wright, 1988) defined trade barriers as any policy action that results in a discrepancy between the world price for a good or service and the value to a nation's producers or consumers (adjusted for transport and marketing costs).

Variable levies are the means used by members of the European Community (EC) to protect domestic prices by imposing levies on imports that represent the difference between the domestic price of the commodity and the presumably lower price in the exporting country. In the case of fruits and vegetables entering the EC, reference prices are used as a basis for setting additional levies. Health and sanitary regulations would seem to be legitimate efforts to protect the health of the population or guard against the introduction of insects or diseases injurious to people or plants, but they are sometimes used simply as a method of restricting imports.

Other nontariff barriers are often a part of domestic policy aimed at protecting vital industries. Whether a particular barrier to trade is legitimate or not can be difficult to determine.

As well as protecting domestic markets through tariffs and quotas governments or industry organizations are sometimes motivated to dispose of greater quantities of farm products abroad than can be sold under normal trade conditions (Bierlen, Blandford, 1984). Increased exports can be obtained in several different ways. Of particular importance to fresh fruit and vegetable marketers are government subsidies to the export industry or the operation of a two price system under which export prices are maintained at lower levels than domestic prices. Under certain conditions the two price system can result in higher returns to the industry without requiring direct government subsidy, as we have seen earlier. The successful operation of such a scheme does, however, require a virtual monopoly, usually government supported. Sale of product abroad for less than in the domestic market is generally referred to as dumping, and along with export subsidies is considered an unfair trading practice.

International negotiations for the reduction in government intervention in agriculture have led to the development of measures called the producer subsidy equivalent (PSE) and the consumer subsidy equivalent (CSE) (Mabbs-Zeno, Dommen, 1989). These are attempts to quantify the effects of government programs on the agricultural sector and consumers in order to have some basis for equalizing the impact of government programs between trading nations.

EXCHANGE RATES AND MONETARY POLICY

Recent fluctuations in the value of the U.S. dollar relative to other currencies has been widely noted. What determines exchange rates, how present conditions developed, and what the future may bring are matters of considerable importance in marketing.

A rate of exchange of one currency for another is, after all, nothing more than a price. We can express exchange rates in either of two ways, depending on which is more convenient. We can quote the value of U.S. currency in terms of the Japanese, for example 200 yen to the $1.00, or the value of the yen in terms of U.S. currency, in this case $.005. Because we tend to think in terms of how much our dollar will buy or how strong or weak it is, we usually quote the exchange rate in terms of how much of the foreign currency $1 U.S. is worth, in this case 200 yen.

In a free market, prices are determined by supply and demand. The supply of and demand for a currency is generated through either current account or capital account transactions. Current account transactions are primarily merchandise trade and tourism. From our standpoint our purchase of Japanese goods or our trips to Japan constitute a demand for Japanese yen, while the purchases of U.S. goods by Japanese on their vacations in this country represents an increase in the supply of Japanese yen to us. Together these transactions result in a balance on current account which contributes to changing the value of one currency relative to the other.

In recent years, U.S. citizens have been importing more and traveling more abroad compared to purchases and travel in the United States by foreign nationals. As a result we have been experiencing a substantial trade deficit. This normally would have led to a weak dollar, but capital flows and government policies both here and abroad have more than offset this situation. The Japanese and people from other countries have been investing heavily in the United States since business prospects and interest rates in real terms (net of inflation) have appeared more favorable compared to the situation in their own countries. To some observers these capital flows, not always in the same direction, will continue to be more important than merchandise trade and tourism in the future in determining the strength of our dollar.

In a free economy the market usually provides a self-correcting mechanism to imbalances in trade. A strong U.S. dollar would normally result in rising prices and costs in countries exporting to this country, while declining U.S. exports would bring about lower prices and costs in this country and an eventual weakening in the relative value of the U.S. dollar. The increased importance of capital flows and of government intervention makes the adjustment process difficult to predict. Government intervention delays the adjustment process.

Following World War II many countries, with the help of the International Monetary Fund, (IMF) attempted to maintain relatively stable exchange rates with the idea that this would encourage international trade through reducing risks. Unfortunately the established rates in some countries became out of line with actual conditions of monetary supply and demand. In the early 1970s the practice of maintaining stable rates was abandoned by many countries, and rates were allowed to float within limits. Policies of different countries with respect to exchange rates differed, as did the effects of those policies.

The Canadian dollar was traded at about par with the U.S. dollar in the early 1970s but the U.S. dollar strengthened relative to the Canadian dollar beginning in 1978. By 1986 the U.S. dollar was worth about $1.39 in Canadian funds, but then declined to an average of $1.23 in 1988 (Figure 15.4). The Japanese continued to manage the value of the yen using monetary and fiscal policies until early 1986 to maintain a weak currency relative to the United States. The value of the U.S. dollar consequently rose from about 210 yen in 1978 to 238 yen in 1985. This benefited Japanese exporters who received more yen for each dollar their products sold for in the United States. But changing Japanese policies resulted in a decline in the value of the yen in 1988 to an average of 128 yen to the U.S. $1.00. This was ex-

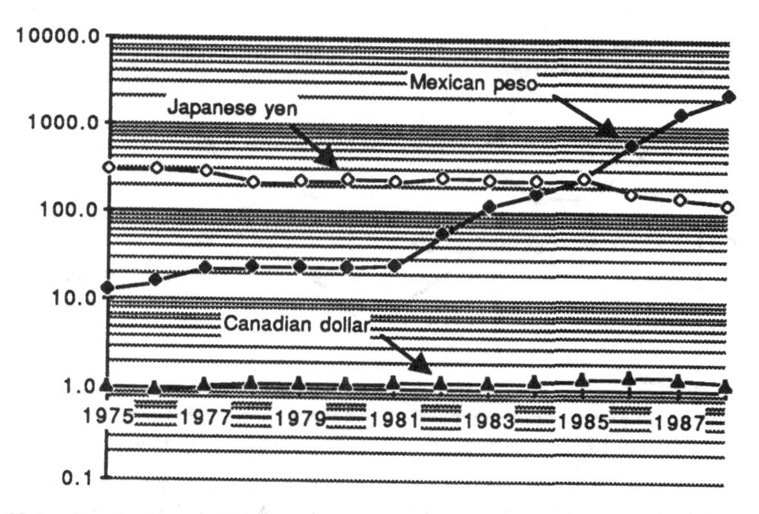

Figure 15.4. Yearly Average Exchange Rates of Selected Currencies in Terms of Units Per U.S. Dollar, 1975–88 (ratio scale). *Source:* Adapted from International Monetary Fund 1989.

pected to stimulate U.S. exports to Japan, and aid the U.S. industries previously suffering from Japanese competition. The effects were delayed, as industries resisted the change. Japanese auto and electronics manufacturers were also apparently willing initially to accept lower returns in order to maintain sales to this country, and U.S. manufacturers were slow to expand exports.

Mexican currency, which had been maintained at 12.5 pesos to the U.S. $1.00 for many years, was not allowed to float until 1976. After a period of adjustment relative stability at about 23 pesos to the dollar was maintained until 1981. Since that time the Mexican peso has sharply declined in value relative to our dollar. The number of pesos one could obtain for a U.S. $1.00 rose from 24.5 in 1981 to 2,273 in 1988. The fact that a U.S. dollar can be exchanged for many more Mexican pesos than formerly would, other things equal, be expected to greatly stimulate Mexican exports to this country. Other things are seldom equal, however, and inflation in Mexico has been much higher than in the United States, so the larger number of pesos obtained from selling vegetables here is not nearly as attractive as it might seem. Even so it appears that the strong U.S. dollar has contributed to increased imports from Canada and Mexico in recent years.

Since changing relative rates of inflation between trading partners must be taken into account as well as exchange rates a value known as the *real exchange rate* is sometimes calculated. To do this the nominal exchange is

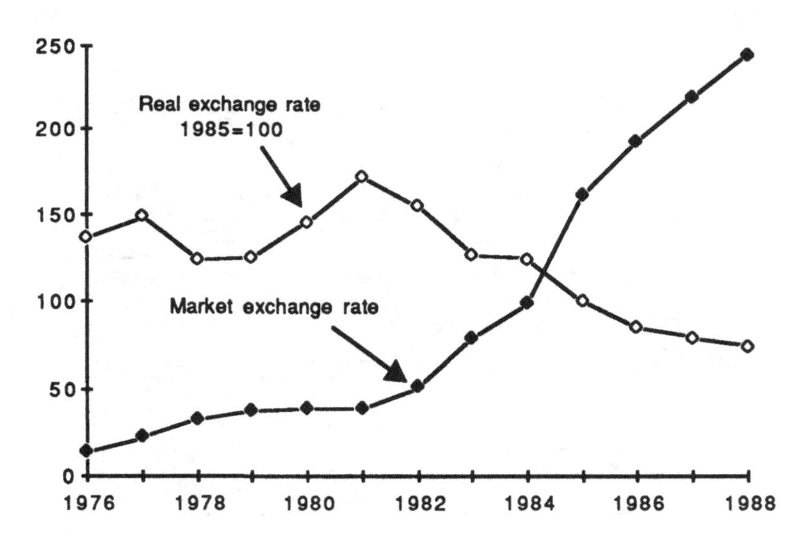

Figure 15.5. Market Rate Exchange in Terms of Pesos per Dollar and Real Rate of Exchange Between Chilean Peso and U.S. Dollar, 1976–88. *Source:* Adapted from International Monetary Fund, 1989.

adjusted by the ratio of the U.S. Consumer Price Index to that of the respective country's Consumer Price Index, measured from a base period such as 1985. In real terms, for example, the purchasing power of the U.S. dollar with respect to the Japanese yen fell from an average of 228 yen per dollar during the period 1981–85 to 139 yen in 1988.

The contrast between changes in the market rate of exchange and the real exchange rate since 1976 is striking in the case of Chile (Figure 15.5). The market value of the U.S. $1.00 in terms of Chilean pesos changed little during the late 1970s, but since then has risen sharply from 39 pesos per dollar in 1980 to 245 pesos in 1988. Inflation has also been rampant, however, so the purchasing power of a U.S. dollar in Chile has actually declined in recent years from an index of 172 (1985 = 100) in 1980 to 73 in 1988. Adjusting to changes like these present a real challenge to firms engaged in international trade.

UNITED STATES TRADE POLICIES

Foreign trade has been used as an instrument of domestic policy in limited and immediate fashion to assist specific industry groups to expand markets or dispose of surpluses. It has also been used with the broader goals of improving the economic well-being of the citizens of this country or helping to reduce world hunger and foster the growth of less developed countries. Trade policy has included unilateral action such as PL 480 under which food donations were made to needy countries, or leadership in promoting multilateral agreements such as the GATT. Various branches of government have sought to encourage trade through U.S. consulates abroad, trade fairs, and other promotional means.

U.S. trade policies and programs are largely based on legislation that provides the President and Congress with authority to undertake certain actions and negotiate agreements with other countries. The Omnibus Trade Bill, 1988, was comprehensive legislation covering a wide range of subjects from general policies to specific actions. The bill detailed U.S. international trade objectives, including an agricultural objective of fostering free and open trade, market pricing, and an end to or lowering of such barriers and subsidies as quotas, tariffs, nontariff practices, including unjustified phytosanitary and sanitary restrictions.

The 1988 bill provided authority for U.S. negotiators to conclude new trade agreements under the Uruguay Round multilateral talks of the General Agreement on Tariffs and Trade. More authority was granted to the U.S. Trade Representative to take action against unfair trade practices and barriers to free trade. The legislation endorsed the Caribbean Basin Initiative and the duty-free status of imports coming from that region. U.S. agricultural export programs and services such as the Foreign Agricultural Service's Cooperator Program, Export Enhancement Program, Targeted Export Assistance and related export programs were continued or increased. The program to monitor pesticides on imported produce was strengthened. Many other programs, studies, and initiatives were also authorized.

Of particular interest to agriculture in recent years has been the Generalized System of Preferences (GSP), the Caribbean Basin Initiative (CBI), Targeted Export Assistance (TEA), and the Trade Agreement with Canada, as well as the continuing negotiations under GATT. Legislation also stipulates the requirements that must be met before industries can obtain relief from what may be considered unfair practices on the part of foreign competitors or their governments.

Generalized System of Preferences

The U.S. Generalized System of Preferences was originally authorized by the Trade Act of 1974, amended by the Trade Act of 1979 and extended in the 1984 and 1988 Trade Acts. Basically it provides preferential tariff treatment to imports of selected products from certain developing countries. The program incorporates safeguards to try to insure that duty free imports under GSP do not dominate our markets or cause hardships to domestic producers. Countries such as Brazil, Mexico, Argentina, Ivory Coast, and Taiwan have benefited under the program shipping products such as sugar, cocoa butter, corned beef, inedible molasses, and unsweetened cocoa. However, there have been objections to the program on the part of some U.S. industries such as California vegetable growers to concessions granted to Mexico.

Caribbean Basin Initiative

The Caribbean Basin Recovery Act of 1983 established the Caribbean Basin Initiative (CBI). This legislation grants tax and trade incentives to countries in the Caribbean and Central America with the aim of promoting regional economic growth. Most horticultural products imported from the Caribbean already entered the U.S. duty free prior to the CBI under the GSP or other provisions of the U.S. tariff code. Dutiable items which enter the U.S. under the CBI are pineapples, orange juice, mangoes (June–August), pineapple juice, roses, macadamia nuts, and limes. Other products which may enter in smaller volume winter and midsummer are cucumbers, onions, spring tomatoes, asparagus, broccoli, mushrooms, avocadoes, and papayas. Some of the Caribbean commodities which showed the greatest increase in imports during the first 4 years of the program were pineapples from Costa Rica and Honduras, other melons from Guatemala and Panama, cantaloupes from Honduras and the Dominican Republic, and celery from Guatemala (USDA ERS, 1988).

Targeted Export Assistance

The 1985 Farm Act contained provisions for providing funds to industry groups or firms that met certain qualifications to enable them to expand sales overseas. This program has already been discussed in chapter 14.

Free Trade Agreement with Canada

The Free Trade Agreement (FTA) with Canada in 1988 is intended to eliminate all tariffs and some nontariff barriers between the two countries by the year 2000. This could have a substantial impact on fruits and vegetables which represent a major part of the agricultural trade between the United States and Canada (USDA ERS, 1988). Canadian horticultural imports from the United States have been five times the value of U.S. horticultural imports from Canada because of climatic differences. In 1986 the U.S. had a trade surplus with Canada of $777 million in horticultural commodities.

The FTA in addition to the gradual elimination of tariffs over a 10-year period would work toward equivalent or harmonization of technical regulations, inspection systems, and personnel training, and also provide for improved communication as changes are made in regulations and standards affecting trade. One of the practices that hinder the flow of fresh fruits and vegetables from the U.S. into Canada are the strict regulations on labeling that Canada imposes. Another is the regulation banning the importation of fresh fruits and vegetables on consignment.

For fresh fruits and vegetables the FTA has special provisions on temporary duties and transshipments (USDA ERS 1988). Both countries will reserve the right for 20 years to apply a temporary duty on designated fresh fruits and vegetables when prices fall below a certain level and domestic plantings have not been increased. The problem of transshipments to take advantage of reduced FTA tariffs is addressed by "rules of origin." Commodities from third countries must be substantially transformed before they can be reexported under the reduced tariffs granted to the FTA partner.

The major Canadian imports of fresh fruits and vegetables from the United States are grapes, oranges, lettuce, tomatoes, and apples (Table 15.1). Most of the produce items entering Canada were subject to import duties although some like oranges and apples entered free. The Canadian duty on some commodities like grapes, lettuce, and tomatoes is only im-

Table 15.1. Major Canadian Fresh Fruit and Vegetable Imports from the United States, Canadian Duty, and Season, 1986.

PRODUCT	VALUE	SHARE OF U.S. PRODUCTION EXPORTED TO CANADA	CANADIAN DUTY	SEASON
	million U.S. $	percent	Can $	weeks
Grapes, fresh	90.0	18	2.21c/kg	15
Oranges, fresh	87.7	11	Free	—
Lettuce, fresh	77.4	9	2.76 c/kg, but not less than 15%	16
Tomatoes, fresh	74.0	9	5.51 c/kg, but not less than 15%	32
Apples	29.5	3	Free	—

Source: Adapted from USDA ERS, 1988.

posed for a certain number of weeks each year. The period is set by the minister or deputy minister of agriculture, not to exceed the specified number of weeks during the 12-month period ending March 31. Some Canadian tariffs have included an ad valorem amount of 10 or 15 percent which could have a significant impact when removed.

Potatoes was the major dutiable produce item imported from Canada, but the value of imported potatoes was considerably less than the value of major produce items exported (Table 15.2). The dates that U. S. seasonal duties change are specified as, for example, for cucumbers and grapes. The U.S. duty on grapes per cubic foot is unusual.

Relief From Unfair Competition

The process by which an industry group may seek relief from unfair competition is also specified in U.S. trade legislation as well as in international agreements. Producers may seek countervailing duties in cases where foreign goods have either been sold in this country at lower net prices than in the country of origin (dumped), or been subject to government export subsidy. If producers can substantiate either of these claims and the government finds that the industry group has suffered economic losses as a result, then additional duties may be imposed equivalent to the amount of the price discrimination or government subsidy. The procedure an industry has to follow to secure relief is expensive and time consuming, and not many groups have succeeded in obtaining relief.

Table 15.2. Major U.S. Fresh Fruit and Vegetable Imports from Canada and U.S. Duty, 1986.

PRODUCT (FRESH OR FROZEN)	VALUE	IMPORTS FROM CANADA AS SHARE OF TOTAL U.S. PRODUCTION	U.S. DUTY
	U.S. $ million	percent	
Potatoes	33.5	1.0	35c/cwt fresh
Apples	18.3	2.2	Free
Carrots	12.4	4.4	lc/lb under 4 in. long 0.5c/lb other sizes
Onions	3.0	1.2	1.75c/lb any time
Cucumbers	2.1	0.9	2.2c/lb 12/1 to 2/28 3c/lb 3/1 to 11/30
Grapes	1.7	1.0	4c/cu.ft. 2/15 to 3/15 free 4/1 to 6/30 6c/cu. ft. any other time

Source: Adapted from USDA ERS, 1988.

INTERNATIONAL ECONOMIC AGREEMENTS

Recent growth in international trade can be traced largely to multinational economic agreements between major world powers that have replaced the unilateral actions and bilateral agreements of the past. These seek to encourage trade rather than to erect barriers to protect local industries. Three major institutions developed shortly after World War II were the International Bank for Reconstruction and Development (World Bank), the International Monetary Fund (the IMF, or Fund), and the General Agreement on Tariffs and Trade (GATT). A regional grouping, the European Community (EC or Common Market), also has had a major impact on international trade.

International Monetary Fund

The IMF was developed to bring some order and stability to the international monetary system and so permit expansion and growth of world trade. The system is designed to deal with the fundamental disequilibrium that can occur as nations develop at differing rates and that may result in surpluses or deficits in balance of payments and major changes in exchange rates. The guiding principles have been consultation, cooperation, adaptability, and flexibility. Under the IMF, exchange rates were maintained at relatively stable levels until 1973. Since that time rates have been allowed to fluctuate or float with the market to a greater degree.

International Bank for Reconstruction and Development

The International Bank for Reconstruction and Development has played an indirect role in the expansion of world trade. The chief function of the World Bank has been to channel capital into development in developing countries which has often led to increased purchases and sales by them on international markets. Dams and irrigation systems, highways and bridges, and other components of the infrastructure financed by the World Bank have contributed to increased agricultural production in many areas.

The General Agreement on Tariffs and Trade

The General Agreement on Tariffs and Trade (GATT) is a multilateral treaty, originally signed in 1948 by 23 nations, that establishes rules for international trade and provides a means for reducing or eliminating both tariff and nontariff trade barriers. Several rounds of negotiating conferences have been conducted by members of GATT. The Tokyo Round, initiated about 10 years ago, dealt with both tariff and nontariff barriers. Special attention was directed to trade problems in developing countries.

The latest round of negotiations, called the Uruguay Round, began in Punto Del Este, Uruguay, and proceeded in Geneva, Switzerland (Mabbs-Zeno, Dommen, 1989). There were 105 nations participating in the talks including most industrialized market countries, most of the less developed countries (LDCs), and several Eastern European countries. Agricultural issues became a focus of GATT, which previously emphasized industrial trade issues. A waiver granted to the United States in 1955 had largely removed agricultural trade issues from GATT discussions, at least until the Uruguay Round.

High farm program costs and expensive surplus commodity stocks in developed countries, particularly the United States and the European Community (EC), gave a strong push to agricultural trade liberalization. The United States, for example, has proposed that in 10 years all agricultural programs that stimulate production or cause overinvestment in agriculture be removed.

The European Economic Community (EC)

The EC is one of the most influential trading blocks in today's world. Formed by West Germany, France, Italy, the Netherlands, Belgium, and

Luxembourg in 1957 by the Treaty of Rome, it enlarged in 1973 with the addition of the United Kingdom, Ireland, and Denmark. Greece, Spain, and Portugal joined later.

The goals of the EC are to promote harmonious development of member nations' economies. To achieve their goals, the EC has fostered free movement of goods, capital, and labor among members, while maintaining a common tariff for the rest of the world, and common economic, social, and to a limited extent monetary policies within the Community.

Special trading relationships have been developed between the EC and several other individual nations and groups of nations. These agreements offer varying degrees of preferential treatment to participating countries. Special agreements have been made with less developed countries and with countries bordering the Mediterranean. The United States, Japan, and Australia are among the few nations that do not have special agreements with the EC. Despite this wide network there has not been a general lowering of protection to Community farmers.

Agriculture has proved to be a very difficult sector to bring into the Community framework. Community farmers have been heavily protected, and social concerns dictated the continuation of protective measures. Community agricultural policy is called the Common Agricultural Policy (CAP). The objective of the CAP has been to maintain internal prices at levels that have been generally above world prices.

Moulton (1983) described the EC policy toward the production and marketing of oranges, which consisted of four programs:

The protection of internal producers (Italy and France) from excessive external competition through a system of seasonal tariffs.

The maintenance of acceptable producer prices by a reference price system and by subsidy payments for oranges diverted to processing.

The encouragement of export marketing by the payment of subsidies for the export of oranges to other EC countries and to third countries.

The improvement of production and marketing efficiency through subsidies for varietal improvements or for modernizing of packaging and storage facilities.

The common external tariff was designed to discourage imports when Italian and French production was at its highest. The tariff rate reached 20 percent during the period October 16 through March 31. The EC granted exceptions to the regular tariff schedule on oranges primarily to benefit Mediterranean producers. The United States and other producers argued that such exceptions violated the provisions of the GATT.

The EC annually determines a reference price for oranges that is based on representative market prices for the past 3 years. When import entry prices fall below the reference price, a countervailing duty is levied against subsequent imports from the offending country until prices are equalized. The reference price program periodically increases import prices above the levy they would otherwise reach for certain suppliers under the tariff structure alone. However, very strict enforcement of phytosanitary regulations effectively insulates Italy from foreign competition at each price level (Moulton, 1983).

TYPES OF FIRMS AND BUSINESS PRACTICES IN INTERNATIONAL TRADE

Export Sales

Buying or selling outside the country requires special knowledge and ability. Information must be gathered on market opportunities and business practices before any deals are negotiated, many government regulations must be met, transportation and handling must be arranged, and special care exercised to be sure that full payment will be received.

Potential market opportunities abroad can be discovered in several different ways. The Foreign Agriculture Service (FAS) of the USDA has developed Product Marketing Profiles that identify leading foreign markets, best selling products, and principal competitors. Country Marketing Profiles outline trade activities for selected foreign countries and their food imports from the United States. The Agricultural Information and Marketing Service (AIMS) operated by the USDA provides trade leads to potential exporters. Agricultural trade offices operated by the FAS around the world promote U.S. farm products abroad and can provide information on business customs, government regulations, and transaction procedures. The FAS sponsors trade fairs in this country and abroad and makes grants to commodity associations to promote exports. In addition to the public agencies many private firms such as brokers or export trading companies are in touch with overseas opportunities. Many exporters make personal visits to potential buyers abroad not only to initiate business but to maintain good relationships. As in domestic sales it is important to develop an export marketing plan.

There are two basic techniques in export sales: direct and indirect selling. Direct selling involves dealing directly with the foreign importer or foreign sales representative and taking responsibility for shipping the product overseas. In addition to foreign sales representatives this includes selling to wholesale distributors or retailers directly. Selling direct may only be feasible for firms with considerable export experience, adequate resources, and well-accepted products.

Using the services of other firms through indirect marketing gives smaller firms with little export experience a way to penetrate foreign markets without getting involved in the complexities of exporting (Nicholas, 1985). An export management company can provide domestic producers with a fast, efficient, and economical way to reach foreign buyers. The export management company undertakes the total marketing function and becomes the export marketing department for a group of noncompetitive companies.

Firms that wish to sell direct may deal with many different types of business (PMA, 1984). There are commission agents that act as finders for foreign firms that want to buy U.S. products, country-controlled buying agents that are government or quasi-government firms empowered to locate and purchase desired goods, brokers that act as the export department for different firms, and export wholesalers or trading companies that purchase U.S. goods for resale in foreign markets.

Foreign freight forwarders facilitate export shipments by acting as an exporter agent in moving cargo to overseas destinations. Freight forwarders can advise an exporter of freight costs, port charges, consular fees, costs of any special documentation, and insurance and handling fees. They can recommend proper packing and arrange for repacking at destination if necessary. When the order is ready to ship, the foreign freight forwarders review the letters of credit if desired, and the packing lists and other documentation. They can reserve necessary space on an ocean vessel, arrange for customs clearance, prepare the ocean bill of lading, and any special documentation required.

Some fruit-exporting cooperatives have well established marketing strategies (Hirsch, 1981). They may use sales representatives to whom they grant various degrees of exclusivity, sales coordinators, or specialized importers.

Practices followed in Britain illustrate how imports are handled in that country (McNitt, 1985). Large grocery chains, called multiples, dominate the British food business. Over half the total food retail sales are made by a few giant enterprises. They do not normally import directly but employ the services of an intermediary, most often one of the large importer/wholesaler/distributors. These integrated firms may either buy from the foreign supplier on their own account, act as commission agents, or both. They take responsibility for customs clearance and usually perform other functions as well such as wholesaling, distributing the imports to the supermarkets' warehouses or retail outlets, and perhaps marketing the product through instore merchandising and sales promotion.

State trading is important in many countries. Exporting to Japan requires an intimate knowledge of trading practices and business customs in that country (Humphrey, 1980).

Merchandise for export generally requires special packaging and marking. Nicholas (1985) provides a useful reference on the packaging and handling of fruits and vegetables for export, and has assembled export specifications for 30 horticultural products. Extra care must be taken to avoid breakage, provide strength without excess weight, protect against moisture, and guard against theft and pilferage. Containers must be adequately marked to provide identification in the language of the destination country and conform to foreign regulations. Detailed information is available from U.S. consulates in major port cities.

Documentation necessary for export shipments is complex. Delivery instructions are given by the shipper or foreign freight forwarder to the carrier. The bill of lading is the most significant document and serves as a receipt for the cargo, an evidence of contract, and a document of ownership. Other documents include the shipper's export declaration, an export license indicating proper authorization to export, a certificate of origin, consular invoices, and inspection certificates. The documents required differ by country of destination and type of transaction.

Apples, pears, and vinifera grapes exported from the United States must meet minimum quality and other requirements established by the Export Apple and Pear Act and the Export Grape and Plum Act and must be officially inspected by the U.S. Department of Agriculture.

Several different methods of payment are used for export sales, depending on the buyer's credit standing, exchange restrictions, and market competition. Cash deposit in advance is the safest method of exporting but is only possible under special conditions. Open account transactions are risky except when made between foreign branches or subsidiaries, and are only used when sellers are dealing with buyers they know well in nearby or established markets.

The export letter of credit offers a high degree of protection. This is essentially a declaration by a bank that it will make certain payments on behalf of a specified party under specified conditions, informing the recipients by letter of these conditions. The sight draft is used when sellers desire to maintain control of the shipment. A bank is designated to act as intermediary and is presented with shipping documents and instructed not to release these to the importer until payment of the draft. The buyer can refuse to honor the draft, but at least the merchandise is not lost. Buyers often ask for credit terms which will allow them to sell the products prior to paying for them. Extending credit clearly increases both the risk of the transaction as well as working capital requirements. In some cases a bank will purchase, or "discount," a draft allowing the exporter to get paid at an earlier date. The exporter may none the less be liable if the importer defaults on payment.

Importing Practices

Importing is generally a lot simpler than exporting, since the foreign exporter takes responsibility to deliver the goods to this country. Products must meet Animal and Plant Health Inspection Service (APHIS) regulations in addition to Customs requirements (Wiser, 1970).

Exporters in developed countries like Canada, New Zealand, the Republic of South Africa, and France that ship fresh fruits and vegetables to this country generally take the initiative in making sales, although importers here may also see opportunities and call for bids abroad. The foreign firm is usually in position to pack and ship a product that meets the necessary specifications for the U.S. market, take care of necessary phytosanitary regulations and documentation, and arrange for the shipment to be delivered to the port of entry. Here the customs broker employed by the receiver will shepherd the shipment through Customs, APHIS, and make sure it meets other regulations.

Most of our fresh fruit and vegetable imports come, however, from less developed countries. Here the U.S. importer often gets involved in foreign production for export to this country in the early stages. Foreign firms may not have the necessary knowledge of U.S. market requirements or the technology of growing, harvesting, grading, and packing. The capital requirements may not be available. Under such circumstances U.S. firms are obliged to become involved in the production process right from the beginning either as sole operator or in some kind of partnership or joint venture arrangement with local producers. As foreign nationals gain experience and accumulate capital they often assume more responsibility in growing and shipping the produce. Members of land-owning families are sent to the United States to study growing and marketing methods here so as to be able to direct the operations at home.

One good example of the evolutionary process the business can go through is the experience with bananas. Originally most of the bananas were grown, harvested, and packed by one U.S. company that owned the land, the packing houses, and the ships that transported the bananas to market. Now there are three major and several minor firms in the business, and a sizable proportion of the bananas are grown on plantations owned by local citizens. There is still, however, close supervision between the local grower and the U.S. importer as there is for Mexican winter vegetables and Chilean grapes.

In the case of the grapes and winter vegetables, and other fresh fruits and vegetables from less developed countries, the importer is in many cases a U.S. shipper obtaining out-of-season supplies to maintain his market.

Sometimes, as in the case of Mexican winter vegetables, the imports compete directly with U.S. production and the interests of the U.S. shipper-importer and the U.S. grower come in conflict.

PROSPECTS FOR FOREIGN TRADE IN HORTICULTURAL PRODUCTS

Future prospects for foreign trade in horticultural products are highly uncertain at this time. The strong U.S. dollar coupled with expansionist farm policies in many other countries led to increased imports, declining exports, and growing concern for greater protection against unfair trade practices. Currently our relatively expansionist monetary policy and low interest rates have led to a decline in the value of the currency which should retard imports and stimulate exports. But a reduction of the federal deficit, if and when it comes, could imply either less public expenditure or more taxes, both of which would reduce the demand for imported goods and contribute to a strengthening of the dollar.

U.S. imports of horticultural products have increased steadily throughout the past decade, which contrasts with exports. Increased imports of fresh fruits are largely due to growth in receipts of grapes and other produce from Chile, and Granny Smith apples from France and several Southern Hemisphere countries. Canada has increased shipments of fresh vegetables and potatoes to this country, and Mexico has maintained large shipments of winter vegetables and increased shipments of late summer and fall produce.

The flow of trade will continue to be greatly influenced by the strength of the dollar, and that depends largely on U.S. monetary and fiscal policies. Reducing the federal deficit would decrease government borrowing which could take upward pressure off interest rates. A decline in the inflow of foreign capital may then help maintain more realistic exchange rates without generating inflationary pressure. Progress in this direction has not been especially rapid, but within the next few years the budget deficit may be brought under control and lead to conditions that could again provide opportunities for increased exports of horticultural products. One problem is that international trade policies often appear to be in conflict with domestic goals and there is difficulty in achieving an acceptable balance.

REFERENCES

Bierlen, Ralph W., David Blandford, 1987. *The Causes of Increased Canadian Exports of Carrots to the United States.* Department of Agricultural Economics, Cornell University, A. E. Res. 87-4.

Goodloe, Carol A., 1988. *Government Intervention in Canadian Agriculture.* U.S.

Department of Agriculture, Economic Research Service Staff Report No. AGES871216.

Hillman, Jimmye S., 1976. *Nontariff Agricultural Trade Barriers.* University of Nebraska Press, Lincoln, Neb.

Hirsch, Donald E., 1981. *Marketing Strategies for Cooperatives Exporting Fruit to Western Europe.* U.S. Department of Agriculture, ACS Research Report No. 4.

Humphrey, Michael L., 1980. *A Guide for U.S. Exporters of Fresh and Processed Fruits and Vegetables to Japan.* U.S. Department of Agriculture FAS M-293.

International Monetary Fund, 1989. *International Financial Statistics Yearbook.* Washington D.C.

Mabbs-Zeno, Carl, Arthur Dommen, 1989. *Subsidy Equivalents: Yardsticks of Government Intervention in Agriculture for the GATT.* U.S. Department of Agriculture, Agriculture Information Bulletin No. 558, January.

McClure, Gail, Sharon Farsht, Editors, 1978. *Speaking of Trade: Its Effect on Agriculture.* University of Minnesota, Agricultural Extension Service, Special Report No. 72.

McNitt, Harold A., 1985. *The British Market for U.S. Food Exports.* U.S. Department of Agriculture, Foreign Agricultural Economic Report No. 210.

Moulton, Kirby S., 1983. *The European Community's Horticultural Trade: Implications of EC Enlargement.* U.S. Department of Agriculture, Economic Research Service, FAER No. 191.

Nicholas, C. J., 1985. *Export Handbook for U.S. Agricultural Products.* U.S. Department of Agriculture, Agriculture Handbook No. 593.

Paarlberg, Philip L., Alan J. Webb, Arthur Morey, Jerry A. Sharples, 1984. *Impacts of Policy on U.S. Agricultural Trade,* U.S. Department of Agriculture, Economic Research Service, International Economics Division, ERS Staff Report AGES940802.

Produce Marketing Association, 1984. *Directory of International Trade.* Newark, Del.

U.S. Department of Agriculture, Agricultural Marketing Service, 1988. *Fresh Fruit and Vegetable Shipments By Commodities, States, and Months,* FVAS-4 April and earlier years.

——Economic Research Service, 1988. *Agricultural Outlook,* March.

——1989. *Fruit and Tree Nuts Situation and Outlook Report,* TFS-252, November.

——Foreign Agriculture Service, undated. *The World is Your Market: An Inside Guide to International Food Marketing.* A reprint of articles from *Foreign Agriculture* magazine.

——1988. *Food and Agricultural Export Directory 1988,* Miscellaneous Publication No. 1459.

——1989. *Horticultural Products.* Foreign Agriculture Circular, monthly.

Venedikian, N. M., G. A. Worfield, 1986. *Export–Import Financing.* New York, John Wiley.

Webb, Alan J., David Blandford, 1984. *The Evolution of Agricultural Trade Policies and Programs.* Cornell University, Agricultural Economics Staff Paper No. 84–30.

Wiser, Vivian, 1974. *Protecting American Agriculture: Inspection and Quarantine of Imported Plants and Animals.* U.S. Department of Agriculture, Agricultural Economic Report No. 266.

Wright, Brian, Study Group Leader, 1988. *Trade Barriers and Other Factors Affecting Exports of California Speciality Crops.* University of California, Agricultural Issues Center.

Chapter 16

Shipping Point Operations and Firms

INTRODUCTION

Operations at shipping point play a very important role in marketing fresh fruits and vegetables. Years ago, as we saw, the principal sources of fresh produce were market garden farms that were clustered around major population centers. Now production has largely shifted to specialized growing areas where products in large quantities are harvested and packed, and shipped often a considerable distance to market. Such shipping points may range in size from a growing area with only a single grower–shipper handling his or her production and that of a few neighbors, to a very large production region with many firms involved in various aspects of shipping operations.

Marketing is, of course, just one part of the total production process. Variety selection, cultural practices, and pest control methods all influence the success of the marketing operation. Marketing per se is often considered to begin at the farm gate or the point of first sale, but growers of many fruits and vegetables destined for fresh market turn over the harvesting to a marketing firm. In such cases marketing can be considered arbitrarily to start with the harvest operation.

Marketing functions performed at shipping point can be divided into two categories. One consists of the physical operations performed on the product such as harvesting, cooling, sizing and grading, trimming, wrapping, packaging, storing, ripening, and shipping. The other includes the associated services such as buying, selling, financing, arranging transportation, market reporting, inspecting, and regulating. The kinds of marketing services and the way they are performed differ for different commodities.

In the total shipping point system the operations that are performed will vary from product to product and place to place. The sequence in which they are done, and the way they are carried out will also differ. Shed packing, where the product is harvested and brought in bulk to a central location for packing, is quite different from field packing where harvesting and packing are carried on at the same location. Yet sometimes both can exist in the same area for the same products. In California, however, there has been a shift toward field packing in recent years first for lettuce then for other vegetables such as cauliflower, celery, and broccoli. Changing technology and costs, especially for labor, and concern for quality as well as for total expense have been the driving forces. There is a continual search to provide as economically as possible a pack of the desired grade and quality with the ability to maintain quality and condition, or shelf life, at retail for at least several days to a week.

Different systems can apparently coexist due to management preferences or different priorities with respect to costs and quality. Some firms hydrocool melons before packing them, while others pack first then cool with forced air. Some apple packers size and grade apples before placing them in storage, while others store orchard run fruit and only size and grade out of storage.

There has been a trend toward doing more marketing operations at shipping point, again a result of weighing the effect on costs and quality. The wrapping of lettuce, the ripening of tomatoes, the packaging of fruit and vegetables in consumer containers has either developed at the shipping level or been taken over there from wholesale firms or from the back room of retail stores. Improvements in methods of storage, handling, and transportation to market have enabled shippers to take advantage of generally lower wage rates and opportunities for greater labor efficiency at shipping point.

PHYSICAL SHIPPING POINT OPERATIONS

Harvesting

Most fruits and vegetables destined for fresh market are harvested by hand, so the availability of harvest labor and the management of labor is very important. Highly mechanized equipment is used to harvest relatively sturdy items like potatoes, onions, and carrots. Various types of harvesting aids are used for some crops to reduce labor costs without sacrificing quality. For some products such as lettuce the harvest equipment technology is available, but use of the equipment is not economically justified at present. The amount of product that is lost or damaged by the machine has to be taken into account. Usually cost and quality considerations will determine conclusively whether a commodity will be hand or machine harvested, but in the case of some commodities such as snap beans the decision to harvest by hand or machine may depend on local conditions and markets.

Workers hand harvesting fresh fruits and vegetables are generally highly skilled, specialize in harvesting certain commodities, work on a piece rate basis, and can earn fairly good wages during the harvest season (Johnson, Zahara, 1977). Crews may be recruited and supervised by a labor contractor, but both government regulation and good personnel practices are requiring growers and shippers to become more involved in the hiring and management of harvest help.

Harvesting most fruits and vegetables for fresh market is generally the responsibility of the grower, although practices vary for different commodities and in different regions. When products such as lettuce and broccoli in California are field packed the shipper performs the harvesting operation. When product is to be delivered to central packing sheds the harvesting is often considered the grower's function, except where specialized harvest crews are required. In Florida the crew used to pick oranges is generally under the supervision and employ of the packer as a service to the grove owner, but also in order to better schedule deliveries and achieve quality standards.

Grading, Packing, and Related Operations

Grading and packing may also include sizing, cleaning, waxing, and trimming (Kader et al., 1985). The grading and sizing may be done in the field at time of harvest as for leafy vegetables or at central packing operations as for soft fruits, tomatoes, apples and many other fruits and vegetables. Equipment used in grading and sizing will vary from simple conveyor belt operations to more complex systems. Gentle handling is achieved through emptying pallet boxes into water.

Computerized equipment that will sort product by weight or volume and separate according to color has been developed. In one color-sorting system commodities such as apples or tomatoes are conveyed into singulator cups and moved through a photo-optical device that measures color by means of multiple viewers that scan the product from opposite sides. The product can then be presorted into four grades according to a program stored in the computer that can be easily adjusted. This system is said to be able to grade and sort up to 1,920 28-pound cartons of tomatoes per hour, and reduces but does not eliminate the need for hand sorting.

Weight sizing is a somewhat similar operation. Each tomato or apple is placed in individual cups for weighing and subsequent discharge. One system can handle up to 8 lanes at a speed of 330 cups per minute per lane, sorting the fruit into as many as 14 weight classes and delivering them to specified packing and filling locations according to predetermined computer instructions. The computer records the number and weight of fruit in each category.

Highly automated grading and packing lines may provide the capability of reducing packing costs but only when used at close to full capacity over a long season (Yolz, Anthony, 1977).

Precooling

Precooling refers to the rapid removal of field heat from freshly harvested commodities before shipping or storage (Hardenburg, Watada, Wang, 1986). Properly carried out, precooling reduces spoilage and helps retard loss of preharvest freshness and quality. Storage rooms and refrigerated trucks or railroad cars have neither the refrigeration capacity nor the air movement needed for rapid cooling. Thus for perishable products precooling is generally a separate operation requiring special equipment or rooms. Precooling is accomplished commercially by several methods: hydrocooling, vacuum cooling, air cooling, and contact icing. Certain cooling methods are appropriate for certain commodities and not for others, and differ in the rate of cooling. The rate of cooling is usually measured by the half cooling time which is the time required to reduce the temperature difference between the commodity temperature and the coolant temperature by one-half.

Hydrocooling is accomplished by immersing the commodity in or spraying it with cold water. Two basic systems are used: a conveyor or flow through system or a batch type system. The water is generally recirculated and for effective cooling needs to be kept cool. Commodities for fresh market that are often hydrocooled are asparagus, celery, cantaloupes, green peas, peaches, radishes, and sweet corn.

Vacuum cooling is achieved by enclosing vegetables in an air-tight chamber and rapidly pumping out air and water vapor. Cooling is accomplished by evaporating water from the product surfaces. Moisture loss in vacuum cooling may range from 1.5 to 5 percent. Vacuum cooling is the standard commercial method for cooling crisphead or iceberg lettuce. Lettuce, because of its vast leaf area is readily adapted to this method of cooling. Other leafy vegetables such as spinach, endive, escarole, and parsley are also well adapted. Other vegetables are sometimes vacuum cooled, especially if prewetted, to reduce the amount of moisture lost from within tissues.

Air cooling is accomplished by either room or forced-air cooling. Room cooling may be used for apples, citrus, and pears in the same refrigerated space used for temporary or long-term storage. Forced-air cooling is more effective, and consists of forcing air through stacks of ventilated containers by producing a difference in air pressure on opposite sides. Forced air cooling usually takes only one-fourth to one-tenth the time needed for room cooling, but still takes twice as long as hydrocooling or vacuum cooling. It is especially effective for strawberries, grapes, fruits, melons, and vine-ripe tomatoes, and is sometimes used for cucumbers, peppers, and cauliflower.

Package icing is an old method that involves placing finely crushed ice within shipping containers. Spinach, collard, kale, broccoli, brussel sprouts, radishes, green onions, carrots, and cantaloupes are commonly marketed with crushed ice in shipping containers. Top icing, placing ice on top of packed containers, is now used primarily as a supplement to one of the principal precooling methods for crated sweet corn, celery, and some other leafy vegetables and for radishes and carrots packed in film bags.

Packaging

The process of packing into wholesale shipping containers and consumer packages varies considerably from one commodity to another and depends on the eventual market (USDA ESCS, 1969). In major growing areas the packing of tomatoes for fresh market is a well organized operation (Figure 16.1) Potatoes used to be sent to market in 50- or 100-pound bags, but now they are generally packed either in consumer packages of 3, 5, or 10 pounds and then shipped in a master container, or are packed to a specified count in a master container. Some fruits, such as apples, may be either packed by hand in layers in cartons, or bagged by machine and packed in shipping containers. As labor costs have increased and equipment has improved, the methods of packing have changed. Tomatoes used to be packed by hand in layers in shipping containers but now the containers are frequently bulk filled by machine.

Equipment is now available that will bulk fill fruits and vegetables in cartons or bags to very narrow weight tolerances. The narrow tolerance enables most containers to meet weight specifications without many substantially exceeding requirements. Such highly sophisticated packing equipment is expensive, can generally handle a large volume of product, and needs to be used over a fairly long season to reduce the overhead costs per unit.

Shipping Containers

The shipping container plays a very important role in marketing since it must hold and protect the product, be easy to handle, stand up under stacking, cooling, shipping, and storage conditions, identify the product and the packer, carry information required by law, and make a favorable impression on the buyer. There have been and still are a tremendous number of different containers in use (Stokes, Woodley, 1974; Bongers, Hillebrand, Risse, 1981). The temptation to save money in local sales on wholesale con-

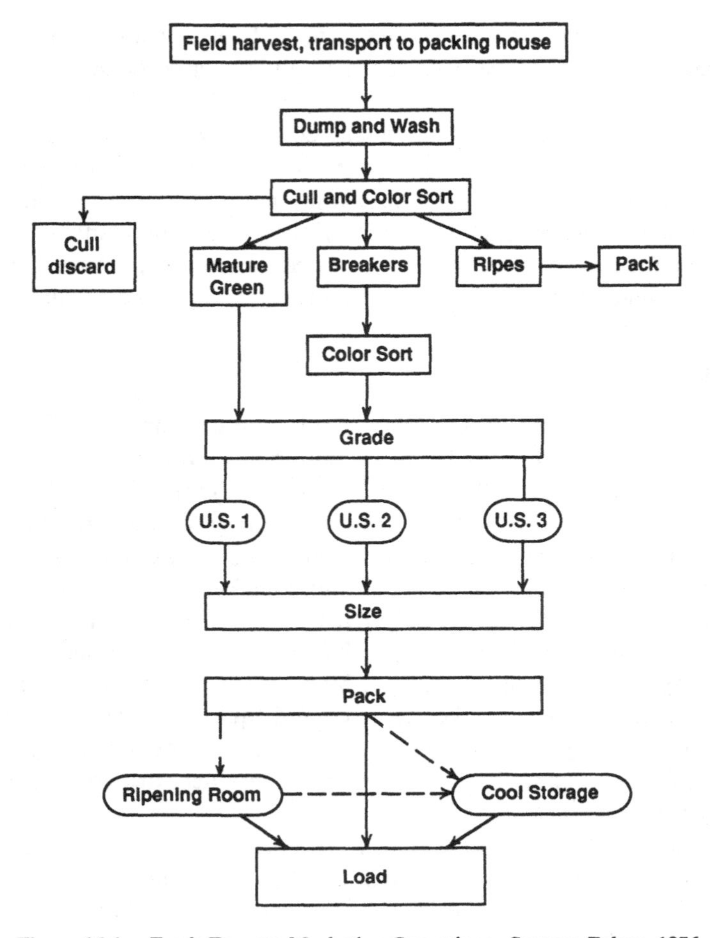

Figure 16.1. Fresh Tomato Marketing Operations. *Source:* Fahey, 1976.

tainers has sometimes resulted in the use of chicken crates, old banana boxes, or used cartons from other shipping areas, with unfavorable effects on future sales. Wooden crates and burlap bags have largely been replaced by fiberboard cartons.

Pallet size has also varied. The use of pallets has increased and pallet size has gravitated toward the 48 × 40-inch dimension (roughly equivalent to 120 × 100 centimeters). As a consequence there has been an effort by the industry to standardize container dimensions to make best use of this pallet size so that different products in different size cartons could be stacked on the same pallet. States such as California regulate the dimensions of shipping containers, but even so there is still a large number of different sized cartons in use.

Storage

Fruits and vegetables differ in terms of optimum storage conditions with respect to temperature, moisture, and atmosphere (Hardenburg, Watada, Wang, 1986). The life of many fruits and vegetables will be prolonged best if held at temperatures close to but just above freezing, while for others a higher temperature will be better. Most products do well under high humidity, but some others prefer lower humidity. Apples primarily, but also a few other fruits and vegetables, will last longer if stored in a controlled atmosphere of lower oxygen and higher carbon dioxide. Simply reducing air pressure in what is called hypobaric storage has been found beneficial for some commodities.

Refrigerated storage is recommended for many perishable commodities. Temperatures must be maintained close to the optimum, and not permitted to vary since fluctuations may cause water condensation. Uniform temperatures should be maintained throughout the storage area.

Relative humidity of the air in storages directly affects the keeping quality. High relative humidity, of 85 to 100 percent, is recommended for most perishable products to retard softening and wilting from moisture loss. This requires providing good insulation, avoiding air leaks, and accurate control of refrigerant temperature.

Air circulation is necessary to maintain an even temperature in a cold-storage room. After field heat has been removed, only enough air movement should be provided to remove respiratory heat and heat entering the room.

Fresh fruits and vegetables are alive and carry on processes characteristic of all living things. The most important of these processes is respiration, which releases energy in the form of heat. Some products have high respiration rates and consequently require considerably more refrigeration than more slowly respiring products. The storage life of products like broccoli, lettuce, peas, spinach, and sweet corn which have relatively high rates of respiration is short; and that of onions, potatoes, and storage cultivars of grapes, which have low respiration rates, is long.

Although refrigeration is the most effective means of retarding spoilage of most horticultural crops, supplemental treatments are often beneficial for certain crops (Hardenburg, Watada, Wang 1986). Supplements include chemical treatments and fumigation, controlled- or modified-atmosphere storage, waxing and surface coating, irradiation, and protective packaging. Products must also be protected from chilling and freezing.

Fruits and vegetables are frequently cleaned, cooled, and conveyed in water as well as treated with aqueous formulations of chemicals after harvest. Examples are washing citrus fruit, apples, and potatoes; hydrocooling peaches; and applying antiscald materials to apples before storage. Low concentrations of chlorine are often used to reduce the buildup of microorganisms in circulating water.

Growth regulators are used to control sprouting of potatoes and onions during storage. Sulphur dioxide has been used to control decay in grapes, and ethylene dibromide (EDB) for disinfection and control of insects. Certain fungicides and bactericides effectively reduce decay of some produce items during storage.

Controlled atmosphere (CA) storage is a technique for maintaining the quality of produce in an atmosphere that differs from air in respect to the proportion of oxygen, carbon dioxide, or nitrogen. The process involves refrigerating an insulated gastight room and controlling the atmosphere within so that it is higher in carbon dioxide and lower in oxygen than normal. The system is now widely used for apples and pears. Optimum proportions of carbon dioxide and oxygen differ for different varieties of apples.

The application of wax or wax emulsion coating to certain perishable products has been practiced for many years. Waxes are used commercially on oranges, grapefruit and other citrus, apples, rutabagas, turnips, mature green tomatoes, peppers, cucumbers, sweet potatoes, cantaloupes, and melons, and to a limited extent on some other products. Waxing improves appearance, and when combined with a fungicide may retard deterioration.

The use of gamma irradiation has been studied for many years for controlling decay, disinfestation, and extending the storage and shelf life of fruits and vegetables. Radiation has been found effective in ridding papayas, mangos, bananas, pineapples, and grapefruit of insect infestations; in inhibiting ripening of papayas, mangos, and bananas, and sprouting of potatoes and onions. Commercial application is limited by the cost and size of the equipment and the uncertainty about acceptability by consumers.

Protective packaging is essential for both storage and shipping. Film wraps that provide a modified atmosphere for individual fruit are under development.

Degreening or Ripening

While storage conditions are primarily designed to retard ripening or maturing there are times when more rapid ripening or degreening is desired (Fahey, 1976; Hardenburg, Watada, Wang 1986). In this situation advantage is taken of the fact that under natural conditions most fruits give off ethylene gas and this hastens the ripening process. Consequently when more rapid and uniform ripening of commodities such as tomatoes, bananas, melons, and some kinds of citrus is desired this can be accomplished through raising the level of ethylene gas in the storage chamber.

Bananas are shipped to market green so they will not soften or sustain serious injury or bruising during handling. Banana ripening is accomplished by introducing ethylene gas into high humidity chambers and adjusting the temperature to extend or shorten the ripening period to meet trade requirements. Under average conditions the ripening period may be as short as 4 days with higher temperatures and extended to 8 or 10 days with lower temperatures.

Shipment may be by truck, rail car, airplane, or ship and the system needs to be designed to protect the product and reduce cost. The use of cartons that stack easily and reduce damage is often essential, and the additional use of pallets and marine or air containers may be worth the cost.

SUPPORTING ACTIVITIES AND SERVICES

Acquiring the Product

Beyond the physical operations performed on the commodity many supporting activities must be undertaken in order to carry out the marketing process at shipping point. The shipper must arrange to acquire the product, finance production if necessary, locate buyers, determine prices, find transportation, absorb risks, and be prepared to guarantee quality.

Shippers acquire product in several ways. Many grow a substantial share of what they pack and sell, often having first been a grower before entering the shipping business. Grower–shippers who have been successful in shipping their own produce frequently find they have the packing equipment capacity and sales capability to handle additional produce from other growers. Shippers who market produce obtained from other growers may buy it outright and handle it on a net return basis. Shippers may pack and sell fresh produce for a fixed fee per carton.

Growers often join together to cooperatively operate packing facilities. The local cooperative packing house may have its own sales staff, employ a broker or sales agent, and be a part of a larger organization through direct membership on the part of the local growers or as a part of a federation of local packing houses.

Selling and Financing

Few shippers are integrated with receivers at destination markets and so must arrange to sell the product. Some retail chain store and institutional firms have buying offices or use brokers located in major shipping regions. Most buyers are located at a considerable distance and must make purchases sight unseen. Chain store people and other major buyers generally have long established working arrangements with a few selected shippers.

Terminal wholesalers are often willing to accept shipments on consignment rather than to purchase the product outright. Consignment sales are risky since the merchant often will not market consigned product as aggressively as purchased product.

Many sales are arranged through brokers or sales agents rather than being made directly. Establishing and maintaining good working relationships with reputable buyers is probably the most difficult part of marketing for many shippers.

Auction markets still provide an outlet for fresh produce in some growing areas, but their use is diminishing. Auctions at shipping point provide the means for buyers to assemble shipments composed of product from several growers. The auction process establishes, discovers, or sets the going market price.

Keeping abreast of market conditions and prices is a major concern. Those responsible for sales at shipping point normally have a network of private and public sources of information. Their day usually begins with calls to business associates locally or in other areas; perusal of telex, fax, or electronic network reports; and a review of USDA Market News or National Agricultural Statistics Service releases.

The burden of financing the product through the shipping operations may fall on different agencies. Growers bear a large share of this function through delayed payment, even when they sell the product outright at harvest. Shippers finance their costs apart from the commodity either internally or through local financial agencies.

Arranging Transportation

Since many sales are made f.o.b. shipping point the buyer is responsible for paying for the transportation to destination, but the shipper often has the duty of arranging for the carrier. This may vary all the way from an annual contract with a truck fleet covering all shipments, to negotiation with independent owner operators on each order. Transportation of individual shipments may be arranged through a truck broker or, in the case of air or water shipments, through a freight forwarder.

Risk Management

There are many risks inherent in marketing fresh produce all through the marketing system. Chief among these at shipping point are the risks of price change, of misrepresentation of the product, and of quality deterioration. These risks can be reduced through appropriate business practices, or their impact shifted from one firm to another, but seldom eliminated completely. U.S. Department of Agriculture market reports and inspection service can help in these areas.

Buyers on major markets often undertake some marketing services at shipping point or have them done for them. Major retail food chains have buying offices at important shipping points. Some food service operations have formed joint buying arrangements to handle procurement at shipping point. Buying brokers at shipping point serve the interests of the distant buyer. There are also specialized firms, known in some areas as "bird dogs," that provide the service of reporting on grower quality for buyers.

TYPES OF FIRMS OR AGENCIES

In many growing areas the grower-shipper provides most of the marketing services described above, while in other areas some services are provided by specialized firms. Separate firms sometimes operate refrigerated or controlled atmosphere storages, or vacuum cooling equipment. Packing-house firms may engage sales agencies to merchandise their products. Brokers may arrange sales for shippers and purchases for buyers or work on just one side of the transaction. Truck brokers and auction firms generally operate as separate entities.

Many different types of firms are found at most major shipping points. The proportion of each type may differ as is the case, for example, at Yakima, Washington; McAllen, Texas; and Salinas, California (Table 16.1). Yakima is a major shipping point for apples, and according to the primary function performed by these firms, the number of shippers about equalled the number of grower-shippers. Yakima has a number of large cooperative and proprietary shippers that handle a major share of the apple business. There were also three firms listed as being primarily in the export business, but none primarily handling imports.

At McAllen, Texas, a shipping point handling onions and other winter vegetables located near the Mexican border, there were relatively fewer shippers and more buying brokers than at Yakima, and four firms classed primarily as importers and one as exporter. Salinas, of course, is a major shipping point for lettuce and other vegetables with three times as many marketing firms as the other locations. Salinas has a heavy concentration of buying

Table 16.1. Number of Fresh Fruit and Vegetable Marketing Firms Operating at Selected Shipping Points By Primary Function, 1986.

PRIMARY FUNCTION	YAKIMA WASHINGTON	MCALLEN TEXAS	SALINAS CALIFORNIA
		number	
Grower-Shipper	18	12	51
Shipper	17	9	35
Buying Broker	1	9	46
Broker	8	5	18
Buying Office	8	5	18
Exporter	3	0	0
Importer	0	4	0
Other	6	4	9
Total	58	48	164

Source: Trade References.

brokers, even considering its size, and no firms primarily engaged in the export or import business.

FORMS OF BUSINESS ORGANIZATION

The full range of types of business organization may be found at major shipping points. Individual ownership is still common among smaller firms, but grower-shippers are often organized as partnerships or family corporations because of the need for more than one person with management responsibility and ownership interest. Publicly traded corporations are less common, but are involved in some areas through the entry of larger diversified firms into the fresh produce business. Cooperatives, mainly grower owned, play a major role in the packing of many products or as sales agencies.

SUMMARY

Operations at shipping point are the key to maintaining quality through the marketing system. Once the product starts to deteriorate the process cannot be reversed. Obtaining top quality fresh fruits and vegetables and then handling them the best way for their particular kind and variety is essential. Most produce items are highly sensitive to their environment. Very sophisticated methods and equipment have been developed to grade, pack, cool, and ship different products in order to maintain the highest quality. Within limits these can be adapted to individual situations. Failure to observe the necessary precautions regarding product handling, however, will ultimately result in lost sales or lower prices.

REFERENCES

Bongers, Anton J., Ben M. Hillebrand, Larry A. Risse, 1981. *Containers in Common Use for Selected Fresh Fruits and Vegetables Exported to Western Europe.* U.S. Department of Agriculture, Science and Education Administration, Marketing Research Report No. 1114.

Fahey, James Y., 1976. *How Fresh Tomatoes are Marketed.* U.S. Department of Agriculture, Agricultural Marketing Service, Marketing Bulletin No. 59.

Hardenburg, Robert E., Alley E. Watada, Chien Yi Wang, 1986. *The Commercial Storage of Fruits, Vegetables, and Florist and Nursery Stocks.* U.S. Department of Agriculture, Agriculture Handbook No. 66 (revised).

Johnson, Stanley S., Mike Zahara, 1977. *Mechanical Harvesting and Packing of Iceberg Lettuce.* U.S. Department of Agriculture Agricultural Economic Report No. 357.

Kader, Adel A., Robert F. Kasmire, F. Gordon Mitchell, Michael S. Reid, Noel F. Sommer, James F. Thompson, 1985. *Postharvest Technology of Horticultural Crops.* University of California, Division of Agriculture and Natural Resources, Special Publication 3311.

Lewis, Edgar L., 1989. *Fresh Vegetable Packing Costs for Six Small Cooperatives.* Agricultural Cooperative Service, ACS Service Report 25.

Stokes, Donald R., Glenn W. Woodley, 1974. *Standardization of Shipping Containers for Fresh Fruits and Vegetables.* U.S. Department of Agriculture, Agricultural Research Service, Marketing Research Report No. 991.

U.S. Department of Agriculture, 1969. *Conversion Factors and Weights and Measures for Agricultural Commodities and Their Products.* Economics, Statistics, and Cooperatives Service, Statistical Bulletin No. 616.

Yolz, Marvin D., Joseph P. Anthony, Jr., 1977. *Operating Costs at Four Potato Packing Plants.* U.S. Department of Agriculture Agricultural Research Service, Marketing Research Report No. 1072.

Chapter 17
Long Distance Transportation

INTRODUCTION

Long distance transportation provides the link in the marketing of fresh fruits and vegetables between shipping point and destination market. The current highly efficient system has enabled receivers to reach out to obtain supplies of perishable products from growing areas half way around the world, and shippers to seek buyers in distant lands. A box of apples arriving on the Boston market today has, on the average, come three times as far as one did only 25 years ago. No longer is distance to market a restriction on the distribution of most fruits and vegetables. Transportation costs are not as important a factor in determining source of supply as formerly. Yet many fruits and vegetables are easily damaged and highly perishable so the costs of faulty transportation go far beyond the transfer charges. Knowledge of the available transportation systems and the ability to select the most appropriate mode of transportation and protective services for each product is still essential to reduce quality deterioration and extend shelf life.

THE DEVELOPMENT OF ALTERNATIVE MODES OF TRANSPORTATION

Our transportation system for fresh fruits and vegetables has gradually evolved from the primitive methods used by early settlers to the highly sophisticated and efficient process in use today (Bole, 1980; Mennem, 1980). Initially produce transportation was mainly by horse-drawn wagons, and the growing of fruits and vegetables for commercial sale was largely restricted to the outskirts of population centers. Market garden farms developed around our towns and cities, and some have persisted to this day. Items like apples, citrus, potatoes, and onions that did not deteriorate rapidly could be moved longer distances, even by boat. Production of these commodities could be carried out in specialized growing areas, and consumers outside those regions could still enjoy these products. Even in the days of sailing ships pineapples were brought from Hawaii, and apples shipped to Europe. The development of toll roads and waterways extended the distance that less perishable fruits and vegetables could be shipped. Aside from a few storeable staples, however, consumers were limited to what could be grown nearby during the local growing season.

The Railroad Era

The development of the railway system early in the last century and the eventual linkup with the growing areas of the West and the South greatly improved the domestic transportation system. The production of less perishable crops was able to shift even further from major markets, and transportation replaced storage for less perishable crops as a means of supplying consumers.

About one hundred years ago the first refrigerated railroad cars, boxcars with bunkers for ice at each end, were developed. Initially these could only be used for shipments over relatively short distances. Soon, however, icing stations were established at about 250-mile intervals along main railroad lines and it became possible to ship perishables like lettuce and soft fruits across the country. The railroads' effective system of handling the transportation of perishables was a major factor in the development of fruit and vegetable production on the West coast and in the southern states.

The railroad system in this country at its peak was highly efficient in moving freight long distances, especially in relation to fuel and labor costs (Hutchinson, 1983). Most costs of railroad operation were fixed or sunk expenses for track, equipment, and right-of-way, and were minimal when the system was being used at capacity. Railroads, however, had operated since 1885 under strict regulation by the Interstate Commerce Commission (ICC) and by their very nature were limited in the services they could provide. The ICC regulations restricted them in adjusting rates to meet changing competition, adding new or dropping existing services, and abandoning unprofitable or extending profitable routes. Being restricted to rail tracks reduced their flexibility and often caused delays in pick-up and delivery. Normal freightcar handling practices could damage delicate commodities. Service became unreliable for although rail shipments could be moved across the country in a week or so the trip frequently took much longer. Railroads were, therefore, vulnerable to competition from over-the-road vehicles as trucks and highways were developed.

Economic deregulation in 1978 enabled the industry to take some advantage of the lower costs of rail shipments for fresh produce, especially for large volumes over long distances. Even more important it has permitted the expansion of intermodal arrangements such as the trailer-on-flat-car (TOFC), the container-on-flat-car (COFC), the road-railer, and by the mid-1980s the double-stack container train.

The Advent of Trucks

The development of truck transportation in the 1910s and 1920s soon captured a share of the shorter distance movement of fruits and vegetables (Mennem, 1980). But unsuitable roads, undependable vehicles, and the lack of refrigeration still restricted truck shipments to a few hundred miles. As the highway system developed and trucks became more reliable the ability of this mode of transportation to meet the needs of the industry became apparent. Competition between trains and trucks intensified, and truckers began to secure a larger share of the fresh produce business.

Improvements in over-the-road equipment in the 1930s and 1940s including the introduction of tractor-trailers and mechanical refrigeration, plus better roads, gave a major boost to truck transportation. But railroads continued to hold on to a significant share of the increasing business, even though hampered by economic regulations. The construction of the Interstate Highway System in the 1960s and later, however, along with other developments in growing, shipping, and marketing enabled truckers by 1980 to secure most of the total volume. Even fuel shortages and increased fuel prices of the 1970s were not sufficient to stem this tide. In recent years trucks have been responsible for transporting about 85 to 90 percent of the volume of fruit and vegetable intercity shipments in this country, while railroads handled 3 or 4 percent. The remainder has been shipped by trailer on flatcar with a small volume by air.

In the mid-1930s intense competition among truckers had led to the economic regulation of most segments of that industry. The hauling of fresh fruits and vegetables was exempted, however. Ease of entry into produce hauling resulted in a proliferation of independent owner-operators and highly volatile transportation rates. The Motor Carrier Act of 1980 has made it easier for truckers to obtain operating authority to haul regulated commodities. This has resulted in a decline in independent owner-operators and an increasing share of the perishable business going to fleet operations that can secure traffic in both directions and keep equipment in use. More than half the west-to-east shipments of produce and shipments from Florida appear to be carried by firms operating fleets and by companies whose principle business is not transportation (Dunham, 1989). Some owner-operators now lease their equipment and their services as drivers to these companies. Competition has intensified, and rates have been held down.

Tractor–trailers can now cross the country in 4 days or less, maintain a dependable schedule, pick up and deliver at several locations, and provide the proper conditions to minimize quality deterioration of perishable products in transit.

Increased Role of Ships

Only a small quantity of fresh fruit and vegetables has ever been shipped by inland waterways, mainly because of the slow speed of barges and boats, but alternatives, except for air, have not been available for ocean transportation. Ocean shipment has consequently been used for many years for a considerable volume of the less perishable fresh fruits and vegetables, especially those that cannot be grown or are in short supply in the eventual markets. For many years the Northeast shipped apples in barrels to Europe, and received pineapples from Hawaii. Bananas came by boat from Central America on specially designed ships with refrigerated cargo holds, operated by the United Fruit Company, that controlled most of the banana business at that time.

Ocean shipment of perishables as well as other cargo was virtually revolutionized by the development of the container in the 1960s (Mennem, 1980). Prior to that time the ocean shipment of packaged goods had been severely hampered by high labor costs of loading and unloading, delays of ships in port, severe damage to product, and considerable pilferage. Dockside labor costs reflected not only low productivity but high wage rates imposed by strong union activity.

Containers, large aluminum and steel boxes 8 feet wide by 8 feet high and 20, 30, or 40 feet long were originally developed for the domestic transportation of cigarettes but lent themselves to ocean shipment through the ease and speed with which they could be loaded and unloaded, the protection they provided the product, and the obstacle they presented to pilferage. Through the installation of self-contained refrigeration systems, containers are well adapted to carry perishables. They can be delivered to point of embarkation and picked up at destination on railroad flatcar or truck trailer chassis.

Use of containers, according to some reports, increased labor productivity in loading and unloading from about 1 ton to over 200 tons per man-hour, and reduced ship turnaround time from over a week to two days or less. The adoption of containers was vigorously opposed by longshoremen in some ports who eventually secured such benefits as a guaranteed annual wage for all union members. Union settlements were particularly costly in cities such as New York, which caused a shift in traffic to other locations. But the savings in labor and ship time and improvements in quality have more than offset union guarantees in most places, and container use has spurred the growth in ocean shipment of perishables.

Shipment by Air

Air transportation as it developed was, of course, considerably faster from point of departure to point of arrival than other modes, and consequently often advantageous for highly perishable items or to catch hot market situations. Larger planes and better equipped terminals have improved the service and reduced costs, but the extra handling and delays along the way have limited the advantages of air transportation compared to truck traffic. The use of containers specially designed to fit the cargo space of the airplane has improved handling and reduced damage.

Air shipment of fresh fruits and vegetables within the United States amounts to about 75,000 tons annually and has not changed much in recent years. California normally ships a wide variety of fruits and vegetables by air to other parts of the mainland as well as to Hawaii and Pacific rim countries. Strawberries are the most important commodity accounting for over half the total volume, followed by smaller quantities of cherries, lettuce, avocadoes, mushrooms, and brussel sprouts. Hawaii ships papaya and pineapple to the mainland on a regular monthly basis.

Intermodal Shipments

The use of two or more modes of transportation to move products to market is not a new phenomenon (Brookes, Byrne, 1979). Rail arrivals at terminal markets have frequently been unloaded at team tracks and hauled by truck to distribution centers. Intermodal operations where the modes of transportation are truly integrated is a more recent phenomenon. This is the case when trailers are loaded on flatcars or when marine containers are hauled on truck trailer chassis to be loaded at packing houses and then after transportation by ship are again hauled over the road to final destination.

Although TOFC shipments have been made for many years, at first the railroads were responsible for the total shipment. With deregulation, under what is known as Plan III, a shipper or trucking firm now can provide the trailers and deliver them at destination, and the railroad is simply responsible for moving the trailer on the flatcar from one point to another.

Shipping produce by tractor–trailer on railroad flatcar part of the way requires considerable coordination between the trucker and the railroad since in most cases these are separate business operations. Loading and unloading facilities must be made available, and shipments carefully scheduled. The shipper provides for local pickup and delivery, and the railroad charges for ramp to ramp service. Shippers wishing to use the service can provide trailers from their own fleet, obtain them from a leasing company, contract for the service or, if available, lease them from the railroad.

As noted earlier, fresh produce shipments by TOFC represent a small but significant proportion of the total. The traditional truck broker that was able to put shippers in contact with independent truckers has been supplemented or replaced by the transport broker or shipper agents that provide TOFC service to shippers. These transport brokers purchase TOFC space from railroads, often under relatively long-term contracts that establish somewhat stable rates for a fixed number of trailers on a train over a several month period. The agent then sells the trailer space to produce packers and other shippers. In many cases, shippers agents own the trailers used and arrange for return hauls, lowering costs. Some agents provide delivery and pickup of trailers at both ends of the haul, and some offer credit to shippers. One problem with Plan III TOFC is the considerable amount of paperwork necessary, and the transport broker is set up to handle this for the shipper or receiver.

Much of the fresh produce coming to western New York from California is shipped TOFC to Chicago then hauled over the road to Buffalo. There is often delay in transferring the trailer from one road to another in Chicago, and spotting the car in Buffalo. The real potential in intermodal transportation has been largely achieved by firms such as the CSX Orange Blossom Special that assembles unit train loads of refrigerated trailers from packing houses in central Florida for transportation to the mid-Atlantic region on a regularly scheduled basis.

One of the problems with shipping trailers on flatcars has been the increased movement to which the product is subjected because of being placed over two sets of suspensions (Ashby, et al., 1987). Proper loading and fastening to the flatcar can overcome part of this problem. The other problem has been the jolting caused by the slack in the coupling between the cars when trains are shunted.

Containers are now being widely used for ocean shipments of dry cargo and perishable products, and systems have been developed to bring containers to dockside and to deliver them to destinations. Containers are also being used to a limited extent for domestic shipments by integrating local pickup and delivery over the road with double stacking on specially constructed deep-welled railroad flatcars. Use of the container reduces some of the motion inherent in carrying trailers on flatcars, while the double stack cars are articulated in sets of five to reduce jolting caused by slack coupling. Loading and unloading equipment and facilities must be available nearby to make this system work. Trailers can still be driven onto flatcars, but the preferred systems for trailers or containers are using either a heavy duty forklift or a crane at a paved loading facility.

Intermodal operations offer the opportunity to take advantage of the best features of each mode such as the flexibility of truck pickup and delivery and the efficiency of long-distance rail hauls. Handling the trailer or container as a unit reduces labor costs, damage, and pilferage, though requiring added coordination between modes.

DIFFERENCES IN FRESH PRODUCE WITH RESPECT TO TRANSPORT CONDITIONS

The transportation of fresh fruits and vegetables represents not only temporary storage but also subjects the commodity to additional handling and possible variation in transit conditions. Individual fresh fruits and vegetables differ greatly in the ideal conditions under which they should be packaged, loaded, and transported in order to minimize damage, reduce deterioration in quality, and preserve shelf life. Even apple varieties differ in the proper temperatures at which they should be stored. Excellent information with respect to optimum shipping and handling conditions for fresh fruits and vegetables may be found in the U.S. Department of Agriculture's Handbook No. 668, *Tropical Products Transport Handbook* (McGregor, 1987) and Handbook No. 669, *Protecting Perishable Foods During Transport By Truck* (Ashby et al., 1987).

Desired transit temperatures vary from close to 32°F (0°C) for commodities such as cabbage or broccoli to 56° to 58°F for bananas and many other tropical products. Some commodities like beets and carrots do best in relative humidities of close to 100 percent, while dry items like garlic and onions hold up well at relative humidities in the range of 65 to 70 percent. Many tropical products are subject to chilling injury, while others are particularly susceptible to freezing injury.

Recommended loading methods also differ. Most commodities are now shipped in fiberboard boxes, although wirebound crates, bushel baskets, bags, bulk bins, or other containers are used for some products. Each type of shipping container requires special care in loading. Bracing is important to prevent containers from shifting, especially for trailers that will be carried on railroad flatcars. Crushed ice is blown over loads of some commodities to maintain humidity as well as reduce temperatures, but care has to be taken that the top ice does not impede air movement.

The increased shipment of mixed loads has brought problems in transportation. Many commodities are incompatible, not only because of differences in temperature and humidity requirements but for additional reasons.

Some commodities such as apples and oranges produce a gas called ethylene which hastens ripening, but other products such as cabbage and lettuce are sensitive to it and so loads with conflicting needs should not be stored or shipped together. Some products may need to be shipped under modified atmosphere. Some products produce odors, and others absorb odors readily. Compatability groups have been developed to accommodate those who feel the need to ship mixed loads by relaxing somewhat the recommended transit conditions for some commodities. Six groups have been identified for fresh fruits and vegetables that take into account temperature, relative humidity, and production or sensitivity to ethylene. Because of their particular characteristics, separate transit and storage is recommended for bananas, citrus, nuts, potatoes, and onions.

Unitized loading is becoming more important. Most unitized loads are on wooden pallets but these take up space and are costly to return or dispose of. An increasing quantity of perishable products are being handled on slip sheets and some as blocked units that require a special push–pull lift truck for handling. Pallet loads are sometimes wrapped in stretch or shrink plastic netting, or tied together with vertical or horizontal strapping and protected with cornerboards. The lack of standardization of pallet and container sizes is still a problem.

TRUCK TRANSPORTATION

Types of Firms and Business Practices

The interstate trucking industry consists in general of two major segments, the private and for-hire operations. Private carriers are firms that transport their own goods and supplies in their own trucks. For-hire carriers with interstate operations may either have operating authority from the Interstate Commerce Commission (ICC) to haul regulated commodities, or they may be limited to hauling exempt commodities such as fresh fruits and vegetables. Operating authority is now easier to obtain, so the distinction is not as significant as it was prior to deregulation.

For-hire carriers consist of either owner–operators with up to about three vehicles, or fleet operations. Truckers without ICC operating authority, generally owner–operators, may haul regulated commodities under a leasing arrangement with a firm that has operating authority, either on a trip by trip basis or for a longer period. Fifty percent of the vehicles hauling produce in Florida in the 1985/86 crop year were owner–operators, 37 percent were for-hire fleets, and 13 percent were private carriers (Beilock, MacDonald, Powers, 1988). The share of total fresh produce hauled by exempt carriers has tended to decline, while that hauled by irregular route common carriers (IRRC) has increased. Many IRRC carriers with refrigerated rigs, operating under ICC authority, haul fresh produce in one direction and solid refrigerated products such as frozen foods in the other direction.

Most fresh fruits and vegetables are hauled in refrigerated trailers, except for a limited number of less perishable items such as onions and potatoes and then only under favorable weather conditions or for short distances. Refrigerated trailers also haul frozen foods, and can be used if necessary to haul products not requiring refrigeration.

Since most fresh fruit and vegetable sales in this country are made on an f.o.b. origin basis the receiver is responsible for arranging and paying for the transportation. The shipper, however, is often asked to take responsibility for arranging for the transportation subject to the receiver's approval. Shippers who use independent owner–operators or small trucking firms generally require the services of a truck broker to line up transportation. In Florida during the 1985/86 season, 66 percent of all produce hauls were arranged through a truck broker, 24 percent were made by direct contact between the receiver and the carrier or between the shipper and the carrier (usually on behalf of the receiver). The remainder were either private carriers hauling their own produce or other arrangements not specified (Beilock, MacDonald, Powers 1988).

The truck broker, in addition to securing trucks, may also negotiate truck rates on behalf of the shipper or receiver. Written contracts of hauls with independent truckers are seldom used. The shipper sometimes must make an advance to an owner–operator to cover operating expenses for the trip. In such cases the receiver then deducts the amount of the advance from the final payment to the trucker, and remits this amount to the shipper. Large shippers now may contract with truck fleet operators for services at a fixed rate for the season and avoid the day to day fluctuations in charges.

Truck Costs and Revenues

The USDA estimated that in December 1988, it cost about $1.20 per mile to operate a refrigerated truck under typical conditions hauling fresh fruits and vegetables over long distances (USDA AMS, 1989). Individual cost items differed for fleet operators compared to owner–operators, but the total was about the same (Figure 17.1). The driver's wage was the largest item of cost amounting to 33 cents for the fleet operation and 36.7 cents for the owner–operator. Fuel costs came next, amounting to 21 cents per mile. Together these two costs accounted for almost half the total. Vehicle depreciation at 14 cents was 3 cents higher for the owner–operator than for the fleet operation, but this was largely offset by lower maintenance charges. Other major expenses were insurance and licenses, tires, and mis-

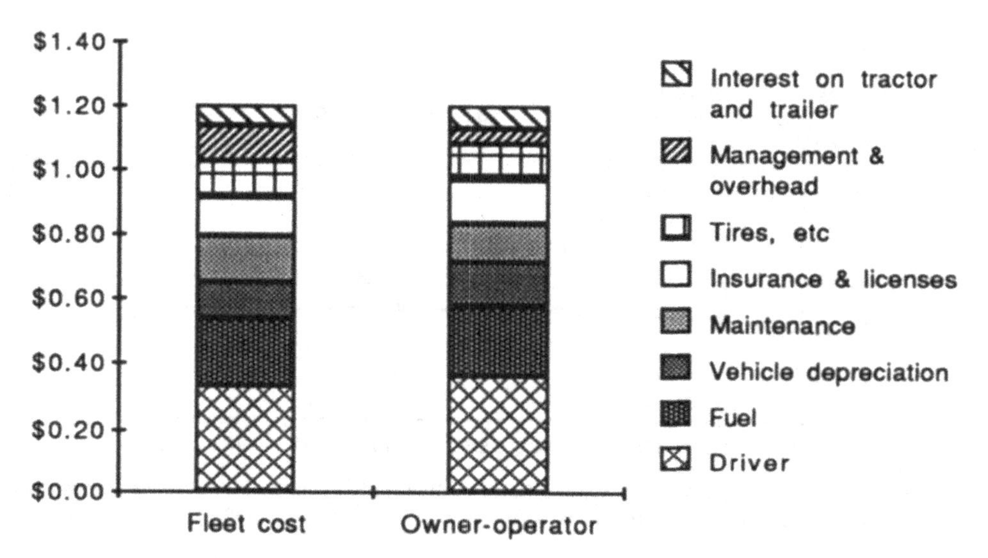

Figure 17.1. Truck Costs per Mile for Fleets and Owner–Operators, December 1988. *Source:* Adapted from USDA AMS, 1989.

cellaneous, management and overhead, and interest on tractor and trailer. Costs may vary widely depending on operations. The USDA cost estimates are based on tractor–trailers averaging 1,400 miles per trip and 140,000 miles per year, and running fully loaded three-quarters of the time.

Costs of truck operation rose sharply during the 1970s but have increased slowly in recent years. According to Dunham (1989) the estimated cost increased from $0.96 per mile in 1980 to $1.18 in 1988, an increase of 22.7 percent (Figure 17.2). The rate charged per box or carton for hauling fresh produce has changed even less. The rate for hauling apples from Washington State to New York City rose from $3.09 per box in 1980 to $3.30 in 1988, an increase of 6.8 percent. This largely reflects the increased loading from 900 boxes of apples per load to 1,000 boxes. Potential increases in rates per box have also been offset at least in part for other commodities by increased loading. The general adoption of larger 48-foot trailers, the use of twin trailers on some limited access highways, and the more careful preparation and packaging of produce has also tended to offset increases in other costs (Wolff, 1980).

Trucking is still highly competitive, and trucking charges between shipping points and markets still vary widely from week to week, depending on the local demand for trucking services and the available supply of vehicles. There is a definite seasonal pattern in the rates reported for hauling fresh

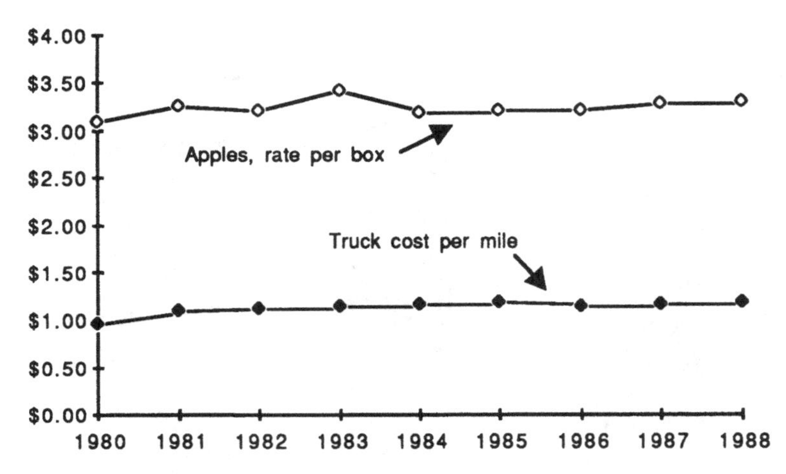

Figure 17.2. Trucking Costs per Mile and Rates per Box for Shipping Apples from Washington State to New York City, 1980–87. *Source:* Adapted from Dunham, 1989.

produce from California to midwest and eastern markets. Heavy demand for trucks during the summer months at a time when equipment is needed in other parts of the country increases the demand; rates consequently rise. In 1988, for example, truck shipments of fresh fruits and vegetables from California in June and July were 2.5 times as great as during the winter (Figure 17.3). Truck rates remained relatively stable from November through through April, then increased as the volume of shipments rose. For fruit and vegetables from the Central San Joaquin Valley to New York City the rate per truck load in 1988 went from about $2,800–$3,000 during the winter to a peak of $4,600–$5,000 in mid-July.

Traditionally truck drivers have unloaded their trailers at destination and their rates have reflected this service. This practice existed when most loads were hand stacked. With the advent of more palletized shipments, however, receivers have sought to have arriving trailers unloaded by temporary help, and to deduct this cost from the transportation charge paid to the trucker. The temporary workers used to unload trucks are called lumpers. Use of lumpers has been a source of friction between truckers and receivers. The Motor Carrier Act of 1980 prohibited receivers from coercing truckers to use lumpers (Hutchinson, 1983). Recent research in Florida has indicated, however, that in markets where lumpers are used the transportation charges tended to be set high enough to more than cover the cost of unloading.

A study of rates charged by truckers hauling produce out of Florida during the 1985/86 season found that several factors helped explain differences in freight rates per trip (Beilock, MacDonald, Powers, 1988). These in-

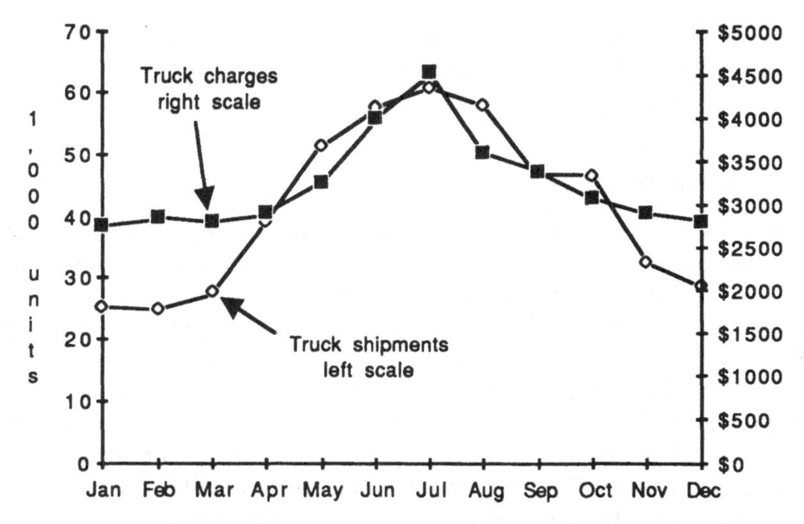

Figure 17.3. Truck Charges per 40,000 Pound Load From Central California to New York and California Fruit and Vegetable Shipments (1,000 cwt) By Months, 1988. *Source:* Adapted from USDA AMS, 1989.

cluded hauling distance, the number of pickups and drops, vehicle size, cargo value and perishability, the destination, and the season of the year. For distances under 100 miles the average cost of an additional mile was $1.10. This declined by $.06 per mile for each additional 1,000 miles total distance beyond the first 100 over the range of the data (up to 3,200 miles). It was estimated that carriers received $50.30 per pickup and $10.48 per drop. Carriers received $0.18 per additional cubic foot of trailer capacity, a difference of $135 per produce load between a trailer 40 feet by 8 feet and one 48 feet by 8.5 feet.

Florida shipper/receivers were apparently willing to pay a premium for the transportation of higher value perishable products, with charges for loads of tomatoes and beans, for example, averaging about $90 more than those of potatoes or celery. Carriers hauling produce up the East Coast received an average of $94 more per load than those headed toward the Great Lakes, and $175 more than those hauling produce directly to the West. During the May/June period of heaviest shipments rates averaged $251 higher than in November, $259 higher than in January, and $224 higher than in March. The availability of refrigeration was not, of itself, associated with higher transportation rates. This was apparently explained by the differentials for perishability, since the unrefrigerated vehicles mainly hauled watermelons or potatoes.

Freight rates were apparently not related to carrier type (owner-operator, for-hire fleet, or private carrier), or by method of arranging for the load (broker, direct contact between carrier and shipper, or direct contact between carrier and receiver).

Equipment and Operating Practices

To maintain product quality and retard deterioration requires careful attention to refrigeration methods and air circulation, and sufficient insulation. The use of appropriate containers and proper loading methods are also extremely important as is pointed out by Ashby et al. (1987) and McGregor (1987). Products differ in their optimum requirements for transit temperature and humidity, and in their compatability in mixed loads.

Mechanical refrigeration units on tractor trailers may be mounted either on the nose of the trailer or under the frame, and will vary according to refrigeration capacity. Top icing is sometimes used to add refrigeration and increase humidity. Top ice should be allowed to melt steadily during the trip otherwise it may crust or freeze and block air circulation.

Air circulation is important to insure that the proper temperatures are maintained throughout the load. The design of the trailer and the method of loading influence the effectiveness of air circulation. In the conventional top air delivery system the ceiling ducts, load patterns, cross-bracing, floor design, recessed-groove sidewalls, and vented shipping containers are all essential for proper temperature control. Some newer trailers and marine containers are now equipped with bottom or under-the-floor forced air delivery systems. Provision must be made for the air to be forced up through the load.

The two types of air-delivery systems require different loading patterns in order to maximize and maintain load cooling efficiency. With the top air delivery system the cold air from the refrigeration exits at the front ceiling, and the air must be at a relatively high velocity to carry it all the way to the rear of the trailer. With bottom-air delivery systems the air from the refrigeration unit is forced down the front bulkhead and through a deep T-rail floor and then vertically through the load. A high static air pressure is maintained under the load to assure a low velocity but steady movement of air through small openings in the load.

RAILROADS

Railroads were hampered for many years by government regulation in the adjustments that they could make to changing economic conditions. There has never been one railroad company that could provide cross-country service, and railroads have been restricted in the other services that they might provide. Deregulation has provided somewhat greater flexibility in setting rates and providing associated services, but this has not enabled the railroads to recapture much of the business lost to truckers except through trailer or container on flatcar service.

Railroads still have a considerable cost advantage compared to over-the-road transportation in hauling large loads of heavy, less perishable products such as potatoes and onions. Delays and uncertain arrival times are not as critical for these shipments. But the increasing prevalence of mixed loads and multiple drop shipments has worked to the disadvantage of the railroads, as has the dispersion of packing sheds at shipping point and wholesale distribution centers at major markets.

AIR TRANSPORTATION

Freight forwarders at major airports are used to arrange air transportation. They can supply the special containers designed to fit the aircraft cargo compartment.

Air transport is also used extensively for many fresh fruits and vegetables imported into this country. Fruits and vegetables from Israel and Holland arrive at Kennedy and other major airports on a regular basis. Most come in the cargo hold of regularly scheduled passenger planes rather than special cargo aircraft, in specially designed containers, but there are some airlines that specialize in freight hauling even though it is an uncertain business.

SHIP TRANSPORTATION

Business Practices and Pricing Arrangements

Ship transportation is arranged largely through freight forwarders since the job is so specialized and complicated. The freight forwarder checks schedules, books space, makes sure the produce is properly loaded at point of origin, and makes sure the documentation is in order. He also collects on the letter of credit, which is important in many foreign transactions. The freight forwarder can quote rates to the shipper but does not negotiate them as does, for example, a truck broker. The negotiation of ocean freight rates is very limited.

Oceangoing shipping firms are generally exempted from antitrust laws. Steamship lines traveling the same route have banded together to form conferences for the purpose of setting schedules, rates, and other shipping rules. Most of the fresh produce that leaves the United States travels on a conference steamship line. The other two methods of produce transportation are on charter and nonconference lines. Few nonconference lines have refrigerated containers, and when a shipper uses a conference carrier he may be asked to sign an agreement prohibiting him from using a nonconference carrier if a conference carrier is available. Shippers can charter a ship for a voyage, but very few shippers are large enough to do this. In spite of the use of conferences to set rates, the ocean shipping industry has suffered periods of overcapacity and low returns that have temporarily benefited exporters and importers.

New technology continues to promise improvement in the cost and quality of container transportation. Improved container design and shipping cartons, atmospheric control, and studies of specific commodity handling conditions are being carried out. Optimum shipping conditions for fruits, vegetables, and even flower bulbs are under study. We will probably see more containers being loaded at packing houses and unloaded at supermarket distribution centers.

TRANSPORTATION COSTS AND PRODUCTION LOCATION

With truck transportation rates approaching $5,000 to haul a trailer load of fresh produce across the country many observers have wondered if this might encourage increased production closer to market. Several factors, however, tend to offset the apparent advantage of the grower near the market. Nearby growers also have transportation costs to market, which for the western New York grower shipping to New York City may run $800 to $1,000 per load. Of more significance is the fact that truck costs from California peak for only a few weeks in the year, and growers and shippers can afford to ship below cost and recover the deficit during other weeks of the season. In some cases distant shippers also have a very favorable reputation for quality and receive a premium in prices over those paid nearby growers that helps to offset the higher transportation costs.

Other factors also tend to work to the advantage of the grower–shipper in major production areas distant from market. Yields per acre may be higher and costs per acre lower, both leading to lower unit costs. A longer shipping season may permit better use of equipment and manpower, as well as preferred treatment by the buyer.

Future increases in transportation costs resulting from rising fuel prices have been thought to represent a future opportunity for nearby growers, but again the extent of that opportunity has often been overestimated. A study of the impact of changing transportation costs on the apple industry found that the profitability of apple production was, however, sensitive to changing costs (Dunn, Garafola, 1986). Results of using an interregional competition model of the U.S. apple industry indicated that changing transportation costs brought moderate changes in total production and consumption levels, but greater changes in the utilization of a region's crop in fresh sales and processed products. Fuel costs and transportation rates role rapidly in the 1970s but have declined since that time. Undoubtedly costs will rise in the future as fossil fuels become exhausted. But the industry in major growing areas has shown surprising ingenuity in increasing efficiency in hauling fresh fruits and vegetables so it is unlikely that nearby growers will be able to count on profitable expansion opportunities simply on the basis of rising transportation costs.

FUTURE DIRECTIONS

There have been some very exciting developments in fresh produce transportation in recent years, and the improvement in efficiency and quality of transportation services will probably continue to take place. But difficult problems and challenging decisions also lie ahead. The tremendous increase in the use of truck transportation over the past 20 years was not envisioned when the Interstate Highway System was developed, and sufficient funds have not been budgeted to adequately maintain roads and bridges. Increased competition between truckers due in part to deregulation has resulted in the failure of many to properly maintain equipment, and in the tendency to ignore safe driving practices with respect to highway speed and rest intervals. The Commercial Motor Vehicle Safety Act of 1986 will eventually require federal driver testing and licensing that may reduce accidents.

Efforts on the part of the federal government to remove state restrictions on truck weights and lengths will probably permit the use of trailers of increasing size and the greater use of twin trailers. Improved trailer design and refrigeration equipment will maintain quality better on longer hauls. Modified atmosphere will come into greater use. The TOFC will enable shippers to combine the flexibility of the truck with the economy of the railroad. Low cost truck transportation, however, has come in part at the expense of more rapid deterioration in our highways and bridges than was originally expected, in the failure of owners and operators to adequately maintain trucks and trailers, and in provisions to insure that licensed operators had adequate training and ability.

Railroads will focus on services they can do best and develop equipment and facilities to perform those functions most effectively. Repair and renovation of equipment may enable them to return to the reliability and speed of earlier days. Through mergers or acquisitions there will be less need for costly delays in transferring cargo from one road and the next. Union work rules are undergoing change to permit no cabooses, fewer changes in crew, less labor, and more efficient locomotives.

Pallets or slip sheets will be used to a greater extent in the future for the loss in cube will be more than justified by the gain in labor efficiency and reduction in damage. Packages and cartons will be modular to fit different commodities on the same pallet for local delivery. Products will be trimmed and packaged to economize in weight and cube.

The container that initially was largely used for ship transportation will be increasingly used for over-the-road or rail transportation, although more cost-effective means of generating electrical power to supply the double-stack container train operations is needed. Containers are now being shipped two deep on railroad flatcars, so that four containers replace the two trailers formerly carried on the flatcar. The container is being and will be used for temporary storage by receivers just as rail cars have been.

There will still be advantages in being close to market, but the transportation industry will introduce improvements that will offset partially if not entirely the rising costs of fuel and labor that would otherwise result in increased costs per carton in the future.

REFERENCES

Ashby, B. Hunt, R. Tom Hinsch, Lawrence A. Risse, William G. Kindya, William L. Craig Jr., Mark T. Turczyn, 1987. *Protecting Perishable Foods During Transport By Truck.* U.S. Department of Agriculture, Office of Transportation, Agricultural Handbook Number 669.

Beilock, Richard, James MacDonald, Nicholas Powers, 1988. *An Analysis of Produce Transportation: A Florida Case Study.* U.S. Department of Agriculture, Economic Research Service, Agricultural Economic Report No. 597.

Boles, Patrick P., 1980. *Trends in Fresh Fruit and Vegetable Transportation 1963–75.* U.S. Department of Agriculture, Economics and Statistics Service, Report No. 1.

Brooks, Eldon, E., Robert J. Byrne, 1979. *Piggybacking Fresh Vegetables—California to the Midwest and Northeast.* U.S. Department of Agriculture, Economics, Cooperatives, and Statistics Service, FCR Report 10.

Dunn, James W., Lynn A. Garafola, 1986. Changes in Transportation Costs and Interregional Competition in the U.S. Apple Industry. *Northeast Journal of Agricultural Economics,* Vol. 15, No. 1, April, pp. 37–44.

Dunham, Denis, 1989. *Food Cost Review, 1988.* U.S. Department of Agriculture, Agriculture Economic Report, No. 615 July.

Hardenburg, R. E., A. E. Watada, C. Y. Wang, 1986. *The Commercial Storage of Fruits, Vegetables, and Florist and Nursery Stocks.* U.S. Department of Agriculture, Handbook 66, Revised.

Hinsch, R. Tom, William G. Kindya, Roger E. Rij, 1983. *Improved Arrival Temperatures in Produce in a Modified Refrigerated Trailer.* U.S. Department of Agriculture, Marketing Research Report No. 1132.

Hutchinson, T. Q., 1983, *Implications of the Motor Carrier Act of 1980.* U.S. Department of Agriculture, Economic Research Service.

McGregor, Brian M., 1987. *Tropical Products Transport Handbook.* U.S. Department of Agriculture, Office of Transportation, Agriculture Handbook No. 668.

Mennem, Gary M., Editor, 1980. *Transportation Policy Primer.* National Extension Policy Task Force Publication No. 4, Oklahoma State University.

Risse, L. A., W. R. Miller, T. Moffitt, 1982. *Shipping Fresh Fruits and Vegetables in Mixed Loads to the Caribbean.* U. S. Department of Agriculture, Agriculture Research Service, AAT-S-27.

Ulrey, Ivan W., 1969. *The Economics of Farm Products Transportation.* U.S. Department of Agriculture, Economic Research Service, Marketing Research Report No. 843.

U. S. Department of Agriculture, Agricultural Marketing Service, 1989. *Fruit and Vegetable Truck Rate and Cost Summary, 1988.*

Wolff, Donald G., 1980. *Study of Methods to Improve Fuel Efficiency in Delivering California Iceberg Lettuce to Market.* California Iceberg Lettuce Research Advisory Board.

Chapter 18

Wholesaling at Destination and Terminal Market Facilities

INTRODUCTION

In the ever-changing world of marketing fresh fruits and vegetables, wholesaling at destination and terminal wholesale market facilities play several important roles. Fresh produce arrives by truck, railcar, ship, or airplane at major population centers across the country from nearby and distant growing areas. Wholesalers, some clustered on central terminal facilities and others scattered throughout the market area, receive this produce. These wholesalers perform essential market services such as ripening or repacking, and resell smaller amounts to many different types of customers.

Chain stores also receive wholesale lots for distribution to their retail outlets and in some cases to other buyers. Information on arrivals and prices in major markets is collected by the Market New Service of the U.S. Department of Agriculture in cooperation with state departments of agriculture. This contributes to collective knowledge of market conditions so essential to the smooth operation of the marketing system.

Initially wholesalers operating on terminal markets were mainly small independent firms specializing in a few commodities and operating at arms length with shippers and customers. Today there are still many small independents, but some are now integrated both vertically, that is having close business relationships with suppliers or retailers, and horizontally, that is being able to provide a wider range of goods and services. The number, size, and functions of these firms have changed over the years as have central market facilities.

The role of the terminal market facility has changed and its importance has diminished in some communities, but not to the extent predicted by many only a few years ago. Marketers need to be aware of the history of terminal market facilities and the structure and operation of destination wholesalers so as to have some insight into the role these wholesalers perform in such markets and may provide in the future.

EARLY HISTORY OF TERMINAL MARKETS

In early days most towns and villages had a central place where local farmers congregated to sell their produce to consumers. As these places grew in size businesses developed whose function it was to obtain fresh fruits and vegetables in wholesale quantities from more distant areas as well as locally, and to provide produce in smaller, or job lots to retailers and other buyers. Such wholesalers tended to cluster together to take advantage of transportation facilities, and for the convenience of buyers who liked to inspect the produce before buying and might need a greater variety of items than was handled by any one firm. With the development of the railroads in the latter part of the last century, access to rail service became an important consideration.

As cities grew and transportation methods improved, the numbers of wholesalers and the volume and variety of products handled increased. Many firms specialized in a few commodities or groups of commodities, in certain types of customers, in the way they did business, or in different marketing services. Early in this century facilities on many central markets became overcrowded and antiquated in terms of current handling and storage methods, yet firms on these markets were reluctant to move for fear of losing business or incurring higher costs.

Experience with the major New York City wholesale market illustrates how difficult it could be for a terminal market to adapt to changing conditions. For many years the primary wholesale terminal market in New York was centered on Washington Street on the lower west side of Manhattan, near where the World Trade Towers stand today. Food peddlers were reported to operate in this area as early as 1800. In 1925, according to Rasmussen (1929) about 85 percent of the fresh fruits and vegetables arriving in the New York metropolitan district went through the Washington Street market. Most of this arrived initially in New Jersey in railroad boxcars that were either floated across the river or unloaded onto trucks for transportation to market.

By the late 1930s the market had grown to the point where it occupied a 25-block area around Washington Street. This street was only 30 feet wide between the curbs and was intersected by 14 cross-streets, each typically 34 feet wide. There were 267 stores selling fruits and vegetables on the market, as well as 200 other businesses providing other services. Local farmers could send merchandise directly by small truck, but rail and ship receipts had to be reloaded from the many piers and 22 railroad terminals scattered throughout the region. Large trucks were barred from the immediate market area because of congestion, and arriving merchandise had to be wheeled several blocks by handcart to stores. In spite of obvious inefficiencies documented by several studies it was impossible to mobilize the necessary resources and elicit support to move the market to a more modern facility until several years later.

Market facilities in other areas may not have been quite as antiquated as those in New York, but efforts to move many of them to more efficient facilities were equally unsuccessful. In some other areas, however, aided by studies conducted by the U.S. Department of Agriculture, markets were able to move to better locations and upgrade their facilities at this time. New markets built during the period between World Wars I and II included the Los Angeles Terminal Produce Market in 1918, The South Water Street Market in Chicago in 1925, the Detroit Produce Terminal, and the Northern Ohio Food Terminal in Cleveland in 1929, and several markets in upstate New York, including the Niagara Frontier Regional Market in Buffalo in 1938.

POST-WORLD WAR II DEVELOPMENTS

The period following World War II brought many changes in the growing and marketing of fresh fruits and vegetables that had a major impact on the operation of firms on organized terminal markets. On many terminal markets business declined, the number of firms shrank, the market for locally grown produce dropped off, quality suffered under existing handling methods, and costs of operation escalated.

Not only obsolete markets but other factors were apparently responsible for these changes. Fruit and vegetable production was shifting to distant specialized growing areas. Truck transportation provided greater flexibility in deliveries to dispersed warehouses. Chain stores were gaining a larger share of the food business and buying more produce direct from shipping areas. Consumption of some fresh fruits and vegetables was declining as consumers shifted to canned and then frozen products.

Opinions differed as to what should be done, if anything, to improve market operations. Some focused on the deplorable conditions existing on many markets and believed there were great opportunities for potential savings and growth with improved facilities. Others felt it was simply a matter of time until there would be little business left for destination wholesalers, and this could easily be handled through existing stores. As it turned out neither viewpoint was entirely correct.

The U.S. Department of Agriculture conducted a major study of wholesale markets across the country in 1958 (Manchester, 1964). The 52 markets in the study varied not only in size but also in other characteristics.

The New York City phase of this study revealed many changes on this market between the mid-1930s and the mid-1950s. Total receipts on the New York market declined from 240,000 carlots in 1931 to about 170,000 in 1958. The number of wholesalers in the New York market dropped from 1,080 in 1939 to 769 in 1958. Less than a quarter of those that had been in business in 1939 were still active in 1958. Almost half the firms interviewed said the outlook for their business was poor.

In 1958 about half the produce coming into the New York City area, including that bought by chains, was sold by firms in the Washington Street area. Wholesalers on Washington Street accounted for 55 percent of sales of all wholesalers in New York, not counting sales to chains. Wholesalers in other organized markets handled 2 to 5 percent of the total. Wholesalers in other parts of the city sold 28 percent of the total. Only one of the 18 retail chains in the area had a warehouse on the Washington Street Market.

At the same time, the U.S. Department of Agriculture reported that about $10 million might be saved annually in direct costs of handling and local transportation if the Washington Street Market was moved to a better location and provided with better facilities (USDA AMS, 1960). Five sites for a new market were evaluated, including rebuilding on Washington Street, expanding the existing Bronx Market, or moving to Maspeth in Queens near the population center of the market, to the Secaucus Meadows in New Jersey, or to Hunt's Point in the Bronx. Even so, no decision was made until several years later.

CHARACTERISTICS AND TRENDS IN WHOLESALE MARKETING

In 1965 the National Commission on Food Marketing (1966), as part of a larger study, interviewed industry leaders in major terminal and secondary markets across the country to determine current characteristics and trends in fruit and vegetable wholesale markets. They noted that terminal market firms served the function of matching producer supplies with buyer demands and performed several other functions such as receiving, storing, and forward physical distribution of products. In addition public and private market information on wholesale receipts, unloads, holdings, and prices in the central terminal markets was employed extensively by producers and shippers as well as terminal firms as the basis for marketing decisions.

The terminal market complex was considered to consist of the full range of firms and physical facilities throughout the market, not just those clustered on a particular street or district. Within this complex there were two distinct entities—the integrated and nonintegrated sectors.

The integrated sector consisted primarily of the food retailing firms. These were the national, regional, and many local food chains that operated integrated warehouse and distribution systems. Not only corporate owned chains but also cooperative and wholesale sponsored chains had developed integrated systems. Such integrated wholesale–retail systems bought largely direct from shipping point, were aided by shippers who performed some wholesaling functions by making up mixed carlots or arranging pool shipments. Under this system the terminal wholesaler was used by them only for fill-in and specialty items.

Integration into wholesaling had freed the retail chains from the need to locate in the terminal market districts, which were usually in the heart of the city in congested facilities unsuited to efficient parking, unloading and loading, and materials handling. Instead they built modern distribution centers with easy access and adequate space generally on the edge of the city.

Even 20 years ago there were indications that firms in the food service market, particularly chain restaurants, were integrating into wholesaling. Chain stores were also beginning to service the institutional market from their warehouses.

The nonintegrated sector of terminal wholesale markets was considered to consist of carlot receivers, merchant wholesalers, jobbers, purveyors, commission merchants, repackers and prepackers, and auction companies. Central market districts composed on nonintegrated firms were still important in many cities, but their importance had been declining, and they had almost disappeared from smaller cities. The growth of the integrated sector had imposed many changes and adjustments.

Before the dominance of the wholesale–retail integrated firms the terminal market wholesalers had tended to specialize in a limited number of marketing functions. This firm specialization and the consequent interdependence was one factor that contributed to the concentration of firms in a central market district. The functions of wholesale firms had tended to become combined, especially in the smaller markets, and this lessened their interdependence. The service wholesalers performing the combined functions of receiving, jobbing, and distributing had developed the best position to serve the growing institutional food market and the shrinking number of independent retail food firms. The clientele being served was changing also, with jobbers who sold to retail firms now servicing institutions. Purveyors who normally served the institutional markets were continuing, however, to restrict their operations to these customers. Some jobbers were also adding processed products to their produce line.

Another change noted 20 years ago was the decline in consignment and auction selling on wholesale markets, replaced by f.o.b. or delivered sales direct from shipping point. Consignment sales were still persisting, but mainly for distressed merchandise or unsold supplies from shipping point firms.

The increase in wholesale-retail integrated operations, their direct buying from shipping point and exodus from the central market district, the combination of wholesaling functions in single firms and the resulting lessening of interdependence had led to many terminal market firms to move their operations away from the central market district. Continued movement away from the central city was expected to continue, with some moves involving the complete relocation of the central market districts.

The expansion of consumer packaging of fresh produce was also considered to be having an impact on the nonintegrated sector of the terminal market. It was estimated that 40 percent of the fresh fruit and vegetable supplies were prepacked either at shipping point or in terminal markets. Terminal market prepackers performed substantial packing services for both chains and nonintegrated firms. The chains were considered likely to increase their own packing services or move toward integrated packing operations. The increase in shipping point consumer packaging for crops such as citrus fruits, potatoes, onions, and carrots was noted, as was the increase in prepacking of vine ripe tomatoes at shipping point and the boxing of bananas by importing firms.

The growth of the integrated sector had not only resulted in a sharp decline in the number of terminal wholesalers but had the effect of concealing data on price and movement for a considerable quantity of fresh fruit and vegetable supplies. This also altered industry pricing procedures. Terminal market statistics no longer played as important a role, and shipping point information carried greater weight in establishing prices.

MARKET FACILITIES AT MAJOR MARKETS

The Green Book published by the National Association of Produce Market Managers (NAPPM) contains information on a large number of wholesale markets operating in the United States and Canada. These range in size, type, and location from smaller markets located in farming areas that assemble products for shipment as well as receive products for local distribution, to very large markets in major cities that only perform wholesale operations. Many markets, particularly in the Southeast, serve not only as wholesale distribution centers but also provide facilities for farmers to bring local produce to sell at wholesale or retail. A few still operate auction markets during the local growing season, or have special facilities for the assembly and shipment of commodities such as watermelons. Ownership varies:

A survey of 44 markets by NAPPM in 1986 found that 19 were state owned, 11 were owned by private corporations, 5 were privately owned, 4 were owned by municipalities, 2 each by public corporations and cooperatives, and 1 by other means.

The major markets throughout the country are primarily if not exclusively wholesale operations serving as receiving and distribution centers for the surrounding metropolitan area. Though their functions are similar they still differ in facilities and operations. A brief description of four of these will illustrate this point.

Los Angeles

Los Angeles has three wholesale produce markets—the Los Angeles Union Terminal Inc. (Seventh Street Market), the City Market Company of Los Angeles (Ninth Street Market), and the Los Angeles Wholesale Produce Market (Central Market). These operate not far from each other in the center of the City.

The Seventh Street Market, originally the major produce market, opened in 1916 and is owned and operated by the Southern Pacific Railroad. The Ninth Street Market is operated by a privately owned corporation formed in 1908 by a group of growers and shippers that broke away from other wholesalers. Extensive renovations were completed in 1987. The Central Market developed during the 1950s as a result of the inability of new firms to find available space in the Terminal or City Markets and was owned and managed by a private real estate firm. New facilities were completed in 1986.

Efforts in the late 1960s to consolidate these markets in one up-to-date food distribution center led to a major study by the U.S. Department of Agriculture (Taylor, et al., 1972). The study found that there were 133 fruit and vegetable wholesalers in the area. These firms handled a total volume of 1.3 million tons of produce annually of which 86 percent was in direct receipts and 14 percent interwholesaler transfers. Wholesalers on the three markets handled 83 percent of the volume, over half at the Terminal Market and most of the rest at the City Market. The study identified 244 of the 538 food wholesalers of all types in the area that might benefit from new facilities, including 114 of the 133 fresh fruit and vegetable wholesalers. Several alternative sites were evaluated, and the proposal included several single-occupancy buildings as well as multioccupancy structures. In spite of apparent savings through consolidating food wholesaling operations at one location no move was made for many years.

Eventually the new Los Angeles Wholesale Produce Market was built. It covers 29 acres, and is exclusively a wholesale operation, shipping to as well

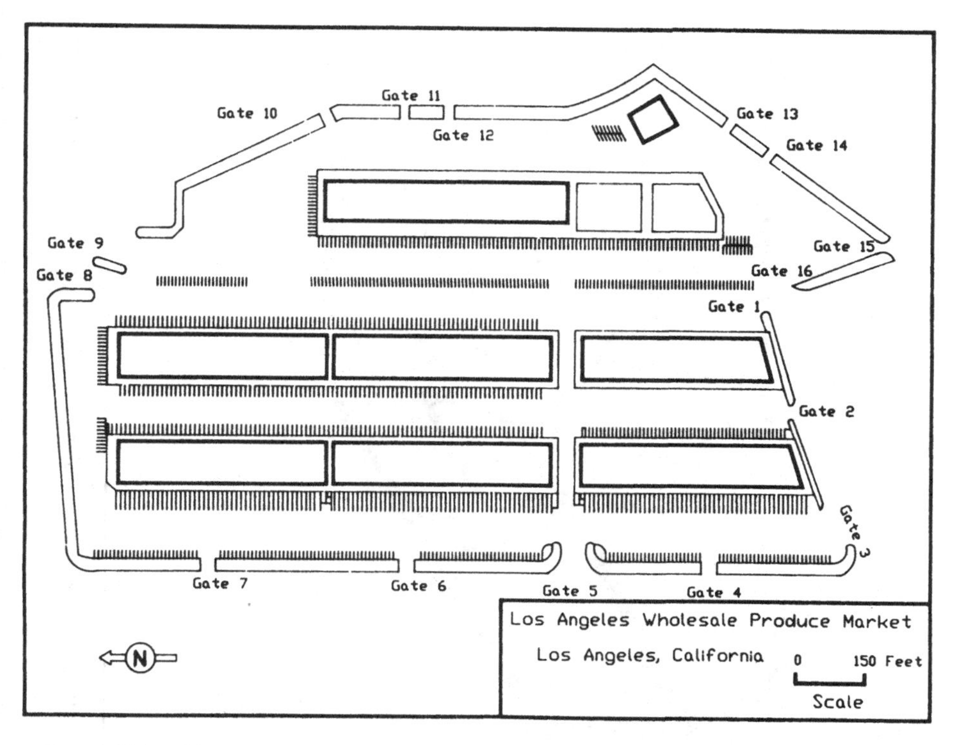

Figure 18.1. Diagram of Los Angeles Produce Wholesale Market. (Permission to use this diagram granted by L.A. Wholesale Produce Market.)

as receiving from nearby and distant areas. It is jointly owned by a real estate firm and the wholesalers of the market. When the new facility opened larger wholesalers in the area quickly moved into it, and their places on the two older markets were taken up by smaller wholesalers and jobbers. The layout of the market and the design of the facilities reflect the needs of the trade to handle fresh produce efficiently into and out of the market (Figure 18.1, Figure 18.2). Wholesalers in Los Angeles, by reason of their proximity to southern California growing areas, ship considerable produce, speciality crops in particular, to distant markets, as well as serving local markets.

New York City

Even with all the changes and uncertainties facing the industry following World War II the conditions on the Washington Street Market were finally

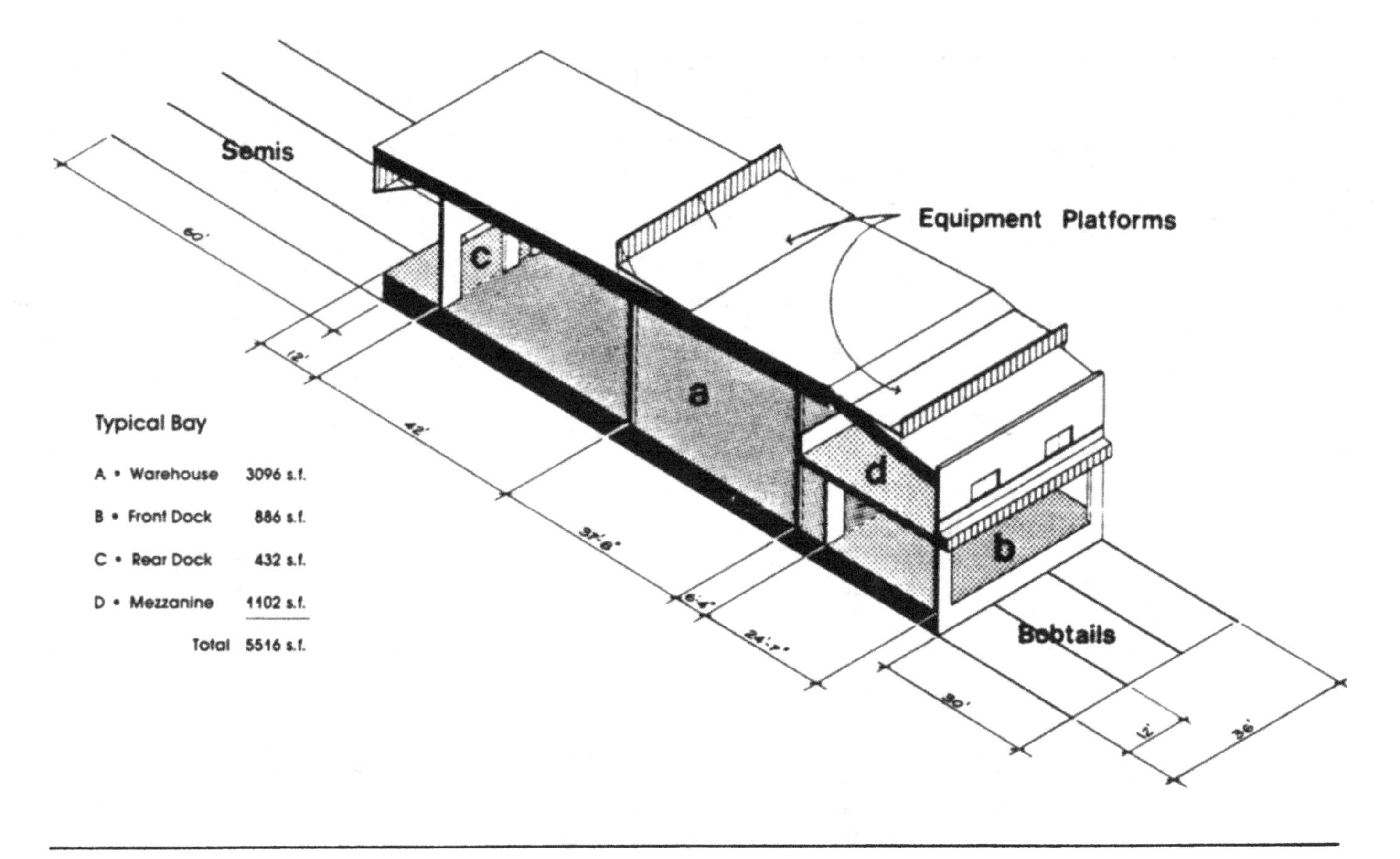

Figure 18.2 Diagram of Typical Bay on Los Angeles Wholesale Produce Market. (Permission to use this diagram granted by L.A. Wholesale Produce Market.)

considered intolerable. The Hunt's Point location for a new market was chosen by the City of New York, and was finally opened in March 1967 after the City required occupants of the Washington Street Market to vacate their premises.

The Hunt's Point Market is typical of many designed by the U.S. Department of Agriculture, although of a larger scale. The original market consisted of 252 stalls arranged side by side along four fingerlike buildings each 1,752 feet long and running approximately north and south. The buildings are separated by 200-foot streets which can easily handle trailer trucks. Office space is provided on the second floors of these buildings, which are connected at the southern end by a second-story bridge. In this bridge are offices of organizations that serve the market, such as the U.S. Department of Agriculture, the Fruit and Produce Association, and the offices of the market manager.

Each stall or store unit is 25 feet wide and 60 feet long, the same size as the old Washington Street market stalls. But at Hunt's Point each stall has a 15-foot rail platform at one end and a 25-foot-wide truck platform at the other. The platforms are at truck- and rail-bed levels, or about 5 feet high. The platforms are covered by marquees to facilitate use in all weather. Ceilings in each stall are 30 feet high to permit palletized stacking.

New York is not only a major population center but owes its dominance in part to being a port of entry. Imports, much of which are reshipped to other markets, account for over 20 percent of total arrivals.

Boston

Boston, like New York City, had several clusters of fruit and vegetable wholesalers, one being at Faneuil Hall and another in South Boston. In 1965 the Boston Redevelopment Authority took over the Faneuil Hall area and the wholesalers were forced to leave. The Faneuil Hall merchants had seen the writing on the wall, however, and with a few wholesalers from the South Boston area they formed a corporation to build a new market. In 1968 they established the New England Produce Center (NEPC) with 128 bays, at Chelsea, Massachusetts. The remaining seven wholesalers in South Boston who had been reluctant to join the NEPC finally also decided to relocate. They established the Boston Market Terminal on 20 acres adjacent to the NEPC and opened there in 1969. The two markets have continued to operate separately. The New England Produce Center had about 50 tenants in 1986 (NAPPM, 1986).

Chicago

The South Water Market, Chicago's main terminal market, has been in the same location for many years. The present facilities were opened in 1925, well before the advent of tractor–trailers and fork lift trucks. There are 166 privately owned units in six long and narrow buildings, occupied by about 100 different tenants. There has been continuing discussion about the possibility of building new market facilities, but it has not been possible to muster support from all the various parties involved. In spite of the antiquated facilities the market continues to handle a substantial volume of produce.

Chicago is a rail center through which much produce passes on its way to other markets. Over half the volume of reported arrivals in Chicago have come by rail, a significant proportion by TOFC. This is much higher than other markets, except for New York.

CURRENT OPERATIONS ON TERMINAL MARKETS

The physical operations involved in wholesaling fresh fruits and vegetables on terminal markets are still in a state of change. The integrated wholesale–retail operations, mainly corporate and voluntary chains, have adjusted fairly well to changing technology and market demands. But many nonintegrated independent wholesalers, especially those on terminal markets, have not been able to take full advantage of modern methods of transporting and handling fresh produce. Facilities designed and built as late as 20 years ago, let alone prior to World War II, have already become obsolete. Uncertainty with respect to the future of the independent wholesaler and the advantages and limitations of clustering on organized markets delays action.

Continuing study by the U.S. Department of Agriculture reveals substantial savings that might be made through providing new facilities on consolidated and organized markets, or rebuilding existing markets. Among recent studies are those recommending new wholesale produce facilities for Dallas in view of proposed urban development in that city's existing wholesale market, a food distribution center for northern New Jersey that would bring together wholesalers handling all kinds of foods, and a relocation and modernization of the Niagara Frontier Terminal Market built in Buffalo in 1938. These document the same conditions that plagued the Washington Street Market for many years, namely inability to properly deal with modern tractor-trailers or to store and handle pallet loads of perishables efficiently.

Firms in the industry have changed, as well as what they do. Direct buying on the part of retail chains and some food service operations has reduced the business of independent wholesalers. What packing of produce in consumer packages is still done is largely performed at shipping point, and modern methods of ripening and handling tomatoes has reduced but not yet eliminated the need for specialized tomato ripening firms on terminal markets. When the Hunt's Point Market was opened in New York, bananas were still arriving in this country on stalks weighing up to 100 pounds each rather than in boxes as they do today, and banana ripening has become more specialized and technologically advanced.

On the other hand some independent wholesalers have been able to take advantage of the expansion of the food service industry to supply full service restaurants, fast food chains, and institutions. The increase in international trade has provided added business for new and existing export-import firms. The introduction and expanded use of exotic and unusual products has led to new opportunities for diversified as well as specialized wholesalers. With so many products available from so many places the produce brokers have flourished.

Not all firms have benefited equally. Most types of firms have declined sharply in number in recent years as many wholesalers, often family owned and operated, hung up the telephone and closed the door. The remaining businesses have in general grown larger, handle more commodities, and provide more services than before.

The adjustments have been especially severe in the smaller markets. Wholesalers in the larger markets have been able to service greater areas and so maintain their volume. On smaller markets, often those serving less than 1 million people, only one or two primary receivers do most of the business, along with a few jobbers who specialize in certain commodities and services.

On large markets like New York and Chicago the terminal wholesalers continue to play a significant role. Hunt's Point is still a very vital market in spite of certain limitations. The market is in a very depressed and dangerous part of the city not convenient for either truck or rail deliveries and shipments. Security and sanitation have been deficient. Many receivers operate several units now, one as many as 17, and could make good use of specially designed single-occupancy buildings. Volume on the fruit auctions declined and they were closed. But additional facilities have been built and are being fully used, and the ownership and operation was recently transferred from the City of New York to a tenants' cooperative that promises to improve security and sanitation.

Types of Wholesalers

Many different types of wholesalers operate at major markets, some on central terminal facilities and others scattered through the metropolitan area. The wholesale operations integrated with retailing or food service generally are located in single occupancy buildings away from the terminal area. Arrival data do not indicate how much or what proportion of the fruits and vegetables received in the market area physically passes through the terminal facility, but probably ranges from one-third to one half. Some types of wholesalers operate primarily on the terminal market, some types both on the market and elsewhere, and some primarily away from the market (Table 18.1).

Receivers, jobbers, and commission merchants were the most numerous type of firm according to primary function in both markets, numbering 98 out of the 174 firms listed in New York City and 75 out of 127 firms in Chicago. Most firms operating primarily under one of these categories probably performed functions in the other two categories also. Receivers handle some produce on consignment just as commission merchants sometimes buy outright. Some receivers function as full-service wholesalers. Others specialize in certain products such as bananas, or in certain customers such as restaurants and institutions.

There were 39 importers or exporters in New York but none in Chicago. Some of these are trading companies that simply arrange sales. Governments of other countries have offices in New York that serve as exporters or importers.

Brokers and buying brokers were more numerous in Chicago than New York, as had been found in the earlier study. Brokers normally represent the sellers, but many also serve as buying brokers. The brokerage business has apparently expanded with the increase in the number of commodities handled and the sources of supply.

Table 18.1. Number and Type of Fresh Fruit and Vegetable Wholesalers in Chicago and New York City By Primary Function, 1986.

	CHICAGO			NEW YORK CITY		
PRIMARY FUNCTION	SOUTH WATER ST.	OTHER	TOTAL	HUNT'S POINT	OTHER	TOTAL
Receiver	17	19	36	34	21	55
Jobber	18	13	31	8	14	22
Commission Merchant	8	0	8	21	0	21
Importer	0	0	0	11	17	28
Exporter	0	0	0	2	9	11
Shipper	1	6	7	2	4	6
Broker	3	12	15	7	3	10
Buying Broker	1	1	2	4	0	4
Repacker	0	0	0	4	1	5
Branch Sales	0	4	4	1	2	3
Other	5	19	24	0	9	9
Totals	53	74	127	94	80	174

Seven shippers had offices and 4 marketing firms had branch sales offices in Chicago, one more in each category than the number of comparable firms in New York. Shippers and sales offices appear to play a more important role in terminal markets than formerly. As well as arranging sales they can advise and supervise merchandising practices. The largest firm on the Hunt's Point Market is a division of a shipper in Salinas, California, that also has an office on the Boston Market.

Repacking was the primary function for five firms in New York but for none in Chicago, although several firms included repacking as one of their secondary functions. The repacking of potatoes, onions, and other fruits and vegetables into consumer packages and ripening and repacking tomatoes has declined in importance, but the preparation of salad vegetables and fruits for salad bars in restaurants and retail stores has increased and several firms now specialize in providing this service.

Other functions in each market included processor, pickler, wholesale grocery, retail chain, and corporate headquarters.

Over half the wholesalers listed for New York City proper operated on the Hunt's Point Market while about 40 percent of those in Chicago were located on the South Water Street Market. In the New York area a few wholesalers are also clustered on the Bronx and Brooklyn Terminal Markets, and in Chicago on the Randolph and South State Street areas. All the commission merchants and buying brokers listed in New York operated at Hunt's Point, as did all Chicago commission merchants at South Water Street. Offices of shippers and marketing organizations tended to be located off the terminal markets in both cities. Most of the brokers were on the organized market in New York but at other locations in Chicago.

DESTINATION MARKET WHOLESALERS' BUSINESS PRACTICES

Wholesalers have many different ways of procuring supplies. Each type of firm combines the various buying methods in different proportions. Direct purchase by phone or wire from shipping point is a most common method for many receivers, as is receipts by consignment. Some produce is bought through shipping-point brokers, or by the wholesaler's own buyer. Commission merchants and wholesalers, of course, rely heavily on consignment, wholesale grocers and receiver–jobbers buy by phone, service jobbers and purveyors use shipping point buying brokers, and service wholesalers, prepackagers, and tomato repackers depend heavily on their own buyers.

When the wholesaler takes title to the produce the usual terms are f.o.b. shipping point (suitable shipping condition) for much of their purchases. Firms occasionally use delivered sale, and a few purchases are made using f.o.b. acceptance.

Many wholesalers have a group of shippers from whom they buy regularly. The vast majority of all direct purchases from shipping point are made by phone. Usually the shipper calls the wholesaler, except when the product is in short supply. First contact by the shipper is more likely for those wholesalers who deal with a group of regular shippers than for wholesalers without a regular source of supply. Prices are often determined without bargaining, but frequently there is some dickering. Promptness of payment varies. In general larger firms have been found to make payment more promptly than smaller ones. A few years ago approximately one-fourth of all purchases were made through brokers, but the proportion may be higher today.

Wholesalers buy at least some commodities on the basis of U.S. grades when buying direct from shipping point. Specific commodities bought on grades most frequently include potatoes, tomatoes, and peaches. Few wholesalers have their own specifications, but rely heavily on the shipper's or packer's brand.

The U.S. Department of Agriculture found a few years ago that primary wholesale handlers operated on a gross margin that averaged about 13 percent of sales. Net income amounted to about 1 percent of sales. Secondary receivers had a gross margin averaging 18 percent of sales, partly due to the margins of 25 percent of sales obtained by prepackagers and tomato repackers. Their net income averaged 2 percent of sales. Salaries and wages for both groups accounted for about half of all costs.

Price making was considered to be a competitive supply-and-demand process that took into account certain objective facts on existing prices, supplies, and weather conditions, but was complicated by imperfect information especially about quality. There were several different price-making systems. The street sale is the most prevalent method of selling in larger wholesale markets. It is found wherever wholesalers' places of business are concentrated in a small area. Wholesalers display samples of their offerings in front of their places of business. Buyers who wish to compare prices and quality walk the market to shop some or all of the wholesalers before making actual purchases. The street sale sets the tone of the market, even though some buyers do all of their business by telephone. Wholesalers tend to conceal the extent of their supplies, and buyers their immediate needs. Prices are determined by the interplay of buyer and seller.

The Chicago carlot track sale was a nearly unique marketing institution. Rail cars, primarily of potatoes but also onions and lettuce, were spotted on the Wood Street railroad yard. There they were opened by receivers for inspection by potential buyers, and after the contents were sold were resealed and shipped to final destination.

The terminal sale is a system used on the Boston Chelsea Market. Samples are taken from cars or trucks and placed on the terminal floor. All buyers are admitted at the opening hour. Most buyers make a circuit of the floor before bargaining with salesmen and placing orders. Usually deliveries are made directly from the car or truck.

In many medium-sized and smaller markets wholesalers are physically dispersed. In such markets prices are typically announced by the sellers; some of them issue price lists. Buyers shop around either in person or on the basis of the price lists. Prices may be adjusted on the basis of sales, although the adjustment is not as prompt as on street markets.

Market News Service reports on prices and volume of shipments from major shipping point markets as well as arrivals and prices at major wholesale markets are used for general information, to follow the market, or occasionally to set prices when settling claims for spoiled or damaged merchandise.

Activity on the Hunt's Point market starts before midnight when buyers from as far away as Boston or Philadelphia may begin filling their needs. Wholesalers have already set out samples of their wares on the truck dock for buyers to inspect before making their purchases. Regional wholesalers and purveyors serving the independent stores and institutional trade start arriving about 1 A.M. Around 3 A.M. some chain store buyers may come to check prices and quality and make fill-in purchases. After 5 A.M. the market is flooded with buyers for specialty fruit and vegetable stores, most of them operated by Korean immigrants. By 8 or 9 A.M. most of the activity is over for the day. Shortly after that the Market News reporter distributes reports of prices on the market that day as well as on other markets across the country and shipments from growing areas. This and other information picked up by the local grapevine will provide the basis for future marketing decisions.

FUTURE PROSPECTS

The imminent demise of the terminal market facility has been forecast for many years, but refuses to happen. The ability of some wholesalers to adapt to changes in the market has rendered this forecast false. The great increase in numbers of produce items, the distance many of them come to market, the proliferation of food outlets, and the high standards demanded in the marketplace are some of the reasons destination wholesalers have survived and in some cases prospered. They have done this by recognizing the needs of their customers, and adapting their businesses accordingly. Many firms have diversified their product line and added new services.

The need to cluster in centralized market facilities does not seem as essential as it once was. Many destination wholesalers in smaller communities have built facilities designed to meet their needs with large coolers and banana ripening rooms adjoining extensive warehousing space. But in the major cities there is still some advantage in being close together. Whether this is because of the need to keep in touch with market developments or serve customers better is hard to say. Certainly it facilitates market reporting. Whatever the need it does appear as though centralized market facilities will continue to serve a useful purpose.

REFERENCES

Manchester, Alden C., 1964. *The Structure of Wholesale Produce Markets.* U.S. Department of Agriculture, Agricultural Economic Report No. 45.

National Commission on Food Marketing, 1966. *Organization and Competition in the Fruit and Vegetable Industry.* Technical Study No. 4 Washington D.C.

National Association of Produce Market Managers (NAPPM), 1986. *Green Book Produce Market Information Directory.* 1354 Rutherford Road, Greenville SC 29609.

Rasmussen, M. P., 1929. *Some Facts Concerning the Distribution of Fruits and Vegetables by Wholesalers and Jobbers in Large Terminal Markets.* Bulletin 404, Cornell University Agricultural Experiment Station.

Taylor, Earl G., Bruce E. Lederer, Jesse W. Goble, Marvin D. Volz, James J. Karitas, James N. Morris, Charles F. Stewart, and John C. Bouma, 1972. *Los Angeles Wholesale Food Distribution Facilities.* U.S. Department of Agriculture, Marketing Research Report No. 966.

U.S. Department of Agriculture, Agricultural Marketing Service, 1960. *Improving Market Facilities in New York City for Wholesaling Fresh Fruits and Vegetables.* Marketing Bulletin No. 6.

Chapter 19

Food Retailers and Retailing

INTRODUCTION

Retailing, in terms of costs, is the most important function in marketing the fresh fruits and vegetables consumers purchase for home consumption. Retailing charges account for about one-third of the amount consumers pay for these commodities, or about 40 percent of the total marketing margin.

The bulk of the retail food business, for fresh produce as well as other foods, is done through supermarkets that now come in several different kinds and sizes. A significant share, however, now moves through other types of stores and these outlets may be important for fresh fruits and vegetables in the future. There has been a gradual change in retail food stores in recent years, not only in their number and size but also in the types of stores and in the products and services they offer. Although the retail food business is dominated by companies that each own and operate many stores, there are still many small firms and individuals that own less than 10 stores and are affiliated with a wholesaler through either a cooperative or a franchise type arrangement.

Retailing is a very specialized operation. Each individual supermarket is a substantial business, often employing several hundred people. Stores are organized into departments, and purchasing and merchandising practices are carefully planned. Measures of performance have been developed by management to keep control of operations.

Specialty produce stores have sprung up in many areas in the past 15 years. These have found a niche in certain markets where they provide a significant outlet for fresh fruits and vegetables, especially items that might be classed as unusual, exotic, tropical, or ethnic.

FOOD STORE NUMBERS, SALES, AND OWNERSHIP

Food Store Numbers and Sales

Food stores are the major source of food for consumers to eat at home, and in 1987 there were an estimated 235,800 such stores with annual sales of $314 billion (Table 19.1). The 26,500 supermarkets constituted 11.5 percent of the total but these stores, with average sales of $8.3 million, did close to 70 percent of the total food store business. Supermarkets were originally defined in 1972 as food stores with more than $1 million in annual sales. Due to inflation this definition has been revised to include only stores with sales of more than $2.5 million in 1985 dollars. Convenience stores were 21 percent of the total number and did 11 percent of the business on average sales of $718,000 annually. Superettes, sometimes referred to as Mom and Pop stores, were still the most numerous with 38 percent of the total number, but did only 13 percent of sales. Specialized food stores had the lowest average sales equal to about $250,000 a year. They did less than 6 percent of the business, but constituted about 30 percent of the number.

The number of supermarkets in this country and their share of the total food business grew rapidly in the post-World War II period until about the mid-1970s (O'Rourke, 1982). In 1977 the number of supermarkets reached a peak of 30,800, and their share of total food store sales topped 70 percent. Since the late 1970s the number of supermarkets has hovered close to 27,000 and their share of food store sales continued at about 70 percent. Sales per supermarket in current dollars have continued to increase, rising from $2.3 million in 1972 to $8.8 million in 1987.

Supermarket Format

The larger food stores with current sales of more than $2.5 million annually (adjusted for inflation) at one time were very similar. Today large food

Table 19.1. Number and Sales of Food Stores by Type, 1987.

	NUMBER OF STORES		SALES		SALES
	TOTAL	PERCENT	TOTAL	PERCENT	PER STORE
	thousands	%	billion $	%	thousand $
Grocery Stores					
Supermarkets	26.5	11.3	$219.2	69.8	$8,272
Convenience Stores	50.0	21.2	35.9	11.4	718
Superettes	89.3	37.9	40.9	13.0	458
Specialized Food Stores	69.9	29.6	18.2	5.8	260
Total Food Stores	235.8	100.0	$314.3	100.0	$1,333

Source: Adapted from USDA ERS, 1989.

stores come in several different types and sizes, and the proportions in each format are constantly changing (Table 19.2). The conventional supermarket is self-service, ranges in size from 3,000 to 30,000 square feet, and has the highest percentage of food versus nonfood items. About 60 percent of the supermarkets fell in this category in 1988 with 43 percent of the total supermarket sales. One trade paper reported that in 1987 the average supermarket had a selling area of 23,775 square feet, stocked 16,500 different items with an inventory value of $556,000. These stores had an average of 8 checkouts and 3,000 square feet per checkout, and 62 percent of these markets were equipped with electronic scanners.

Superstores are larger than conventional supermarkets, with an average size of 35,000 square feet (USDA ERS, 1989). They carry some prescription drugs, have generic and specialty product areas, and some self-service bulk foods. These were just over one-fifth of the total, but had more than 30 percent of the sales. Warehouse/limited assortment stores have strong price appeal, stock a limited number of product brands and only the most popular sizes, and often cover less than 10,000 square feet. About 12.5 percent of the supermarkets were in this category with the same percentage of sales.

Combination food and drug stores ranged in size from 35,000 to 45,000 square feet, and had more product variety than conventional supermarkets. Their nonfood items amounted to 25 to 35 percent of sales, and they carried prescription drugs as well as nonprescription drugs and general merchandise. In 1988 they had average sales estimated at $16.5 million per year. They constituted 4.6 percent of the number of supermarkets, but had 8.6 percent of the sales.

Superwarehouse stores were from 50,000 to 140,000 square feet in size, carried primarily food, some health and beauty aids, had a low general

Table 19.2 Supermarket Numbers and Percent of Sales By Format, 1988

FORMAT	NUMBER		SALES		SALES PER STORE
	TOTAL	PERCENT	TOTAL	PERCENT	
	thousand	%	billion $	%	thousand $
Conventional	15.7	60.3	$98.7	42.8	$6,287
Superstore	5.4	20.8	69.5	30.1	12,870
Warehouse/Limited Assortment	3.3	12.5	28.8	12.5	8,727
Combination Food and Drug	1.2	4.6	19.8	8.6	16,500
Superwarehouse	0.2	1.4	9.0	3.9	22,500
Hypermarket	0.1	0.4	5.1	2.2	51,000
All formats	26.1	100.0	$230.9	100.0	$8,847

Source: Adapted from USDA ERS, 1989.

merchandise emphasis, and had strong price appeal. About 1.4 percent of supermarkets fell in this category, and with average annual sales of $22.5 million they together accounted for about 3 percent of all supermarket sales. The hypermarket, a relatively new development introduced from Europe, is a huge one-stop shopping supermarket that brings together a broad variety of food and nonfood products in a single store, and may range up to 200,000 square feet in size. There were about 100 of them in this country in 1988 with annual sales averaging over $50 million.

Supermarket Ownership and Organization

Supermarkets may be owned in large numbers by major companies, or in smaller groups or ones or twos by individuals or family corporations. Supermarket chains are generally defined as having 10 or more stores under the same ownership. Individuals or corporations that own fewer than 10 stores often voluntarily either band together cooperatively to obtain their merchandise or become affiliated with a wholesaler. The cooperative or the wholesaler may in these cases also provide management and merchandising advice. Few remain completely independent. Stores owned by both corporate chains and voluntary chains and independents range in size, with corporate chains tending to have more larger stores.

Most chain supermarkets are operated by large publicly owned corporations such as Kroger, Safeway, American Stores, and Winn-Dixie. In 1986 the sales of the 20 largest food retailers reached $106.7 billion, or 38.3 percent of total grocery store sales. In many markets, however, the four largest retailers do a much higher share of the business. These large retailers also operate wholesale distribution centers, often one to every 40 to 100 stores. Some of these companies are U.S. owned and operate in other countries, and several that operate in this country are now foreign owned.

Many of the smaller chains are privately held corporations. Most operate their own distribution center or centers, but some contract with independent wholesalers for fresh produce.

Some supermarket corporate chains also service stores owned individually or in small numbers through an affiliated or franchise type arrangement. These stores in most cases, though, are serviced by a cooperatively owned wholesale operation or through affiliation with an independent or nonintegrated wholesaler. Corporate chain retailers may also supply food service operations like hospitals, universities, and hotels.

Convenience Stores

In the past 10 years the only major type of food store to increase in numbers has been the convenience store. Convenience stores have been defined as small grocery stores selling a limited variety of food and nonfood products, and typically open extended hours. Many operate as chains. These increased in number by about 50 percent between 1977 and 1987, a time when supermarkets, Mom and Pop stores, and specialized food stores were all declining in number. Sales increased by 4 times in current dollars, while supermarket sales doubled, Mom and Pop stores increased by one-third, and sales of specialized food stores went up 75 percent. The growth in sales now seems to have leveled out. Many convenience stores are located in self-service gasoline stations and a large proportion of their sales are in nonfood items.

Convenience stores are distinguished from Mom and Pop stores by presenting a standard format as part of a chain (Leed, 1983). Convenience store chains are operated by gasoline companies such as Atlantic Richfield or food companies such as Southland Corporation that operates 7-Eleven Stores. The produce they carry is generally limited to a few staples like potatoes, onions, lettuce, tomatoes, apples, oranges and bananas. Both their cost of goods and their costs of operations per item handled are substantially above those of supermarkets, so their prices are correspondingly higher.

SUPERMARKET PERFORMANCE

Operating Results

Supermarkets, like other retail stores, operate on a gross margin which is the difference between gross sales revenue and the cost of goods sold. For example, a store with annual sales of $10 million might have spent $8 million for the goods it sold. The gross margin in this case would be $2 million, or 20 percent of sales. The $2 million is what is available to pay the employees, the rent, the utilities, taxes, and other expenses including interest on borrowed money, and hopefully to have something left over as profit for the owners. For comparison and analysis the gross margin in dollars is usually expressed as a percent of sales.

Retail supermarket gross margins vary from chain to chain and store to store. In 1986–87 a representative group of 32 supermarket chains had an average gross margin of 24.3 percent of sales (Table 19.3). Labor or payroll expense was by far the largest item, accounting for more than half the total. Profits after taxes averaged 1.45 percent of sales. Among this group those in the upper quarter in terms of profitability had slightly higher gross margins but significantly lower total expenses per dollar of sales. In contrast, those in the lower profitability range had about average expenses per dollar of sales but significantly lower gross margins. Margins and profits can vary widely from year to year and by type of store. Warehouse or economy stores can operate on a slightly lower gross margin and may get down to 18 percent or so, saving the customer 4 cents on the dollar. Convenience stores require a larger gross margin to recover expenses and return a profit, and may operate on margins of 30 to 40 percent.

Supermarkets are organized into departments according to the types of goods sold. Sales and margins vary from department to department. Management sets prices to obtain certain gross margins, recognizing that the actual realized gross margin will fall short of the margins anticipated on the basis of markup on cost because of losses or price reductions due to damage in handling, deterioration in quality, pilferage, or changes in market conditions. Potatoes bought at 50 cents a bag and priced at $1.00 (100 percent markup) will not normally return a 50 percent gross margin.

The proportion of food and nonfood sales in supermarkets varies considerably. On the average food sales accounted for 75 percent of total sales in 1988. Produce departments tend to have the highest gross margins among the food sections except for perishables such as the service deli and instore bakery (Table 19.4). Dry grocery and meat departments generally have the lowest. Each department's share of total food sales also varies a great deal, with produce generally accounting for about 12 percent of food and 8 or 9 percent of total store sales. Although produce sales are lower than those of

Table 19.3. Supermarket Gross Margins, Expenses, and Earnings, as Percent of Sales, 32 Retail Chains, 1986–87.

		PROFITABILITY RANGE	
	AVERAGE ALL FIRMS	LOWER QUARTER	UPPER QUARTER
Gross Margin	24.33	22.61	24.66
Expenses			
Payroll	12.67	12.66	12.09
Property (rentals, depreciation, repairs)	4.01	4.12	3.96
Supplies	0.88	0.99	0.86
Utilities	1.24	1.49	1.47
Other	3.61	3.20	2.84
Total	22.41	22.46	21.22
Net Operating Profit	1.93	0.15	3.43
Other Income (cash allowances, etc.)	0.76	0.31	0.73
Total Net Income	2.69	0.46	4.16
Income Taxes	1.24	0.20	2.11
Net Earnings after Taxes	1.45	0.26	2.32

Source: Adapted from McLaughlin, Hawkes, 1987.

Table 19.4. Supermarket Food Department Performance, 1988.

DEPARTMENT	AVERAGE PERCENT GROSS MARGIN	PERCENT OF FOOD SALES	PERCENT OF FOOD $ GROSS MARGIN
		percent	
Meat, Fish, and Poultry Fresh and Cured	18.9	23.2	18.5
Produce	32.8	11.9	16.4
Dairy	20.8	8.2	7.2
Frozen Foods	24.5	7.1	7.3
Other Perishables Deli, Baked Goods, etc.	39.0	13.3	22.0
Dry Grocery(Food)	19.2	36.3	28.6
Total Foods	23.6	100.0	100.0

Source: Supermarket Business, 1989. (Permission granted)

other major departments the high gross margin tends to result in total dollar gross margins comparable to departments with greater sales.

The differences in department margins probably reflect differences in costs, or at least management's concept of cost. In retail store operations as in many other businesses some costs can be directly identified with products or departments, but some costs have to be arbitrarily allocated. The wages of produce department employees and the costs of packing supplies can be charged directly, but there are many overhead or general store costs such as rent, utilities, manager's salary, and checkout or frontend operation that have to be allocated in some way. Allocating rent on the basis of space used, and checkout expense on the basis of department sales, may seem reasonable but is arbitrary and may not reflect costs. The low margins on meat are probably due to the high dollar volume relative to the space required and number of employees, plus the fact that meat is highly competitive and narrow margins tend to draw shoppers. The produce department takes up a lot of space per dollar of sales and requires a lot of labor compared to groceries. Traditionally the produce department has had high margins, but some firms are recognizing the drawing power of the department, and shaving margins hoping to pick up sales in other departments.

Retail Produce Pricing and Margins

Pricing produce at retail is truly an art, and not just a mechanical process of marking up a fixed percentage over cost. Retailing costs and margins play a part, but there is also the psychology of pricing on the 9s or other odd numbers as well as pricing in multiples. Having to take into account competition, merchandising strategy, the drawing power of certain specials, and maintaining store image are what makes the process an inexact science.

Retailers may follow different paths to reach to the same conclusion—that the dollar margins over cost of goods sold must be sufficient to pay expenses and provide an adequate return on investment. If one could forecast sales and expenses one could simply calculate margins as a percent of sales and then determine the markup necessary to produce those margins. In fact, however, margins and sales are jointly determined and so adjustments must be made continually to arrive at the desired results.

The aim for produce departments in most supermarkets is to operate on a realized gross margin of from 30 to 50 percent. This would imply a markup of 50 percent over cost, providing there were no markdowns or losses. There is bound to be a certain amount of "shrink," or difference between markup over cost and realized gross margin. So allowance for shrink as well as a gross margin target must be considered when setting retail prices.

There are still significant differences in sales of particular fruits and vegetables across the country, and produce margins do vary considerably from store to store and week to week for the same items. A survey in 1988 by the Food Marketing Institute (FMI) and the Produce Marketing Association (PMA) found the median or mid-50 percent value to be most valid since other measures were biased by the fact that retailers often use key commodities in promotions at no gross margin. Using this measure gross margins varied for major commodities from about 25 percent for strawberries to over 40 percent of tomatoes and cucumbers in both conventional supermarkets and superstores (Table 19.5).

Commodities with the lower gross margins tended to be specialty items like strawberries, mangoes, mushrooms, and cantaloupes which could have higher shrink but also which tend to be bought on impulse and consequently have a relatively elastic demand. Commodities with higher gross margins like apples, potatoes, tomatoes, and cucumbers are, with the possible exception of tomatoes regarded as hardware and are more staple items with a more inelastic demand. The average realized gross margin for potatoes and apples and other staples of 40 percent or more is often the result of using a standard markup of 75 to 100 percent, and then putting the item on special every few weeks.

On the average the gross margin on produce in conventional supermarkets of 32 percent was 1.5 percent higher than in superstores. The larger stores may have been able to achieve greater economies because of their size, or may just have had different pricing strategies. Retailers do have

Table 19.5. Conventional Supermarket and Superstore Gross Profit Margins on Selected Fresh Fruit and Vegetables, 1988.

ITEM	CONVENTIONAL SUPERMARKETS	SUPERSTORES
	percent	
Strawberries	26.5	25.5
Bananas	30.9	30.6
Mangoes	34.6	30.8
Mushrooms	33.4	32.8
Cantaloupes	35.0	34.3
Watermelon	35.2	35.5
Lettuce	37.9	38.2
Apples	39.2	39.5
Potatoes	39.9	39.6
Tomatoes	40.8	41.2
Cucumbers	40.7	41.7
All Produce	32.1	30.6

Source: Adapted from Tomes, 1989.

different pricing strategies. There did seem to be a consistent difference between conventional supermarkets and superstores in the way individual items were marked up. Superstores had lower gross margins on the perishable impulse items and larger margins on the staple demand items.

In many supermarket chains the order form, or at least the invoice accompanying the shipment, specifies not only the merchandise cost but also the suggested retail price and expected gross margin. The invoice accompanying the order may also contain a summary of the total cost of the shipment item by item, the suggested retail price, the corresponding retail value, an allowance for shrink, and the anticipated gross margin. Actual sales revenue will be used to determine the realized gross margin both in dollars and as a percent of sales.

EFFICIENCY IN STORE AND PRODUCE DEPARTMENT OPERATION

The accurate determination of the profitability of store and department operation requires taking inventory and allocating all costs and revenues, can be a tedious and time-consuming process, and may only be done on a quarterly basis. Food stores operate on a weekly cycle, however, in a highly competitive business and a lot can change in three months. For purposes of management control and evaluation there are a series of factors that can be easily calculated that retail management tend to watch to monitor profitability and detect change or the need for remedial action. These performance indicators are used to detect problems or identify opportunities for improvement, by making comparisons between this week and last week or the same week last year for individual stores, or between stores in the same organization.

Data used to develop performance indicators in addition to daily and weekly sales by department are measures like cost of goods sold, customer count, and hours of hired labor or number of full-time equivalent employees by department. Knowledge of store characteristics such as the square feet of selling area, the hours of operation, and the number of checkouts enables the calculation of indicators of store performance such as weekly sales per checkout, sales per employee hour, total gross margin, store sales per hour, sales per square foot of selling area, and average transaction per customer. Produce department performance indicators can also be calculated such as department sales as a percentage of store sales, sales per employee hour, realized gross margin dollars and percentage of sales. All these measures give an interim indication of how well each store is doing in terms of attracting customers, effectively merchandising the various items, and achieving labor efficiency.

Retailers monitor the number of customers attracted to the store as a measure of advertising and past performance, and sales per customer as an indicator of current merchandising practices. Sales per checkout and per square foot of selling area are measures of the effective use of store facilities, and sales per employee hour an indicator of labor efficiency. In 1987 supermarkets were reported to have average weekly sales of $22,700 per checkout, $88.90 per employee hour, $7.64 per square foot of selling area, $1,498 per hour of operation, and $15.95 per customer transaction. There was considerable variation in these numbers between stores of different sizes and types, and probably even between stores of the same size reflecting differences in market conditions and management.

Department sales as a percent of total store sales and department realized gross margins are key factors in following department performance. Care must be taken not to rely too heavily on any one indicator, however, for this might be to the detriment of the business.

Electronic scanning equipment has been introduced in many supermarkets. In 1987 62 percent of supermarkets had scanners installed at the checkout (USDA ERS, 1989). When a package with a Universal Product Code (UPC) printed on it is passed over the scanner the code number is read and transmitted to a computer in the store where the name of the product and the price is identified, added to the customer purchase, and recorded on the sales slip. Price changes can be made easily by simply changing the price in the computer. Stores with scanners can simply put prices on shelf labels rather than marking each item, at considerable savings in labor. Many localities, however, require stores to mark prices on each item, or at least a large percentage of them believing this protects the consumer. The computer hooked to the scanner can also be programmed to provide other information that might be useful in market research on a specified product such as the day of the week and time of day the item was purchased, other related items that were purchased at the same time, and other information.

MERCHANDISING AND HANDLING PRODUCE AT RETAIL

Department Layout

Operating a successful produce department in a retail food store requires considerable knowledge and skill. There are so many different fruits, vegetables, and floral items, and each has specific temperature and moisture requirements for optimum holding conditions. Each product also has its own particular uses and methods of preparation. The great variety of color and texture provides excellent opportunities for creative merchandising.

Produce merchandising starts with the store layout. Produce is usually first in the shopping pattern, perhaps because so much produce is bought on impulse, produce sets the tone for the store, and the quality of the produce department is often what draws customers to a particular store. The layout of the display area is generally similar whether the customer traffic pattern is clockwise or counterclockwise. Many produce departments consist of a wide aisle lined with display cases, many refrigerated. In between will be islands of produce items in display cases, pallet boxes, or on other structures. Like items are often grouped together, such as salad vegetables, soft fruits, citrus, and apples. Staple or demand items like potatoes are often placed so that customers must move past impulse items like avocados to reach them.

Retailers are introducing new layouts in larger stores, Wegmans, a regional chain in upstate New York, has opened several new supercombination stores that feature the produce departments. Customers entering a 100,000-square-foot store in Greece, New York, are greeted with an impressive display of fresh fruits and vegetables (Figure 19.1, bottom left). Displays are grouped around preparation areas where clerks can observe what is going on and to whom customers can direct their questions. Produce is part of a section of the store featuring fresh foods that also includes a deli-

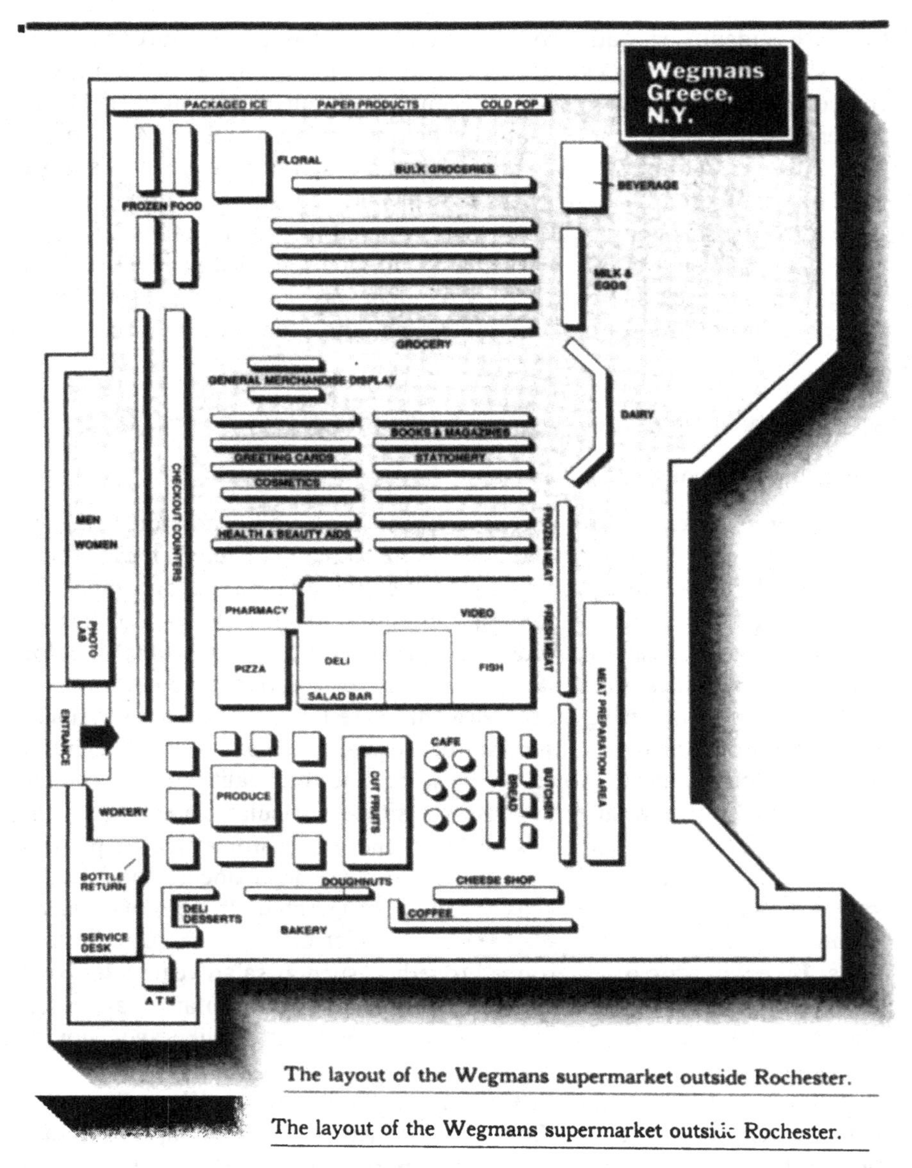

The layout of the Wegmans supermarket outsiü Rochester.

Figure 19.1 The Layout of Wegman's Supermarket Outside Rochester, New York. (Permission to use this diagram granted by Wegman's Food Markets, Inc.)

catessen shop, bakery, salad bar, cheeses, fish and meat that takes up about one-third of the total store area. A small cafe enables customers to eat something they have bought or sample a free cup of coffee. On their way in to the produce department shoppers pass a small Chinese restaurant called the wokery.

A technique called customer flow analysis is sometimes used to study department layout and introduce improvements (Leed, German, 1985). This consists of plotting on a plan of the department, as inconspicuously as possible, the movement of many customers through the shopping area recording individually the route they took, the merchandise they examined, and the purchases they made. Examination of these flow patterns can often indicate areas that are being overlooked, bottlenecks in the movement, displays that might be made more accessible, and other possible improvements. Time spent observing customer shopping in a self-service operation is seldom wasted.

Space Allocation

The allocation of space to different items is not an exact science. Generally the more space allocated to a product the more sales will result, but as more space is added the additional sales decline. Since space is limited the added space given to one product reduces the space available for other products. In theory space should be allocated to each product so that the additional net revenues resulting from giving any one item a little more space is the same as for all other items. In this case, any change in the space allocated to individual items would result in a decline in department gross margin or net revenue. A similar challenge is the decision on how many and which items to carry in the department, given the increasing number of different items being offered to the trade.

There are several different measures used by management to allocate space and decide what items to carry (Leed, German, 1985). Gross sales is sometimes used but this ignores the costs of goods and the costs of retailing. Gross margin is frequently used but this still ignores the differences in handling costs and shrink between products. The concept of Direct Product Profit (DPP) has been introduced and is being tested by some retailers. This is an attempt to identify all costs, both direct and indirect, associated with marketing each item, to calculate the resulting profit, and use this information to choose items to handle and to allocate space.

One problem with looking at items individually is that it neglects any consideration of the overall effect of the total department on the store image, the attraction it may hold for potential customers, and the effect on sales per customer. Stores often like to feature, for example, certain categories of items like organic produce, salad greens, oriental vegetables, or tropical fruits and must carry many of these at low margins in order to achieve the desired effect.

Merchandising in Bulk Versus Prepackaged

There is a choice in merchandising most products as to whether they are displayed in a package like a bag or an overwrapped tray, or are simply set out in bulk. Even strawberries, which are mostly offered for sale in pint or quart baskets, are sometimes displayed in bulk. The degree to which produce is displayed in bulk or packaged varies considerably from store to store. Some retailers choose to display most of their produce items in bulk, others offer mostly prepackaged, and some take an intermediate position. There are regional differences in the extent of bulk and prepackaged merchandising. On the West Coast much of the produce is displayed in bulk, while in the East there is a higher proportion of produce prepackaged. The practice also varies between communities in many market areas according to consumer preferences.

There are advantages and limitations to both merchandising practices. Much produce can now be bought already prepackaged, but some items still have to be packaged in the back room of the store which adds to costs in terms of labor, space, and supplies. Unless the preparation area is moved out into the store selling area as it has been in some stores the department employees packaging the produce are not available for customer service. Produce packaged and weighed in advance presents no problem for the cashier and speeds the checkout process. With proper rotation of packages shrink is low.

Many customers like bulk displays so they can choose which items to buy and how many. Bulk produce saves packaging expense either at the store or prior to receipt. Provision must be made for weighing and pricing the purchase either in the department or at checkout. If the produce is to be weighed and priced at checkout then the cashier must be able to identify each item, distinguishing cucumbers from zucchini and McIntosh apples from Empires. Cashier training may be minimized when merchandising in bulk by such practices as pricing all red apples the same, all green apples the same, and all yellow apples the same. This can, of course, restrict pricing opportunities.

Research at Cornell a few years ago found that a combination of prepack and bulk displays sold more apples than either method alone, but this dual system is not feasible in many situations. There are profitable produce departments that sell mainly prepackaged produce, but the shift in the past seems to have been toward bulk merchandising. With more branded items, more precut salad greens, more interest in monitoring sales with the UPC, and more households where both spouses work outside the home, the growing interest in nutritional labeling and suggested uses, the need for packaged produce will probably continue and may increase.

Merchandising techniques favored in the trade include building bulk jumble displays, developing color contrasts, and using ribbon displays with mirrors to give a mass effect. Neat and well-positioned price signs give the department a pleasing overall appearance. Nutritional information can also be provided advantageously.

RETAIL STORING AND HANDLING OF PRODUCE

Many retail supermarkets receive shipments of fresh produce from a distribution center two or three times a week, and larger stores may receive a tractor-trailer load daily. Merchandise is usually delivered at night so that displays may be built in time for the morning rush of business. Cartons and crates, possibly stacked on pallets, are offloaded into either a refrigerated holding room or the nonrefrigerated storage area or some into both. The special requirements with respect to temperature and humidity of different products generally cannot be taken into account. Lettuce, tomatoes, and bananas may be held under the same conditions.

Refrigerated display cases are usually provided for perishable produce, and some retailers follow the practice of sprinkling leafy vegetables with a mist from time to time. Care has to be taken not to pile the displays too high and thus impede cooling. Frequent rotation and culling is necessary to preserve quality and reduce spoilage.

Studies of produce loss generally uncover considerable deterioration at the retail level. This is not entirely the result of poor retail handling practices, since deterioration is often only discovered at this stage, and retailing is also the end of the line where deterioration is most likely to occur. There is still, however, considerable opportunity for many retailers to extend the shelf life of their produce and reduce shrink through better handling practices.

SUPERMARKET CHAIN BUYING PRACTICES

Supermarket chains generally have a produce staff located at the distribution center who are charged with buying produce, filling orders from the stores, and setting up pricing and merchandising programs. Chains with more than one distribution center may have a headquarters produce staff located at one of the centers. These offices procure produce in different ways, sometimes at the same time. A few larger chains have buying offices in major production areas such as the Salinas Valley or the San Joaquin Valley. These offices can follow the market and the normal variation in quality for high volume products from these areas.

Supermarket chains obtain a large volume of their produce by buying direct from shippers in distant production areas. While some shopping around is often done to get the feel of the market, most supermarket buyers have established close relationships with their major suppliers over the years. Often they have visited the shippers' packing houses or met with them at industry conventions. Consequently the shipper is usually aware of the buyers' specifications, and the buyer has some knowledge of the shippers' quality compared to the competitors. Written specifications are seldom considered necessary.

Many chains, especially smaller- to medium-sized operations, rely heavily on brokers. A few years ago it was believed that the role of brokers would decline as the number of firms involved in shipping and retailing declined. Yet brokers continue to play an important role in produce procurement, apparently in part at least due to the increased number of items being stocked and the worldwide hunt for good sources of supply. Brokers call suppliers for quality and price, negotiate the terms of sale, arrange the purchase and the transportation, see that settlement is made, and on the infrequent occasions when differences of opinion arise between buyer and seller the broker may help arbitrate disputes.

The practices of retail chains with respect to their use of destination wholesalers and terminal markets also differs. Many chains depend on local wholesalers for product to fill in shortages, and sometimes for small volume specialty items. Some chain buyers will occasionally walk the terminal market to check on prices and quality, and also to compare produce quality from different shippers and growers. They are always on the lookout for reputable new suppliers and additional items with customer appeal.

Supermarket chain buying practices with respect to locally grown produce also differs. Some chains actively purchase local produce in season, featuring it in their merchandising. As with shippers in major producing areas, chains tend to work consistently with the same small group of growers from year to year, occasionally adding a new one as an older one drops out or is let go. Chains that have a reputation for paying favorable prices and settling promptly generally have a waiting list of local growers willing to supply produce. Buyers expect local suppliers to provide consistently high quality, adequate volume, suitable packaging, and prompt service to fill unexpected needs, not necessarily in that order of importance.

Produce, even locally grown items, is generally delivered to the distribution center and assembled into orders for the stores. Produce is often received in mixed loads even from distant shipping areas. Chain management generally prefers that produce moves through the distribution centers so that quality may be checked and costs controlled. Direct store delivery is sometimes authorized for highly perishable items such as sweet corn or strawberries when grown close to the store location, or for bulky items like potatoes when the transportation costs would be substantially reduced. Direct delivery from distant growing areas may be more feasible for larger stores in the future.

Produce management usually tries to ship to stores just what the store has requested, but sometimes the total store orders exceed that amount of the item on hand and the shortage must be allocated equitably among stores. Conversely the total store orders may fall short of the distribution center inventory, and in the case of perishable products an excess allocation is often made to all stores. Supermarkets in the Northeast that order produce from California, Florida, and Mexico usually must do so at least a week in advance of the time it is needed. The advent of computers has greatly aided the ability of produce management to forecast store sales by referring to the same period in previous years, and predict movement item by item.

FRESH FRUIT AND VEGETABLE SPECIALTY STORES

The U.S. Department of Agriculture has estimated that specialty produce stores in 1987 had sales of $2 billion (USDA ERS, 1989). This is not large in total, but represents a considerable increase in recent years. Most of their sales are in fresh fruits and vegetables. In contrast to fresh produce sales by convenience stores, which may exceed specialty stores in total sales volume, the sales by these stores consist of a wide variety of items rather than just a few staples. Many of these stores feature items that appeal to particular ethnic groups or come from certain parts of the world such as tropical fruits or oriental vegetables. Unlike farm roadside markets, from which some have developed, these stores operate all year and buy all or most of their produce from wholesalers or shippers.

Many specialty stores have opened in major cities, catering to young people with full time jobs, Stores that started out just carrying fresh produce have expanded and added salad bars and other perishable items like dairy products.

Experience in New York City illustrates how important these specialty stores can become in some markets. In the early 1970s a few immigrants from South Korea bought or opened specialty produce stores on Manhattan Island, and later arrivals followed suit. By the mid-1980s there were estimated to be about 1,200 produce specialty stores operated by people of South Korean origin, mainly in Manhattan. Almost every block in residential neighborhoods on the island has a Korean produce store. Most of these carry produce from around the world, and provide a salad bar stocked with unusual and exotic items. They differ from other local stores in black and hispanic neighborhoods in that their customers are not of the same ethnic background, they do not offer credit, and they limit their sales to produce and other perishables.

Korean merchants in New York obtain most of their supplies from wholesalers on the Hunt's Point Market. Initially they purchased individually, but in recent years they have obtained supplies cooperatively or through Korean operated wholesale houses. They have a Korean Produce Association that looks after their interests on such matters as dealing with the city on zoning or sidewalk obstruction. They provide a significant share of the business for New York wholesalers.

Many of the South Korean immigrants who operate these businesses are well educated and in some cases professionally trained. They came to this country without much capital and with little knowledge of English, and so have been unable to continue their trade or profession here. A produce store could be opened without substantial financial resources, sometimes with assistance from relatives already in the business. Family members could contribute the help needed to operate over long hours. Many families through diligence and hard work have been able to send their children to colleges and universities and achieve other goals. Whether ownership and operation will extend into the next generation appears unlikely, although difficult to predict.

REFERENCES

Leed, Theodore W., Gene A. German, 1985. *Food Merchandising Principles and Practices.* Lebhar-Friedman Books, 3rd Edition.

——, 1983. *Convenience Store Merchandising.* Cornell University, Ithaca, New York.

Marion, Donald R., 1977. *Supermarkets in the City.* University of Massachusetts Publication No. SP-102.

McLaughlin, Edward W., Gerard F. Hawkes, 1987. *Operating Results of Food Chains 1986-87.* Cornell University, Ithaca, New York.

O'Rourke, A. Desmond, 1982. *Growth Characteristics of Major Retail Food Chains 1963-1977.* Washington State University, Research Bulletin XB 09707.

Price, Charlene, Doris J. Newton, 1986. *U.S. Supermarkets: Characteristics and Services.* U.S. Department of Agriculture, Economic Research Service, Agricultural Information Bulletin No. 502.

Supermarket Business, 1989. September, pp. 74.

Tomes, Barb, 1989. Vital Signs Good at Retail. *The Packer.* Vance Publishing, February 11.

U.S. Department of Agriculture, Economic Research Service, 1989. *Food Marketing Review, 1988.* Agricultural Economic Report No. 614.

Chapter 20

The Foodservice Industry

INTRODUCTION

Sales of food to eat away from home reached $210 billion in 1988, according to U.S. Department of Agriculture estimates (USDA ERS, 1989). This equaled 45 percent of total food expenditures, up from 26 percent in 1960. About one-third of all meals and snacks are now eaten out. The increase in the foodservice market has been attributed to rising household income, increases in two-income and single-parent households, and demographic factors like the aging of the population. The kinds of foods consumed away from home and their source is changing. Supermarkets and convenience stores are entering the market by offering fast food and limited menu carry-out service, and occasionally customer service on premises.

Foodservice operations can be broadly divided into commercial or noncommercial services. Commercial operations consist of separate eating places such as fast food outlets and restaurants, as well as eating places associated with other businesses such as hotels, department stores, or recreational facilities (Linstrom, Putnam, 1986). Noncommercial service is provided through many different types of outlets such as educational institutions, hospitals, and correctional facilities.

Unfortunately no official statistics are available on the amount or value of fresh fruits and vegetables moving through foodservice channels, nor the margins or costs of marketing them through these channels. One might assume that this market is more important for some items than others. Full service restaurants and fast food places have probably increased the demand for baking type potatoes and take a substantial share of these items, but not fresh apples. Salad bars and prepared salads require substantial quantities of lettuce and other vegetables, some of which like broccoli and cauliflower at one time were not considered salad vegetables. Many foodservice operations, particularly noncommercial operations like hospitals and correctional facilities, still rely heavily on processed fruits and vegetables in the belief that by this means they will obtain uniform quality and reduce storage, handling, and preparation costs (Van Dress, 1982).

A conservative estimate would indicate that at least one-quarter of the 72 billion pounds of the fruits and vegetables available for fresh consumption in this country moved into the foodservice market in the late 1980s. This 18 billion pounds of fresh produce cost foodservice operators about $6.4 billion. The costs of handling, storage, preparation, and serving varied widely depending on the type of foodservice operation. On the basis of an average gross margin of 60 percent, typical of the industry, the value added by foodservice operations to the fresh produce would amount to $9.6 billion, giving a final value of $16 billion. In terms of value added the food service business therefore exceeds that of any of the other major marketing sectors, shipping point operations, transportation, destination wholesaling, or retailing.

Restaurants and fast food places dominate the commercial foodservice business. Many restaurants are locally owned and managed, but most fast food operations belong to chains. Franchises are an important form of ownership for fast food chains. Foodservice operations obtain their supplies in different ways according to the type of business. Foodservice buying practices and the firms supplying them are in a state of change.

FOODSERVICE ESTABLISHMENT AND SALES

In 1988 there were estimated to be almost 732,000 foodservice outlets in this country ranging from restaurants and fast food operations to educational institutions and plant and office buildings (Table 20.1). Almost 60 percent of these were classed as commercial feeding places, the remainder as noncommercial feeding places. The total number of outlets has continued to rise.

Sales of meals and snacks, excluding nonalcoholic beverages, amounted to $193 billion in 1988. From 1981 to 1988 the total number of foodservice establishments increased 7.9, while sales increased 56.5 percent. Restaurants and fast food outlets, each with over 125,000 establishments, each also recorded over $60 billion in sales. Together they had 63 percent of the total foodservice market. In terms of both number of establishments and sales the fast food outlets increased faster than the restaurants and lunchrooms between 1981 and 1988. Cafeterias declined in number, but made a modest gain in sales. Other commercial feeding establishments fared differently.

Noncommercial feeding was provided by many different types of firms and organizations. Educational institutions ranging from elementary schools to universities were the most numerous and together had the most

Table 20.1. Foodservice Establishments and Sales, 1988, and Percentage Changes Since 1981.

	ESTABLISHMENTS		SALES	
	NUMBER IN 1988	PERCENT CHANGE SINCE 1981	AMOUNT IN 1988	PERCENT CHANGE SINCE 1981
	thousands	percent	billion $	percent
Commercial				
Separate Eating Places				
Restaurants, Lunchrooms	125.0	2.4	60.5	55.8
Fast Food Outlets	127.9	12.4	60.6	92.2
Cafeterias	4.3	−30.8	3.0	17.8
Total	257.2	6.3	124.1	70.2
Lodging Places	26.3	11.4	8.0	35.0
Retail Hosts	54.0	−7.1	5.5	43.4
Recreation, Entertainment	36.7	7.8	4.5	80.8
Separate Drinking Places	37.1	−19.1	0.8	−27.1
Commercial Feeding Total	411.3	1.9	142.9	65.7
Noncommercial				
Education				
Elementary, Secondary	89.7	−3.5	9.0	10.8
Colleges, Universities	3.5	7.9	4.7	51.5
Other Education	3.2	10.8	0.3	86.2
Total Education	96.4	−2.7	14.0	23.0
Plants, Office Buildings	17.3	9.5	7.8	38.0
Hospitals	6.9	−2.3	7.0	31.0
Extended Care Facilities	31.9	28.1	6.0	46.9
Vending	3.5	−5.2	3.8	20.6
Military services				
Troop Feeding	1.2	−12.4	2.1	28.2
Clubs, Exchanges	1.9	−21.8	0.7	35.9
Total Military Services	3.1	−18.4	2.8	30.0
Transportation	0.6	−16.8	2.8	70.4
Associations	19.0	−0.7	1.8	40.7
Correctional Facilities	7.3	4.4	1.5	96.6
Child Daycare Centers	96.9	49.9	1.0	63.3
Elderly Feeding Programs	20.0	54.9	0.9	64.9
Other	17.7	12.2	49.8	61.5
Noncommercial Feeding Total	320.4	16.7	49.8	35.1
Grand total	731.7	7.9	192.7	56.5

Source: Adapted from USDA ERS, 1990.

sales. Colleges and universities showed the greatest gain in numbers and sales. Extended care facilities increased substantially in numbers and volume. The fastest growing segment of the noncommercial foodservice operations in terms of sales were correctional facilities. Next in order of growth were child daycare and elderly feeding programs, although these still represent small volumes of sales.

Average annual sales per restaurant and lunchroom still exceeded those of fast food places in 1988, but only by a small margin (Figure 20.1). Average annual foodservice sales of hospitals and extended care facilities combined were slightly less, while the average sales of educational institutions including elementary schools and universities was considerably lower.

FOODSERVICE ORGANIZATIONS

Foodservice organizations range from large chains to individually owned and operated coffee shops and restaurants. Large franchise operations such as McDonalds and Burger King and several others dominate the fast food business. There are many restaurant chains with either company owned or franchised operations like Bennigan's or Pizza Hut, while many other restaurants are individually owned and operated. McDonald's is by far the largest fast food operation with total sales in 1985 of $11 billion, reportedly more than the next two competitors, Burger King and Wendys, combined.

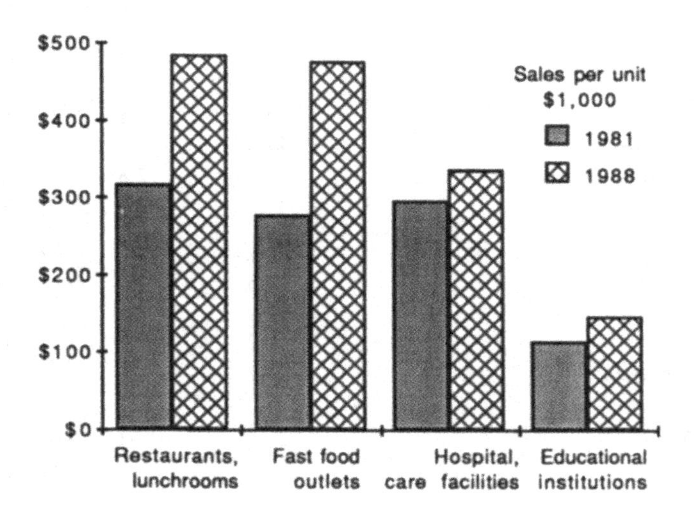

Figure 20.1. Annual Average Sales per Establishment for Restaurants and Lunchrooms, Fast Food Outlets, Hospitals and Extended Care Facilities, and Educational Institutions, 1981 and 1988. *Source:* Adapted from USDA ERS, 1989.

With 7,400 stores in this country and 2,000 abroad, another McDonald's is said to open somewhere every 15 hours.

Food conglomerates have been entering the fast food business in recent years. The Pillsbury Company owned Burger King as well as several other smaller fast food and restaurant chains. PepsiCo, Inc., which already owned Pizza Hut and Taco Bell, acquired Kentucky Fried Chicken in 1986. Hotel and motel chains like Holiday Inn and the Sheraton organization are major factors in foodservice.

When the first McDonalds opened in 1955 it offered only hamburgers, french fries, and milkshakes at the counter to go. Drive-through business is now substantial, and tables are provided in solaria. Burgers are still the main item, but the menu has broadened to include chicken and fish as well as prepared salads and many other items. Breakfasts are available with special selections. Other chains have followed suit or developed their own specialties, and many have salad bars. Fast food outlets featuring hamburgers are still the most numerous, but others favor chicken, roast beef, pizza, Mexican food or some other specialty.

Restaurants are much more diverse than fast food outlets. They vary in the type of menu offered and in the quality and cost of the meal. Restaurants now specialize in foods of many different countries, reflecting not only our diversity in ethnic origin but also the range in tastes accompanying higher income and educational levels and increased foreign travel. Some restaurants offer meals at little more than the cost at a fast food outlet, but others can and do charge many times that amount.

Franchise Operations

Franchising is very common in foodservice, especially in fast food operations and some specialty restaurant businesses (Van Dress, 1983). Franchising enables the parent company to expand its operations with limited capital investment, and turn over part of the risk and management responsibility to the local owner. Hiring and training the labor force is an important part of the business, and this can best be done or supervised by someone with an immediate interest in the operation. The franchise enables the independent owner, the franchisee, to enter the food service business with limited experience. Initial capital requirements vary according to the franchise, and may be upwards of several hundred thousand dollars. Franchise operations are closely monitored and sometimes severely restricted by the franchisor.

The franchisee makes an initial investment for the right to conduct business under the franchise logo, and usually pays a royalty and advertising fee to the parent firm based on a percentage of gross sales. The parent company provides management advice and supplies, and the franchisee agrees to maintain specific uniform products, charges, services, and practices in the operation of the business. The parent company often continues to own and operate some outlets, and a franchisee may acquire many outlets and become a fairly large operation. Franchise outlets operate very much like company owned chain stores in purchasing as well as in merchandising practices.

Selecting a store location that will prove to be profitable is an important decision. Some fast food chains spend considerable effort in choosing a site for a new franchise, while others simply follow the leader. Clustering together seems to benefit each business, rather than being separately spread out over the market area.

TRADE BUYING PRACTICES

Fast Food Chains

The development of fast food chains introduced a new dimension into food purchasing and preparation in that every outlet under the same banner is expected to provide customers with identical products, no matter where located and who the operator. Rigid specifications are established for all purchases including food, and in some cases suppliers must obtain approval from a central quality control office before being able to negotiate sales with a local franchisee. The franchisees purchase from a list of approved sources. Specifications are stringent, often including where the product must be grown and designated internal qualities.

Purchases of fresh fruits and vegetables by fast food chains have increased since the time when the major call was for a leaf of lettuce and a slice of tomato in a hamburger. The addition of salad bars and prepared salads has added considerable volume and variety. Wendy's International was reported some time ago to purchase 200,000 cases weekly of 35 different produce items. The addition of a new item such as the stuffed baked potato introduced a few years ago can also have a major impact on the market. The specifications for 80 count (average weight 10 ounces) russett burbank potatoes caused a significant increase in the price differential between large- and medium-sized baking potatoes.

Initially the components of the salad bars were obtained individually by the chain, but today specialized firms have sprung up to provide salad bar ingredients. Some organizations handle their own produce procurement going directly to major shippers. Ready-to-go salads have been prepared on site, but there are now firms that specialize in providing packaged salads. In some case these may be subsidiary organizations or independent operations. The industry is still in a state of flux, but salads and salad vegetables are now being supplied by specialized firms located in major regions of the country.

While other fast food chains were introducing and promoting salad bars McDonald's was developing and market testing prepared salads. The company had been experimenting with prepared salads since 1977, but only after extensive market testing did they introduce them nationwide 10 years later (Western Grower and Shipper, 1989). In support of this product line a whole new supply system was developed. Stores receive the raw ingredients ready to use, with the exception of eggs and tomatoes which require individual store slicing.

KGM Harvesting Company, a division of Coronet Foods Inc. and McDonald's largest supplier/processor, grows lettuce year-round exclusively for McDonald's. KGM harvests lettuce throughout California and Arizona depending on the growing season, and supplies about 60 percent of the lettuce used for McDonald's salads and sandwiches. The lettuce harvest is based on sales projections that take into account weather conditions and store promotions. In 1988 McDonald's used about 120 million pounds of lettuce, or 60 million heads. Individual store sales averaged 900 salads a week, and each head of lettuce made 4 salads. KGM cut about 175 acres of lettuce a week, harvesting about 15,000 heads per acre and leaving about 25 percent in the field that failed to meet quality standards.

Following harvest the lettuce is promptly cooled and then shipped to one of McDonald's 10 salad processing plants strategically located around the country. At one of these, the Freshpack, Inc., plant in Louisville, Kentucky, lettuce and other fresh vegetables are processed and shipped to 1,500 stores, some as far as 300 miles away. Processing lettuce involves chopping, washing, drying, and packaging it in $3\frac{1}{2}$ pound bags each holding the equivalent of 4 heads. To meet McDonald's quality standards the lettuce must be used within 10 days. The lettuce is cut, shipped, and delivered to the processor within 5 days. Processing takes about 1 day, so the store can frequently sell the product well within the quality time limit.

Uniform quality that meets product specifications is the absolute requirement in most organizations. Price is secondary. The raw material cost is a small part of the final price except when shortages occur. Then purchases are restricted very little by rising prices.

Hotels, Restaurants, and Institutions

Large restaurant and hotel chains obtain much of their supplies in volume directly from the shipper as do fast food organizations, but independent operations buy largely from wholesalers who may specialize entirely in this type of business or feature this service as part of their total operation. Wholesalers who service hotels and restaurants are called purveyors, and they feature product of the quality, size and uniformity needed by their customers to maintain control over portion size and appearance. For some hotels and restaurants the quality must be excellent, for other institutions top quality may not be as important. Purveyors are full-service wholesalers who assemble orders and deliver to customers.

In the past, foodservice wholesalers supplying dry groceries in most cases did not carry fresh fruits and vegetables. Now more grocery foodservice wholesalers are adding fresh produce to their line. Some foodservice wholesalers operate at many different locations, and do a very large volume of business. Food manufacturers as well as other types of firms have entered the foodservice wholesaling business.

Foodservice wholesalers that started with frozen foods or dry groceries have expanded into fresh fruits and vegetables, adding these products gradually in individual markets as the opportunity arises. Sysco Corp., the nation's largest foodservice wholesaler, which started in 1969, ventured into produce in the early 1980s. In 1989, according to one trade source, Sysco's sales of fresh fruits and vegetables accounted for about $500 million of total sales of more than $7 billion (Glynn, 1990). Sysco operated about 80 food service distribution centers nationwide. In early 1990, 3 years after setting up a produce buying operation in the Salinas Valley, Sysco was preparing to begin produce distribution in southern California.

Kraft Inc. branched into foodservice operations in 1976 and today ranks second in foodservice wholesaler sales. The Foodservice Group had 25 distribution centers before 1985 and acquired 21 independents, some with multiple sites, from 1985 through May 1988. Of the 50 operations across the country about half offered fresh produce in mid-1988. Kraft Foodservice maintains national programs with selected suppliers who must be flexible, must ship to many regions, and must be able to meet Kraft's quality specifications. Kraft produce promotions generally run two weeks and so require consistent pricing which has been difficult to develop because of extreme fluctuations. Kraft has an operations manual that establishes specifications and operations guidelines for more than 50 produce items. Local managers are still given the flexibility to buy as needed. The company provides educational materials, additional recipes, and point-of-sale material from commodity boards and works with many of the larger shippers. Kraft distributes products to all segments of the foodservice industry.

Sysco and Kraft are very large businesses but still do only a small share of the business. In 1988 it was estimated that the five largest foodservice wholesalers did less than 15 percent of the total, leaving considerable room for others (USDA ERS, 1989).

Many foodservice companies are in the process of developing private labels to identify product packed to their specifications. Sales to many institutions, especially government operated institutions like correctional facilities and mental hospitals, are made on the basis of bids covering a specified period of time such as six months or longer. The specifications for the bids are generally based on U.S. government grades such as U.S. No. 1 potatoes or U.S. Fancy for apples, and deliveries may require an inspection certificate or be subject to government inspection.

The popularity of fresh fruits and vegetables has extended to educational institutions and most other inhouse feeding operations, but not as yet to many publicly operated institutions. The costs of the fresh produce and the labor required in preparation make these products largely prohibitive in low budget operations.

REFERENCES

Glynn, Mike, 1990. Starting Fresh in L.A. *The Packer.* Vance Publishing, January 13.

Linstrom, Harold R., Judy Jones Putnam, 1986. Building a Foodservice Database. *National Food Review.* U.S. Department of Agriculture, Economic Research Service, NFR-32, Winter.

—— 1987. Foodservice Trends. *National Food Review.* U.S. Department of Agriculture, Economic Research Service, NFR-37 Yearbook.

U.S. Department of Agriculture, Economic Research Service, 1989. *Food Marketing Review 1988.* Agricultural Economic Report No. 614, August.

—— 1990. *Food Marketing Review 1989.* Agricultural Economic Report.

Van Dress, Michael G., 1982. *The Foodservice Industry: Structure, Organization, and Use of Food, Equipment, and Supplies.* U.S. Department of Agriculture, Statistical Bulletin No. 690.

Van Dress, Michael G., 1983. *Dining Out: Separate Eating Places Keep Customers Happy, Suppliers Busy.* U.S. Department of Agriculture, Information Bulletin No. 459.

Western Grower and Shipper, 1989. McDonald's is Major User of Iceberg Lettuce WGA Service Corp. February.

Chapter 21

Direct Marketing By Farmers to Consumers

INTRODUCTION

Some fresh fruit and vegetable growers are able to market their products directly to consumers, bypassing typical marketing channels. The quantity of fresh fruits and vegetables marketed this way, about 3 to 5 percent of total marketings, is small. But for growers of some commodities in some areas and for some consumers it may be very important (Henderson, Linstrom, 1980; O'Rourke, 1980; LeVeen, Gustafson, 1978). The principal ways that farmers market directly to consumers are through farmers' retail markets, pick-your-own operations on the farm, and farm roadside markets (Tracy, Dhillon, Varner, 1982). Gift baskets and other items by direct mail, and mobile wagons to campgrounds or apartment complexes are some of the other ways.

Another marketing channel that has increased in importance in recent years and which bypasses the normal trade routes, is the selling of specialty products by growers directly to retail stores, restaurants, and institutions. While this is not strictly farmer-to-consumer direct marketing, the management and marketing skills needed and the intensive labor requirements are similar to those required for direct marketing and will be included in that category here.

Direct marketing is certainly not a new phenomenon, and was the major marketing channel in the early days of this country. As farming prospered, however, most farmers felt they should concentrate on growing farm products rather than on becoming involved in the time consuming and unfamiliar work of retailing. The number of farmers selling directly to consumers declined, especially in the period of agricultural expansion following World

War II. But in recent years several factors have contributed to bring about a greater volume of direct sales.

The increased interest consumers have shown in fresh fruits and vegetables has also led them to appreciate the taste and flavor they might enjoy from locally grown products such as tomatoes or fresh picked sweet corn (Jack, Blackburn, 1984). The day in the country to pick strawberries or cherries could be a pleasant family outing as well as an opportunity to purchase fruits or vegetables in volume at reasonable cost. A visit to a farmers' market would provide contact with a grower who could explain production practices and even offer assurances regarding pesticide use. Specialty stores and exclusive restaurants might seek out or accept unusual or high-quality items that could be grown locally.

Commercial farmers who found their primary enterprises declining in profitability or their wholesale markets softening, turned to direct marketing to obtain additional revenues (Moulton, 1979). This was a way to use family labor, in some cases made available by mechanization of the major farm operations. Other rural residents also began raising crops or livestock products for local sale either as full-time work or to add a second source of income as so many families were doing in other ways.

Marketing direct to consumers or to specialty stores and fancy restaurants takes special skills and abilities on the part of the marketer, and often also a favorable location with respect to land resources and local markets. This form of marketing, which went through a period of rapid expansion in the 1970s, has now reached the mature stage where further growth will be limited and existing operations will face intense competition. The conditions necessary for success have consequently become most stringent.

FARMERS' RETAIL MARKETS

Fruit and vegetable growers still sell at wholesale and sometimes at retail on some public wholesale markets, but the practice of retailing direct to consumers at community markets one day or so a week has flourished in recent years. Numerous new markets have been established across the country. This method of marketing has a special appeal for part-time or small scale farm operations, although large commercial farmers sometimes participate too. Urban residents are especially attracted to these markets.

Establishing a viable farmers' community market has to be a group or community effort (German, Deckers 1979; Kenyon, Bell, Edgar 1978; Sabota, Courter, Archer, 1980). Markets may be organized by the growers themselves cooperatively, by a Chamber of Commerce, by Cooperative Extension, or by other community organizations. Steps in developing a farmers' market are first to identity interested growers, perhaps 20 or 30, find a good location, obtain the necessary permits and permission, and hold an organizational meeting to determine support and to identify leadership. A set of rules and regulations must be developed, fees set, insurance liability and state and local regulations checked, and responsibilities for management and supervision assigned. The opening date must be agreed upon, the event publicized as widely as possible, and guidance provided to vendors on merchandising and pricing.

Community markets are often located in residential areas or near office buildings. In such cases providing parking may not be important, but parking space is essential where that is not the case. Rules and regulations are needed especially with respect to who is eligible to sell on the market and whether sales are to be restricted to products produced by the vendor. Many vendors on a new community market may not have had experience in selling direct, and so may need guidance in setting out their wares and dealing with customers.

Markets are usually held once a week at a particular location for 3 to 5 hours, and some farmers go to more than one each week during the season (Sullebarger, 1983). Saturday morning is a popular time, but markets are also held on weekday mornings or evenings. Farmers' out-of-pocket expenses to attend a market are usually small, but the major cost is the time it takes. Two people from each farm are often necessary to make sales. If sales are not brisk the time spent may be costly unless there are no other productive alternatives. Regular attendance by the vendor during the season is important to develop good relations with customers since many return each week, but this requires a steady flow and good variety of produce to sell.

Greenmarkets of New York City

The Greenmarkets of New York City are a special group of community farmers' markets. Their history illustrates how such markets can be successfully organized and developed largely through the foresight of one man, and even against the advice of presumed experts. In 1976 after careful study the Council on the Environment of New York City, a privately funded citizens' organization in the Mayor's office, opened its first farmers' market (Sullebarger, 1983). Ten years later there were upwards of 22 Greenmarkets in New York, most of them seasonal but 3 operating all year. Saturday was the most popular day, followed by Tuesday. Over 100 farmers participated, some attending more than one market each week. Farmers selling at Greenmarkets may take in several thousand dollars daily. Some farmers drive over 200 miles to sell at a Greenmarket. Greenmarket staff have estimated that about 45,000 people shop at Greenmarkets each week. Greenmarkets have been credited with introducing many New Yorkers to products and to qualities they had not known before.

To qualify to sell at a Greenmarket one must be a farmer, and must be willing to comply with certain specific regulations. In making application farmers must specify the crops they plan to grow and sell at the market, and must sell only homegrown products unless given express permission by the market manager to purchase other locally grown products for sale. Greenmarket personnel visit every farm to make sure growers are complying with regulations. Vendors are selected to provide variety to the products offered at each market. In addition to fruits and vegetables other items produced on the farm may be sold such as plants and flowers, jams and jellies, cider, breads, eggs, meat, and cheese.

Vendors must arrive at the market site at least a half hour before the market is open to the public, must exhibit a sign giving the name and location of the farm, and must post all prices in a visible location. Perishable produce must have been freshly harvested. Vendors are encouraged to accept Food Stamps, and also to turn over unsold produce at the end of the day to volunteers who deliver it to soup kitchens, shelters, and food pantries for the poor.

The standard fee for one day for one space 12 foot wide and 16 to 24 feet deep varies from $20 to $35 according to the market location. Farmers may rent from one-half to 3 spaces at a time, depending on their volume of business. There is a waiting list at most Greenmarkets. When new markets are opened or vacancies occur at existing markets the applications are screened by the Greenmarket staff who follow specified guidelines.

To obtain sites for Greenmarkets in New York City permission must be secured from many different agencies and individuals, and validating the vendors and managing the markets is a time consuming business (CENYC, 1986). The Council on the Environment employs a small staff year round and seasonal help during the summer and fall to administer and manage the markets. Vendor fees provide only part of the cost. Additional operating funds are obtained from private foundations and other sources with the belief that Greenmarkets help preserve local agriculture, provide New York City residents with produce of better quality and greater variety, and generally help to improve the environment in the City.

PICK-YOUR-OWN

Pick-your-own or U-pick is a form of direct marketing that may have started as a salvage operation but now is important in its own right. For many farms pick-your-own is a subsidiary enterprise, but some farms are organized entirely around this method of marketing. Some consumers depend on securing a major share of their fruits and vegetables this way, preserving them later in the home. Popular crops for pick-your-own are small fruits such as strawberries, blueberries, raspberries, and grapes; tree fruits such as cherries, apples, and peaches; and vegetables such as snap beans, tomatoes, and green peas (Courter, 1979). Many other fruits and vegetables have been sold on a pick-your-own basis.

Characteristics of a commodity that favor selling by pick-your-own and explain the popularity of strawberries are that it requires substantial hand labor to harvest, the maturity or ripeness is easy to recognize, the plant or tree is not likely to be excessively damaged in the picking process, it is easy for consumers to use, prepare, or store, and that it has an elastic demand. Some crops that do not have all these characteristics such as sweet corn, cucumbers, and asparagus can and have been successfully sold on a pick-your-own basis but only with careful management.

Customers may travel considerable distances to pick their own crop if there are no intermediate sources of that commodity, and have been known to drive 50 to 100 miles in search of specific items. Studies have found, however, that most customers for strawberries come from within a radius of 20 miles, and this information can be helpful in determining the potential sales volume based on an estimate of expected per capita sales and population data (Courtner, 1982).

Customers come to pick-your-own farms to obtain better quality and to save money (Hungate, Watkins, 1978; Pelsue, 1983 and 1984; Rossi, Fabian, 1980). They choose one farm over another for ease of picking, and cleanliness and efficiency of operation. Small differences in prices are not a factor in most cases. Pick-your-own farms advertise in many ways, but radio and newspaper are usually the most important. Some maintain a mailing list and use direct mail to alert customers to the start of the season, and recorded telephone messages to provide up-to-date information on picking conditions. "Word of mouth" is often reported to be the most important way customers first learn of the farm, and repeat customers are essential. Many customers are older and tire soon so anything that makes picking easier and safer such as high yields, dwarf trees, and convenient layout will result in greater sales per customer.

Managing a pick-your-own operation requires organizing the traffic flow, establishing parking areas, providing for check-in details, transporting and supervising customers in the field, and handling payment at the check-out area (Ginder, Hoecker, 1975; Courter, 1979). Whether to have customers park centrally and provide transportation to the field or have them drive to the picking area, whether or not to permit children to pick, and whether to sell by weight or volume are decisions that need to be made. Whatever the system it should be operated so that customers do not have to wait around to check in or out, and can pick a large volume of good quality product in a relatively short time.

Setting the right prices at pick-your-own is very important, especially if there are several competing operations in the same trading area. Price wars are disastrous for all concerned, but being priced out of the market can be disastrous too. Arrangements on prices between vendors are not only illegal but usually unworkable. Factors to consider in setting prices are last year's prices, prices for other crops this season, competitors' prices, and prices on wholesale markets or in supermarkets. Costs need to be considered, but generally are only relevant as they may affect the supply. Once set at what appears to be a reasonable amount, pick-your-own prices ae usually maintained at the same level throughout the season. Well-run operations with distinctive or unique features are often able to hold prices moderately higher than those of competitors without losing customers.

A major challenge in pick-your-own marketing is to match customers to the production of the crop, or to provide some other means of balancing supplies and sales. Problems can arise in marketing a crop like strawberries when warm moist weather brings on production more rapidly than expected, often in the middle of the week when customer traffic is low. Additional advertising or an alternative market may be necessary to avoid losing valuable sales at this juncture. Lowering prices is a poor solution, and bypassing a field may be better than having customers pick overripe fruit for the rest of the season.

Evaluation of the marketing operations may require keeping track of gross sales per acre, the number of customers daily, and sales per customer as well as determining the profitability of the operation. Daily and weekly records from previous years with notations as to weather and market conditions can be helpful for planning purposes. Attempts should be made to gauge the effectiveness of advertising expenditures.

Dealing with large numbers of customers on the farm raises all kinds of questions regarding legal liability for damage and accidents (Uchtmann, Archer, 1981). Careful insurance planning can avoid problems later.

FARM ROADSIDE MARKETS

Selling farm produce to consumers at the farm is certainly not new, but has increased in recent years. Whether this trend will continue is difficult to predict. Low prices for farm products have encouraged farmers to do more of the marketing. The increased number of rural nonfarm people, the greater concern for nutrition and health, and tendency of many supermarkets to still feature appearance and not taste and flavor, and the rising costs of supermarket operation all encourage consumers to increase purchases directly from farmers. Roadside markets provide an alternative that meets the needs of many consumers.

Farm roadside markets vary all the way from a card table on the front lawn offering surplus garden produce on the honor system to elaborate buildings containing modern refrigerated display cases, fluorescent lighting, shopping carts, and checkout systems comparable to chain supermarkets. Most farm roadside markets fall somewhere between these two extremes, with a simple building set back from the road fronted or flanked with parking spaces and featuring the fruits or vegetables produced on the farm, perhaps supplemented with other products bought locally or at a nearby wholesale market. Many just operate seasonally, while others stay open all year. Some start out as family projects selling only products grown on the farm. Later they may decide to purchase additional items to attract more customers and provide greater volume. Some eventually give up growing produce entirely, and become a fruit and vegetable specialty store rather than a farm market.

When launching a farm roadside market it is usually better to start small with plans for expansion rather than make a major investment in new facilities immediately (Beierlein, Connell, 1986). A market represents a large fixed cost in terms of facilities and staff since it generally must be operated long hours for six or seven days a week. Sales are difficult to predict, although traffic patterns, trading area population, and the numbers, volume, and quality of already existing places where consumers can buy fresh produce can provide some guide.

Building design and layout of a new market is important, but parking space and exterior traffic flow can be even more crucial. Most people come to farm markets by car, so restricted parking facilities can severely limit sales. Customers are reluctant to come to markets when getting off the highway and back on again is hazardous, and parking spaces are difficult to find. When parking is restricted to the space between the building and the highway there is often little room for expansion, whereas if the parking area flanks the building it can often be easily extended.

The operation of a farm market has many similarities to that of a supermarket produce department except in the general impression to be given the customer, although even this difference is becoming less evident. Farm markets generally try to project an image of natural wholesomeness, fresh flavor, economy, and rural values. But layout, customer flow, and merchandising and pricing policies all play a vital role in the success of the market just as they do in supermarket operation. Daily and weekly sales records compared to the previous year are significant indicators of performance, but customer count and sales per customer can reveal strengths or weaknesses and suggest changes. Observations of customers on the parking lot and in the market can often suggest ways to improve efficiency. Sometimes there may have to be trade-offs in projecting the farm image or using modern equipment, such as replacing baskets with shopping carts or the cash box with a computerized cash register.

As the market grows there may be a need for more sophisticated management tools. Market operators often face decisions whether to add products or product lines to attract more customers and increase sales per customer. Evaluation of the expected or actual contribution to overhead of major products or product lines can be helpful in making such decisions. This does require keeping track of sales by item and costs of goods sold. If decisions regarding what to grow on the farm and what to buy or the choice between alternative outlets for the farm production arise, then it may be useful to set up separate accounts for the market and charge that operation for the farm produce provided to it at the going wholesale prices. In this way the profitability of the market can be determined apart from the farm.

Well-trained, courteous help is even more important on a farm market than in a supermarket (Beierlein, Connell, 1986). Workers are expected to be knowledgeable about all phases of the farm operation and all the products they sell, helpful and polite at all times, as well as maintaining displays and keeping everything in top quality condition. Effective organization of the human resources and careful attention to hiring and supervision can pay major dividends. Here is where the contribution that family members can make will seldom be equalled. Not only the productivity of each worker in terms of sales per labor hour or some other measure is important, but also the quality of customer service which is hard to quantify. Repeat business is the key to success in most farm markets, and helpful friendly clerks can be as important as product quality and value in bringing customers back.

Few markets can grow, or even survive very long, on word of mouth alone (Dalton, Andrews, 1979; Pelsue, 1980a, 1980b). Some form of paid advertising as well as gratuitous promotion will be necessary. Well placed signs can catch the transient trade and bring in new customers, and are generally the most cost effective. Newspapers and circulars are widely used to announce the beginning of the season or the impending harvest of a major crop, providing directions and hours of operation. Radio can help especially if the need for immediate business is urgent. Information about customers, existing or potential, can be used to choose media and develop the advertising program. Mailing lists, contests, license plates, or special surveys are some ways one can learn about customers.

Good promotion can sometimes be obtained at very little cost through articles in the local newspaper, interviews on radio, or a short report on a television station. This may take some planning and ingenuity to obtain, but the payoff is high.

Sometimes there are opportunities for cooperative promotion programs. Farm markets in California and farm wineries in New York have banded together to publicize a trail of markets, wineries, and related businesses such as bed and breakfast lodgings and restaurants. Roadside market operators may find, contrary to first impressions, that a cluster of markets featuring different products and adequately promoted could attract more customers than a single market did before.

MARKETING SPECIALTY CROPS TO STORES AND RESTAURANTS

Rising income levels, the growing interest in unusual and different fruits and vegetables, the satisfaction of eating out in fancy restaurants, and the enjoyment many find in preparing special meals at home have opened up a market for locally grown specialty items. On the supply side, ways have been found to grow these specialty crops in places where they have not been grown commercially before, using carefully chosen varieties and such techniques as plastic tunnels.

New varieties of traditional fruits and vegetables as well as those of tropical or oriental origin have been greatly in demand. Eggplants, peppers, and tomatoes of different colors and shapes have become popular as well as unusual varieties of leaf lettuces and small scale vegetables such as baby carrots. The demand for fresh herbs of different kinds has been expanding in many markets.

Upscale restaurants are the major markets for specialty crops in California and New York. One report from California indicates that restaurants there want freshness and quality, organic produce grown without pesticides, and specialty products not available from local wholesale markets (Gibson, 1987). Some individual farmers serve a group of restaurants on a regular delivery schedule, others work with specialized wholesalers. Specialty crop growers in California, as in some other parts of the country, hold a "Tasting of the Summer Produce" to introduce restaurant buyers to growers and their special items.

The specialty items do not have to be red onions, yellow peppers, cilantro, baby bok choy, or Napa cabbage. Juicy flavorful tomatoes can be a specialty item in some major markets. In 1983 a New Jersey resident grew a few tomatoes and sold them to restaurants in lower Manhattan at the rate of about 10 cases a week (*The New Yorker,* 1987). In 4 years he had expanded to 60 accounts, including many of the better known New York restaurants. Sales reached 500 cases of produce a week, most of this purchased from local growers. Tomatoes were still the backbone of the business, but in eight different varieties ranging in size, shape, and color. Other products varied from sweet corn to squash blossoms.

Serving the restaurant and retail store trade with specialty products is a demanding job. Product must be of top quality and available as needed, sometimes in small quantities. Costs of operation are high, but then so are prices since the cost of the food is not the prime consideration in the operation of these businesses. Dependable service is what these buyers require, and they will seek out sources that can provide it and usually pay them well.

REFERENCES

General

Henderson, Peter L., Harold R. Linstrom, 1980. *Farmer-to-Consumer Direct Marketing in Six States.* U.S. Department of Agriculture, Agriculture Information Bulletin No. 36.

Jack, Robert L., Kitty Lou Blackburn, 1984. *Effect of Place of Residence on Consumer Attitudes Concerning Fresh Produce Marketed Through Direct Farm Markets in West Virginia.* West Virginia University, Agricultural and Forestry Experiment Station Bulletin 685.

LeVeen, E. Phillip, Mark R. Gustafson, 1978. *The Potential Impact of Direct Marketing Policies on the Economic Viability of Small Fruit and Vegetable Farms in California.* Giannini Foundation, University of California, Research Report No. 327.

Moulton, Curtis J., 1979. *Marketing Strategies for Locally Grown Produce in King County.* Washington State University, College of Agriculture, Research Center Circular 614.

O'Rourke, A. Desmond, 1980. *The Role of Direct Marketing in Washington Agriculture.* Washington State University, College of Agriculture, Research Center Bulletin 0890.

Phelps, Joel B., R. Brian How, 1981. *Planning Data for Small Scale Commercial Vegetable and Strawberry Production in New York.* Cornell University, Department of Agriculture Economics, A. E. Res. 80-20.

Toothman, James S., 1981. *Sales Revenue and Selling Costs in Farmer-to-Consumer Marketing, A Report of Fifteen Case Studies in Pennsylvania.* The Pennsylvania State University, Department of Agriculture Economics and Rural Sociology, A. E. & R. S. 153.

Tracy, Marie H., Pritam S. Dhillon, Michael C. Varner, 1982. *Economic Comparison of Direct Marketing Alternatives for Fresh Vegetables in New Jersey.* Rutgers University, Department of Agricultural Economics and Marketing.

Farmers' Retail Markets

Council on the Environment of New York City (CENYC), 1986. *Annual Report.* New York City.

German, Carl L., Mary R. Deckers, 1979. *The Feasibility of Establishing a Farmers' Market in Delaware.* University of Delaware, Cooperative Extension Service, Extension Bulletin 115.

Kenyon, David, Jim Bell, Tom Edgar, 1978. *Planning a Farmers' Market.* Virginia Polytechnic Institute and State University, Department of Agricultural Economics Publication 776.

Sabota, C. M., J. W. Courter, Roberta Archer, 1980. *Establishing a Community Farmers' Market.* University of Illinois, Department of Horticulture, HM-4-80.

Sullebarger, Beth A., 1983. *New York State Farmers' Markets: Key to Rural and Urban Preservation* New York State Council on the Arts, New York Department of Agriculture and Markets.

Farm Roadside Markets

Beierlein, James G., Cathleen M. Connell, Editors, 1986. *Managing for Success: A Manual for Roadside Markets.* Pennsylvania State University, Department of Agricultural Economics and Rural Sociology.

Dalton, M. M., R. A. Andrews, 1979. *Marketing Agricultural Products in New Hampshire IX. Sales Patterns in Roadside Markets.* University of New Hampshire, Agricultural Experiment Station, Research Report No. 74.

Pelsue, N. H., Jr., 1980a. *Consumers At Vermont Fruit and Vegetable Roadside Stands, Part I.* University of Vermont, Agricultural Experiment Station, Research Report.

——1980b. *Consumers At Vermont Fruit and Vegetable Roadside Stands, Part II.* University of Vermont, Agricultural Experiment Station.

Pick-Your-Own

Courter, J. W., 1979. *Pick-Your-Own Marketing of Fruits and Vegetables.* University of Illinois, Department of Horticulture, HM-1-9.

——1982. *Estimating the Trade Area and Potential Sales for a Pick-Your-Own Strawberry Farm.* University of Illinois, Department of Horticulture, HM-6-82.

Ginder, Roger G., Harold H. Hoecker, 1975. *Management of Pick-Your-Own Marketing Operations.* University of Delaware, Cooperative Extension Service.

Hungate, Lois S., Edgar P. Watkins, 1978. *Ohio Customers and Pick-Your-Own Produce.* Ohio State University, Cooperative Extension Service, Leaflet 325.

Pelsue, Neil H. Jr., 1983. *Consumers at Vermont Apple Pick-Your-Own & Roadside Stands.* University of Vermont, Agricultural Experiment Station Research Report 35.

——1984. *Consumers at Pick-Your-Own Strawberry Markets in Vermont.* University of Vermont, Agricultural Experiment Station, Research Report 39.

Rossi, Daniel, Morris S. Fabian, 1980. *Socioeconomic Characteristics, Purchasing Patterns, and Attitudes of New Jersey Pick-Your-Own Customers* Rutgers University, Department of Agricultural Economics and Marketing, AE 378.
Uchtmann, D. L., J. T. Archer, 1981. *Direct Marketing By Farmers to Consumers: Some Legal Implications.* University of Illinois, Cooperative Extension Service, Circular 1195.

Specialty Crops to Restaurants and Retail Stores

Gibson, Eric, 1987. A Slice of the Pie: Direct Restaurant Sales Offer High Profit Potential. *The Grower.* Yance Publishing Co., October.
The New Yorker, 1987. Tomato Bob. *The New Yorker* October 12.

Part IV
Epilogue

Chapter 22
Future Prospects

As we approach the millenium we naturally feel the urge to take stock of where we have been and try to visualize what the years ahead will bring. Rightly or wrongly we tend to consider conditions in our economy to be linked to the decades. To many the 1960s was a period of upheaval and unrest, the 1970s one of energy shortage and inflation, and the decade of the 1980s was dominated by supply-side-economics and expanding business activity. Attempts are now widespread to predict the events and developments that will characterize the 1990s.

Given the tremendous changes that have occurred in our political, economic, and social life in recent years, most of them completely unforeseen even shortly before they happened, one might doubt the wisdom of spending time trying to forecast the future. Within a very short period, on a global scale, we saw the breakdown of the centralized political and economic system in many countries of Eastern Europe. Coincident with this came the sharp decrease in military tension between East and West, and the subsequent reduction in ground forces and in anticipated armament expenditures. Unexpected relaxation in apartheid suddenly occurred in South Africa. The rapid progress of the 12 nations of western Europe toward the achievement of the Common Market by 1992, the first cautious steps by the United States and Canada toward economic union under the Free Trade Agreement, and the continued economic expansion of Pacific Rim countries are just some of the major international economic developments of recent years.

Scientific progress, both physical and biological, has proceeded at a rapid pace. Who would have thought a few years ago that in such a short time personal computers would become so widespread, so powerful, and so capable of performing so many different tasks. And even more surprising was the proliferation of the fax machine almost overnight. In remarkable cooperation between the sciences the biologists now are able to employ equipment developed from advanced physical principles to explore the opportunities provided by biotechnology and genetic engineering.

Private business organization and operation has not been immune to the changing times. Through mergers and acquisitions the large conglomerates grew to extend their activities into new products and services as well as into foreign markets. Satellite firms were spun off, and some of them retired publicly held stock and reverted to private ownership. More recently the collapse of the junk bond market has signaled a change in direction, and a return to greater industry specialization has become the order of the day.

One important lesson we should have learned from the last few years is that trends cannot be depended on to continue indefinitely, and the ability to adjust to change is still extremely important. And emerging from the industrial upheaval is a new breed of manager trained in business principles, in human relations, in financial analysis, in computer technology, and in marketing principles.

If these new developments have been so difficult to forecast why then is there such a great interest in attempting to predict the future or to identify emerging trends? In these days of fast moving action in the political, economic, social, and industrial arenas managers find it essential to be able to take these external factors into account in planning strategies and tactics for their particular business. Popular sayings reflect this attitude. In relation to trends "It is a lot easier to paddle with the current than against it" or "You cannot put the toothpaste back in the tube". Good managers pride themselves on being able to handle any emergency, but have a horror of the unexpected. Don't let me be taken by surprise, is all they ask. So it is important to keep an eye out for the future, even though the crystal ball is clouded. But caution is needed to avoid being unduly influenced by fads, while keeping attuned to the early recognition of underlying trends.

THE BIG PICTURE

We can divide what we see happening that will affect the future into two categories. One of these we might call primary or basic developments, or what John Naisbitt (1982) called Megatrends. These, for example, are the events in eastern Europe, the progress with microchips, gene mapping, or the collapse of the junk bond market to name only a few and not in order of importance. These are to be distinguished from the secondary effects these happenings may have on our particular industry or how they may influence or be incorporated into our individual business.

Comprehending and coping with change is thus a two-stage process. We need to recognize incipient trends or recent developments that are likely to continue to flourish and increase in importance. Naisbitt does this in part by searching the columns of newspapers across the country for items that collectively begin to foretell a groundswell. Here the bottom up trends emerge, as distinct from the top down fads. In the 1980s we recognized that we were in an information society of global proportions, dominated by high tech–high touch. Now Naisbitt and Aburdene (1990) are pointing to new directions such as the emergence of free market socialism, the privatization of the welfare state, and the age of biology. All these have significant implications for the fresh fruit and vegetable marketing industry.

The business climate of the 1980s was generally conducive to development and expansion. On the national scene there certainly were minor interruptions and irritations to cope with. But the new products, the new services, the new competition, the new technology could all be handled readily enough. At home problems with pesticides flared up, but seemed to soon subside. In the international arena obstacles to trade appeared briefly, but the general trends favored expansion. New markets opened up in the Pacific Rim and western Europe.

INDUSTRY CHANGES

Conditions during the 1980s were such that the basic trends continued to strengthen with almost monotonous regularity. New sources of supply appeared in developing countries. The market for new specialty and exotic products seemed insatiable. Larger and more sophisticated grading and packing equipment reduced costs and product deterioration. Better packaging materials and storage techniques added to shelf life to the benefit of all concerned. Handling and transportation from ocean shipment to distribution center operation improved markedly. Supermarkets got bigger, and produce department layout enveloped customers with a tantalizing selection of fresh items. Eating out grew apace, and wider and wider menu selection provided opportunities to offer more new products and services.

Changing business organization and operation also exhibited predictable trends. Apart from supermarket chains, the traditional family firm providing only one or a few services at one stage in the distribution system either took on new responsibilities, expanded into new territories, or faced dissolution. And no longer did one have to be born into the fresh produce business to compete effectively in the marketplace. In shipping, transportation, and wholesaling the management and marketing expertise gained from experience with other products was being applied to an increasing extent in this formerly exclusive arena.

The new scientific technology produced a stream of predictable benefits. Documents are now faxed around the world on a routine basis, which will doubtless lead to video presentations of products and packs before long.

Machines that do chemical tests will soon measure taste and flavor. Not only will minute quantities of many contaminants be able to be detected by a single test, but the test will be completed so promptly that the flow of product to market will not be interrupted.

Likely Uncertainty

The process of management, at least with respect to planning for the future, is much easier and likely to be more successful when relatively stable conditions exist and changes are occurring in a fairly predictable pattern than when disruptions occur and well established trends reverse themselves. But this is what may happen, and certainly needs to be a major part of the planning process.

How does one plan under uncertainty? There are some fairly obvious general principles. When concerned with uncertainty with respect to supplies, markets, or costs the rules are to diversify, to stay as flexible as possible with few long term commitments, to keep options open, and to develop an advance warning system through internal as well as external communication networks. The real challenge is to apply these principles under specific circumstances. Expert management becomes the key to success, or even survival, when faced with uncertainty. The decision to be ready to adapt to change is not one to be made lightly, for it obviously involves certain restrictions on current actions that could reduce current revenues. Some managers will argue that major changes are highly unlikely and will continue to operate as before. Others will look back over history and note the significant turning points in the past and take these into account in future operations and decision making.

The single most important change that affected the whole produce industry was the reversal of the trend by consumers toward processed fruits and vegetables in favor of fresh that took place in the early 1970s. Few recognized it when it happened, and even fewer had predicted it. A change of that magnitude is not likely to happen again in the near future, but there is still a possibility. Other changes have had a major impact on smaller segments of the industry. The introduction of the marine container revolutionized ocean shipping, as the building of the interstate highway system and the improvement in tractor–trailer design changed truck transportation. The expansion of food service wholesalers formerly handling frozen foods and groceries into produce was as big a change as the entry of grocery wholesalers serving retail stores into the produce business 50 or so years ago. Products gain and lose favor with the consumer, as round white potatoes have lost relative to the baking type and fresh oranges lost out to frozen orange juice.

Just as the need to adopt new technologies, go to new sources, or develop new markets must now be done more rapidly to survive under today's environment, so has this had a differential impact on firms of different sizes and forms of organization. The largest retail chains had difficulty adapting to change compared to some of the medium sized firms. Very small companies found themselves at a disadvantage whether they were packing and shipping, transporting, wholesaling, retailing, or serving food.

Future Challenges

Marketing management, whether they be operating at shipping point or dealing directly with consumers, will face severe challenges. Regardless of all the new technology the product is still a perishable agricultural commodity, subject to all the problems in production and distribution that that entails. And these problems will intensify, at times more rapidly and at other times less rapidly, than the means of dealing with them become available.

Sourcing will definitely be a key to success. Traditional sources may face increasing pest pressures, while means of dealing with them could be severely restricted. Supplies from some sources may become more variable in quantity and quality from season to season as we experience periods of excessive or insufficient moisture or times of unusually depressed or elevated temperatures. Surpluses may continue to bedevil growers most years, but shortages will occur unexpectedly as they have in the past. Only now, faced with the inelastic demand from food service operators and the necessity to maintain a brand franchise in retail stores, the imperative of a constant supply of a uniformly high quality is so much more important.

Along with the challenge of obtaining new and existing products from new sources has come that of dealing with the information explosion. What has already happened will only intensify. An increasing flood of data will be generated by supermarket scanners, public and private market research organizations, satellite weather systems, national and international crop reporting services, and other means not even visualized at this stage. All these facts and figures will be instantaneously transmitted around the world by electronic computer networks and fax machines as well as more traditional methods such as teletype and telephone. Without some selection and filtering mechanism computer hard disks will overflow and workplaces buried under paper.

The process of recognizing what information is important and knowing how to use it will be the key to survival. Otherwise the result will be desperation. The ability to identify what is important, what it means, and how to act on it will be essential. This ability, like many, will be part art and part science. Some managers may have a natural flair for recognizing significant information and knowing how to use it. Most will need to be trained in the principles and tools of business management and marketing.

Progress in biotechnology is the third area that will dominate the fresh produce industry in the coming decade. Biotechnology has had a long gestation period, just as did personal computers. Progress in biotechnology, unlike that for personal computers, has been delayed by some who oppose such developments on the basis of social or environmental considerations. Hayenga (1988) found on the basis of a survey of leading biotechnology firms that new products can be expected to come at an increasing rate in the decade ahead. Among these will be biopesticides for plant insect and disease control, and genetically engineered crops such as herbicide-resistant, insect-resistant, virus-resistant tomatoes. Value added plant characteristics will include textural and taste changes in fresh vegetables. Hayenga concluded that consumers' reception of products coming from biotechnology could be critical to their success or failure.

The importance of obtaining products from the best sources, expertly handled and packed, coping with the information explosion, and adjusting to changes in the market as they occur will be some of the significant challenges of the coming decade to those involved in fresh fruit and vegetable marketing. Successful firms will be those who can draw on management ability and technical skill based on rigorous training and continuing education. A feel for the business will continue to be important, but will have to be supplemented with formal classroom and laboratory experience. More time will have to be set aside to stay abreast of new developments. By this means the 1990s could become the Decade of Fresh Produce.

REFERENCES

Hayenga, Marvin L., 1988. *Biotechnology in the Food and Agricultural Sector: Issues and Implications for the 1990s.* University of California, Agricultural Issues Center, Issues Paper No 88-5, September.

Naisbitt, John, 1982. *Megatrends: Ten New Directions Transforming Our Lives.* New York: Warner Books.

—— Patricia Aburdene, 1990. *Megatrends 2000: Ten New Directions for the 1990s.* New York: William Morrow.

Index

Index

Advertising and promotion
effective programs, 225
expenditures on food, 218
farm roadside markets, 335-336
inherent problems, 226
pick-your-own, 332
Agricultural Labor Relations Act, 61
Agricultural Marketing Agreement Act, marketing order provisions, 172
Air transportation, current situation, 279
Alar
estimated effects of negative publicity, 195-196
reported use and effects, 192
Alternative agriculture, 196-199
profitability comparisons, 198-199
Animal and Plant Health Inspection Service, 182
Apples
arrivals at major markets, 100-101
deflated prices, index of, 146
grades, sizes, and packs, 100
imports, changes in, 232
packing and storing, 100
price trend over time, 142
production and utilization, 98-100
retail prices and marketing margins, 104
shipping costs, 276
shipping point prices, 101-104
varieties, 98
Arrivals at major markets, markets and data coverage, 42-43

Baja California, northern, 71-72
Bananas, imports, changes in, 232
Biological control, 198
Biotechnology, impending developments, 348
Bird dogs, role at shipping point, 263
Blue Book, The
information source, 91
mailing address, 162
Brand advertising and promotion
requirements for success, 224
Sunkist experience, 223
to consumers, 223-224
types of shippers involved in, 224
wholesale and packer brands, 222-223
Brokers at shipping point, operations defined, 152

CAADES, Mexico, 70
California, 53-63
agricultural geography, 54-56
drought, impact of, 59
pest control problems, 62-63
pesticide monitoring program, 191
seasonal labor, 59-61
shipments of fresh fruits and vegetables, 54
urbanization, impact on farming of, 62
water requirements and costs for crops, 58
water sources for irrigation, 57
Canada
imports from, changes in, 233
recent exchange rates, 237

Canada (*cont.*)
tariffs on U.S. imports, 242
trade agreement with U.S., 242–243
Capper-Volstead Act, importance to cooperative marketing, 164
Carryout and home delivery, 19
Census of Agriculture
information, source of, 80
limitations of data, 80
Census of Wholesale Trade, 86
Center for Produce Quality, 200
Changes in fresh produce marketing
challenges and opportunities, 346–347
recent developments, 345–346
Chile
grape tampering, 192
imports from, changes in, 232
Commercial foodservice operations, establishments and sales, 320–321
Communication and coordination, problems in marketing, 89
Comparative advantage. *See* Location of production: theory of comparative advantage
Competitive advantage. *See* Location of production: competitive advantage
Conference rates, set in ocean shipping, 279–280
Consignment sales
decline on wholesale markets, 288–289
importance in price discovery, 132
need for special regulations, 158–159
Consumer goods and services, U.S. market for, 3–5
Consumer packaging
expansion in fresh produce, 289
merchandising alternative, 314–315
Containers, marine
impact on ocean shipping, 269–270
shipping, 257–258
Continuing Consumer Expenditure Survey, 15
Controlled atmosphere storage, 260
Cooling methods, postharvest, 256–257
Cooperative marketing, 163–170
current status, 169–170
features of successful operations, 168
history of fruit and vegetable, 163–165
income tax exemption, 167
market conditions favorable for, 168
patronage dividends, 166
strengths and weaknesses, 167–168
Cooperative marketing principles
methods of operation, 165–166
ownership and control, 165
pooling costs and returns, 166
Council on the Environment of New York City, operation of Greenmarkets, 330–331
Cranberry scare, impact on sales, 181
Crop production and disposition. *See* Market information
Crop production estimates, controversy regarding, 120
Crop Reporting Board, availability of estimates, 120
Customs brokers, role in importing, 249

Demand and supply relationships, estimation problems, 146–147
Destination wholesale markets
commission merchants, 153
corporate chains, 85
integrated wholesale-retail firms, 85
nonintegrated wholesale firms, 85
types of firms involved, 153
voluntary retail chains, 85
wholesale receivers, 153
Dietary guidelines, 206–209
Direct marketing by farmers to consumers, 328–339
farm roadside markets, 333–336
farmers' retail markets, 329–331
pick-your-own, 331–333
specialty crops, 336–337

Economic Research Service, source of reports, 126
Electronic information service
availability, 90
information carried, 125
Electronic scanners at checkout, 310–311
Environmental Protection Agency, 181
regulatory actions, 186–187
reregistration procedures, 187
special reviews, 187–188
European Community, 244–246
Common Agricultural Policy, 245
orange marketing policies, 245
reference prices, 235

Exchange rates, 236–239
real rate of exchange, 238
Expenditures per person on fresh fruits and vegetables
by household characteristics and location, 30–31
changes associated with increased income, 32
market entry and expenditure level effects, 32
separate effect of individual characteristics, 31–35
simulated for different household characteristics, 34
Export Apple and Pear Act, 248
Export Grape and Plum Act, 248
Export sales
assuring receipt of payment, 248
firms involved and business practices, 246–247
how handled in Britain, 247
state trading in, 247
subsidized, 235

Farm production, value added in farming, 80
Farm roadside markets
advertising and promotion, 335
features of successful operation, 333–334
organization and operations, 334–336
Farm size distribution
California, 81–82
New York, 82
U.S., 81
Farmers' retail markets, establishment and operation, 329–330
Federal Food, Drug, and Cosmetics Act, 181
Federal Insecticide, Fungicide, and Rodenticide Act
cost–benefit analysis of pesticides, 195
early history, 181
Federal State Inspection Service, 90
FIND/SVP, The Information Clearing House, Inc., The Fresh Produce Market, 36
Florida, 63–66
frost damage, 65–66
geography of farming, 64–65
plant disease and insect pests, 66
shipments of fresh fruits and vegetables, 63–64
Florida, University of, consumer attitudes on food safety, 194
Food and Drug Administration, 181
analyses of pesticide residues, 189–190
Food consumption variation, identifying relation to separate effects, 18–19
Food disappearance changes
beef, chicken, eggs, 13
coffee, soft drinks, fluid whole milk, juices, 14
future prospects, 119–121
lettuce, frozen potatoes, American cheese, 13
national diet, 10–14
Food disappearance per capita
fresh and processed fruits, 24
fresh and processed vegetables, 26
fresh fruits and melons, 25
fresh vegetables and potatoes, 27
individual differences, 29
major food groups, 9–10
potatoes fresh and processed, 28
Food expenditures
by household characteristics, 15–17
changes 1960–87, 7
for away-from-home use, 6
for home use, 5–6
individual differences, 14–19
projections by commodity group, 21
projections for at-home use, 21
projections for away-from-home, 21
variation by household size and income, 15
Food labeling, regulations on nutritional, 211
Food market, U.S., 1–22
Food Marketing Institute, consumer attitudes on food safety, 193
Food preferences
groups with similar, 17
identifying and explaining reasons for, 17
Food safety. *See also* Pesticide use
consumer attitudes toward, 193–194
Foodservice buying practices
fast food chains, 324–325
hotels, restaurants, and institutions, 325–327

Foodservice industry
commercial and noncommercial operations, 319–321
establishments and sales, 320–322
fresh fruit and vegetable market, 319–320
value added in foodservice, 88
Foodservice operations
franchises, 323–324
sales per unit, 322
specialized wholesale firms, 86
types of firms involved, 154, 322–323
wholesale suppliers, 325–326
Foreign Agriculture Service
export sales assistance, 246
source of reports, 127
Foreign market development
cooperator program, 220
export incentive program, 220
Targeted Export Assistance, 221
Franchises, common in foodservice, 323–324
Free-rider problem, marketing order experience, 172
Freight forwarders, foreign, functions performed, 247
Future prospects, 343–348

General Accounting Office, study of FDA pesticide regulation, 190
General Agreement on Tariffs and Trade, 244
Generic advertising
characteristics of promotion, 219
federal programs, types, 219–221
funding, 218–219
Generic advertising and promotion
federal marketing orders, 220
foreign market development, 220
legislation for research and promotion, 219
state programs, 221–222
Grade standards
examples for different crops, 121
needs for, 121
principles followed in developing, 122
USDA policy for uniformity, 121–122
Grades, grading, and inspection, 121–123
services at shipping point and terminal market, 122
Grading and packing
shipping point operations, 255–256
tomatoes, 258
Grading and inspection terms defined, 122
Grapes
imports, changes in, 232
tampering with Chilean, 192
Greenmarkets in New York City, organization and operation, 330–331
Gross margins
importance in retailing, 308
typical for selected fruits and vegetables, 308–309
Grower-shippers
large operations, 83
methods of operations, 83
operations defined, 152

Household expenditures, allocated by function, 4–5
Hunt's Point Market
market activity, 299
market facilities, 293

Immigration Reform and Control Act, effect on seasonal labor, 61
Import purchases, procedure followed, 249
Imports of fresh fruits and vegetables
by major country of origin, 41–42
for major commodities, 40–41
Information management, essential for future success, 347–348
Integrated Pest Management, 197
Integration, in wholesale marketing, 288–289
International Bank for Reconstruction and Development, 244
International Monetary Fund, 237
International trade, 231–251
exchange rates, 236–239
future prospects, 250
recent changes, 233–234
relations with European Community, 245–246
trade barriers and export subsidies, 234–236
International trading companies, role in export marketing, 247

Interstate Commerce Commission
  granting of truck operating authority, 273-274
  railroad regulation, 267

Japanese yen, recent exchange rates, 237

Kraft, Inc, foodservice operations, 326

Lemons
  price cycles, 142-143
  prorates, effects of change in, 177-178
Lettuce
  McDonalds sourcing, 325
  price fluctuations week to week, 138
Life style and food market, major changes over time, 10-12
Location of production
  factors to consider, 50
  future changes, 50-51
  historical changes, 37-38
  influence of transportation costs, 280-281
  reasons for particular locations, 49-50
  regional specialization, 38-39
  theory of comparative advantage, 49
Location of production. *See also* Sources of supply
  competitive advantage, 49
Los Angeles Produce Wholesale Market, terminal facilities, 291-292
Low Input Sustainable Agriculture, 198

Market analysis and outlook, 125-127
Market information, 117-128
  crop disposition, 119-120
  crop production, 119-120
  fruit and tree nut production estimates, 119-120
  storage stocks, 119
  vegetable production estimates, 118
Market News Service, 90
  annual reports, 124
  dissemination of information, 124-125
  source of reports, 126
  terms defined, 123-124
Market, U.S., for fresh fruits and vegetables, 22-36
  changes since World War II, 23-28
  increases since early 1970, 24-25
Marketing bill for food, 7-9
  by commodity group, 9
  by functions, 8
Marketing orders, 171-178
  benefits and costs, 177
  economic impact, 176
  free-rider problem, 172
  historical background, 171-172
  market support activities, 175
  order administration, 175
  procedures to initiate, 175
  provisions of federal, 173
  supply management provisions, 173-174
  types of regulations, 172-175
Marketing systems, 92-114
  apples, 98-104
  changes in firm organization and structure, 89
  diagram for U.S. fresh, 78
  oranges, 92-98
  similarities and differences for individual commodities, 113
  tomatoes, 104-113
  U.S. fresh fruits and vegetables, 74-91
  U.S. system communication and coordination, 89
McDonalds
  lettuce sourcing, 325
  menu development, 323
Megatrends, insight into the general business climate, 344-345
Merchandising produce at retail
  bulk compared to prepackaged, 314
  customer flow analysis, 313
  department layout, 311-312
  space allocation, 313
  Wegmans Food Markets, Inc., 311-312
Mexico, 66-72
  Baja California, northern, 71-72
  export costs, 70-71
  export practices, 70-71
  geography of farming, 67-70
  imports from, changes in, 232
  recent exchange rates, 237
  shipments to U.S. of fresh fruits and vegetables, 66-67
  Sinaloa region, 70-71
Motor Carrier Act of 1980, impact on fleet operations, 268

National Agricultural Statistics Service, 90
source of reports, 126
National Association of Produce Market Managers, information on wholesale markets, 289
National Commission on Food Marketing, wholesale marketing study, 287-289
National Research Council
pesticide residue report, 191
pesticide study conclusions, 201
Recommended Dietary Allowances, 205-206
report on diet and health, 208-209
National Resources Defense Council, pesticide use report, 191-192
Nationwide Food Consumption Survey, 29
Noncommercial foodservice operations, establishments and sales, 320-321
NutriClean, pesticide residue testing, 199-200
Nutrition and health
advertising, impact on consumer awareness, 210
consumer awareness, 209-211
FDA surveys on consumer awareness, 210
Nutritional labeling
fresh produce exemption, 212
health claims defined, 214
medical and health claims, 215
nutrient composition sources, 213
problems with fresh produce, 212-213
processed foods requirements, 212
Produce Marketing Association program, 213-214
use of generic data, 213
use of U.S. Recommended Daily Allowances, 212
Nutritional quality and marketing, 204-216
Nutritive composition of fruits and vegetables, contribution to our diet, 205-206

Ocean shipping, business practices and pricing, 279—280
Office of Technology Assessment, pesticide use recommendations, 199
Oranges
arrivals at major markets, 94
grades, sizes, and packs, 94
packing and sorting, 94
production and utilization, 93-94
retail prices and marketing margins, 97-98
shipping point prices, 95
varieties and types, 92
Organic farming, 196-197

Packer, The
mailing address, 127
Produce Availability and Merchandising Guide, 45
Perishable Agricultural Commodities Act, 158-162
common violations, 161
complaint procedure, 160
trust provisions, 161
unfair trade practices, 160
who is covered, 159
Personal disposable income, U.S. allocation by expense category, 4
Pesticide use, 179-203
alternative agriculture approaches, 196-199
economic impact of withdrawal, 195-196
grower records, 199
history of, 180-181
industry response, 199-201
information and education programs, 200-201
monitoring use and compliance, 188-191
National Research Council report, 191
Office of Technology Assessment, 199
problems evident, 180
public policy, 201-202
quantities used, 182-183
registration, 183-184
reregistration and cancellation, 185-188
tolerances, 184-185
trade association and recommendation, 201-202
Pick-your-own, organization and operations, 331-333
Population projections, 19-20
Potatoes
changes in sources at eastern cities, 48
price cycles, 143
prices on the New York market, 135
seasonal variation in prices, 139-140
year to year variation in prices, 141
Price differences
geographic location, 136-137

origin, variety, quality, and pack, 134–136
stages in the marketing process, 137
Price discovery and establishment, 132–134
auction markets, 133
electronic marketing, 134
wholesale markets importance, 132
Price establishment, retail supermarkets, 307–308
Price theory
competition and monopoly, 131–132
demand schedule, 129–130
elasticity of demand and supply, 130
equilibrium price, 131
price flexibility, 131
supply schedule, 130
Price variation, 137–144
cycles, 142–144
seasonal variation, 139–140
trends over time, 141–142
year to year, 140–141
Prices relative to other commodities, 144–146
correlation with other fruits and vegetables, 145
deflated prices, index of, 145–146
Private label advertising and promotion. *See also* Advertising and promotion
foodservice, 225
retailer, 224–225
Produce Agency Act, current application, 158
Produce Business, mailing address, 127
Produce Marketing Association
mailing address, 127
nutrition marketing program, 213–214
Produce News, The, mailing address, 127
Produce specialty stores
Korean merchants in New York City, 317–318
sales, 317
Producer Subsidy Equivalents, 235
Produce tampering, Chilean grapes, 193
Prorates
criticized, 177
impact of changes, 177–178
marketing order provisions, 174
Quality control regulations, marketing orders, 173
Railroad transportation
current situation, 278–279
development era, 267–268
Interstate Commerce Commission regulation, 267
Reclamation Act of 1902, 57–58
Recommended Dietary Allowances, 205–206
Red Book, The, 90
mailing address, 162
Reference prices, European Community, oranges, 245–246
Residue testing, pesticides, 199–200
Retail food stores
convenience stores, 304–305
numbers and sales of, 302
supermarket format, 302–303
supermarket numbers and sales by format, 303–304
Retailers and retailing, 301–318
Retailing
sales by type of firm, 88
sales by type of store, 88
types of firms involved, 153–154
value added in retailing, 87
Sales agents, operations defined, 152
Seald-Sweet Growers, early history, 164
Shipments of fresh fruits and vegetables, U.S.
by major commodities, 40
by major states. 39–40
Shipping point firms
cooperatives, 83
forms of organization, 264
types of firms, 83, 152, 263–264
Shipping point operations
degreening or ripening, 260–261
grading and packing, 254–255
harvesting practices, 254
packaging, 257
physical functions, 82–83
precooling, 256
storage, 259
systems approach, 253–254
value added in shipping, 82
Shipping point practices
acquiring product, 261
arranging transportation, 262
selling and financing, 262
Shipping Point Trends report, 124
Sources of fresh fruits and vegetables, 37–52

Sources of supply, 53–73
arrivals of 7 leading commodities, 44
by state or country of origin, 43
seasonal and regional differences, 44–49
The Packer Produce Availability and Merchandising Guide, 45
United Fresh Fruit and Vegetable Supply Guide, 45
Sourcing, key to future success, 347
Specialty crops
future prospects, 35
marketing by growers, 336
Specialty wholesalers, establishments, employees, sales volume, 86–87
Storage, methods at shipping point, 259–260
Sunkist Growers, Inc.
advertising to consumers, 223
current status, 169
early history, 164
Supermarkets
chain buying practices, 315–317
gross margins by department, 307
numbers and sales by format, 303
operating results, 305–306
ownership and organization, 304
performance measures, 310–311
retail produce pricing, 307–308
storing and handling methods, 315
Supply and disposition of fresh fruits and vegetables, U.S., 75
Supply management
marketing order provisions, 173–174
marketing programs, 147–148
requirements for success, 148
Surgeon General's Report on Nutrition and Health, dietary guidelines, 208–209

Targeted Export Assistance, expenditures on programs, 221
Tariffs, 234–236
Canadian, on U.S. imports, 242
U.S., on imports from Canada, 243
Terminal market facilities
Boston, 293
Chicago, 293–294
Los Angeles, 290
New York City, 292
Terminal markets
current operations, 294–295
early history, 285–286
future prospects, 299–300
post World War II developments, 286
street sales, 298
types of wholesalers, 295–297
wholesalers' business practices, 297–299
Tomatoes
arrivals at major markets, 107–108
changes in sources on Boston market, 47–48
grades, sizes, and packs, 106
imports, changes in, 232
marketing system for fresh, 107, 108
packing, storing, and ripening, 106–107
production and utilization, 105–106
retail prices and marketing margins, 112–113
shipping point operations, 258
shipping point prices, 109–112
source of arrivals at Boston and San Francisco, 46–47
source of arrivals at St. Louis, 45
types and varieties, 104–105
Trade associations, PMA and UFFVA, 127
Trade barriers, 234–236
Trade papers, mailing addresses, 127
Trade policies, U.S., 239–243
Canadian Trade Agreement, 241–242
Caribbean Basin Initiative, 240
Generalized System of Preferences, 240
Omnibus Trade Bill, 239
Targeted Export Assistance, 241
Trade practices
minimizing disputes and losses, 161–162
sources of information on, 154–155
summary of, 154–155
Trade references
Blue Book, The, 157
information available, 158
Red Book, The, 157
Trade regulation and dispute resolution, 158–162
Trade terms and definitions
acceptance or rejection, 156
point of sale, 155–156
time of shipment, 157
Trailer-on-flatcar shipments
coordination necessary, 270–271
quality problems, 271
Transportation, long distance, 266–283
air shipments, 270

costs and production location, 280
differences in produce requirements, 272-273
early development, 266-267
future directions, 281-282
intermodal shipments, 270-272
major modes, 84
ocean shipping, 269
railroad era, 267-268
trailer-on-flatcar and container shipments, 270-272
trucking development, 268-269
types of firms involved, 152-153
value added in transportation, 84
Truck charges
influence of different factors on, 276-277
seasonality in cross-country shipments, 276-277
Truck Rate and Cost report, 124
Truck transportation
costs and revenues, 274-278
equipment and operating practices, 278
fresh regulation exemption, 268
types of firms and business practices, 273-274
Types of marketing firms, sources of definitions, 152

U.S. Department of Agriculture, dietary guidelines, 207
U.S. Department of Health and Human Services, Surgeon General's Report on Nutrition and Health, 208-209
U-pick, *See* Pick-your-own
Uniroyal Chemical Company, 192
United Fresh Fruit and Vegetable Association
mailing address, 127
Supply Guide, 45
UNPH, Mexico, 70

Value added in marketing fresh produce, 77
at home market, 79
away-from-home market, 79
disposition of U.S. farm production and imports, 77-78
sources of produce for home use, foodservice, and export, 77
Vance Publishing Company, Fresh Trends 1990, 35-36

Washington Street market, early history, 285-286, 287
Wegmans Food Markets, supermarket layout, 311-312
Wholesale marketing
integrated sector, 288
nonintegrated sector, 288-289
value added in wholesaling, 85